EVIL AND THE GOD OF LOVE

EVIL AND THE GOD
OF LOVE

BY

JOHN HICK

Lecturer in the Philosophy of Religion
in the University of Cambridge

MACMILLAN
London · Melbourne · Toronto
1966

MACMILLAN AND COMPANY LIMITED
Little Essex Street London WC 2
also Bombay Calcutta Madras Melbourne

THE MACMILLAN COMPANY OF CANADA LIMITED
70 Bond Street Toronto 2

ST MARTIN'S PRESS INC
175 Fifth Avenue New York 10010 NY

PRINTED IN GREAT BRITAIN

To Hazel

CONTENTS

vii

Contents

Contents

Contents

PART IV

A Theodicy for Today

CHAPTER

XIII. The Starting-point

XIV. Moral Evil

XV. Pain

XVI. Suffering

XVII. The Kingdom of God and the Will of God

PREFACE

THE fact of evil constitutes the most serious objection there is to the Christian belief in a God of love. It is also probably the hardest objection to write about. For in this field it is equally disastrous to say too little or too much. On the one hand, it will not do to remain passively silent in the face of so grave a challenge to one's faith — a challenge that was bitingly summed up in Stendhal's epigram, 'The only excuse for God is that he does not exist' ! The enigma of evil presents so massive and direct a threat to our faith that we are bound to seek within the resources of Christian thought for ways, if not of resolving it, at least of rendering it bearable by the Christian conscience. But, on the other hand, in seeking to justify the ways of God to man, one is inevitably tempted to extend faith's dim sense of a hidden divine purpose and sovereignty into an open map of providence such as could be available only to the Creator Himself. It is this almost inevitable pretension of theodicy to a cosmic vantage point that provokes the thought that any solution to the problem of evil must be worse than the problem itself!

In this dilemma I have felt compelled to risk saying too much rather than enjoy the shelter of an inoffensive but unhelpful agnosticism. For the world has the right to hear Christian voices, however inadequate, on the theological problem of evil; and the recent Christian literature on the subject, although extensive, contains major gaps. There are some important historical works : Friedrich Billicsich's three-volume *Das Problem des Übels in der Philosophie des Abendlandes* ;[1] the first and longer volume of the late Père A. D. Sertillanges' *Le Problème du Mal* ;[2] and R. A. Tsanoff's *The Nature of Evil* ;[3] as well as Charles Werner's slighter but excellent *Le Problème*

[1] Vienna : Verlag A. Sexl, 1936–59.
[2] Paris : Aubier, 1948–51.
[3] New York : The Macmillan Co., 1931.

xi

du Mal dans la Pensée Humaine.[1] But these books, especially
the larger ones, offer a series of articles in chronological order
on the work of individual thinkers without attempting to distinguish types of solution (though here Werner's book is an
exception) or to trace persisting patterns of thought on the
subject. However, a typology emerges quite clearly from the
history of Christian thought, and I have presented it here in
terms of the two divergent responses to the mystery of evil
which I have called the Augustinian and Irenaean types of
theodicy. One of the lessons of these historical chapters is
that there has been continual interaction between theodicy
and theology. The alternative ways of thinking about God
and evil are connected with alternative ways of thinking
about several other topics, such as the fall of man, the nature
of sin, providence, redemption, predestination, heaven, and
hell.

Having thus opened up the resources of past Christian
thought, one should be willing to use them in formulating
a contemporary response to the challenge of evil to the
Church's faith. This has been done on Augustinian lines by
Père Sertillanges in the posthumous second volume of his
book; and again more recently and in a fresh way by Austin
Farrer in *Love Almighty and Ills Unlimited.*[2] However, my own
reflections have led me along the other, Irenaean, path of
Christian theodicy. There are numerous contemporary versions of this in relatively brief compass, but it has not recently
been set forth at length or in a way that takes account of contemporary philosophical criticisms. I have therefore tried in
Part IV to formulate systematically a way of thinking about
the mystery of evil that is basically Irenaean in character,
although with modifications and corrections suggested by
study of the Augustinian tradition. The attempt is addressed
both to Christian believers and to the agnostics, sceptics, and
humanists who are the more characteristic citizens of the
intellectual world of today. These chapters do not seek to
demonstrate that Christianity is true, but that the fact of evil
does not show it to be false: those who have some degree of

[1] Lausanne: Libraire Payot, 2nd ed., 1946.
[2] Garden City, N.Y.: Doubleday, 1961; London: Collins, 1962.

Christian faith should not abandon it in face even of this agonizing problem, nor should those who lack Christian faith rule it out on this account as a possibility for themselves.

One who is invited to read a fairly lengthy book, even on so momentous a subject as this, might well wish to know what sort of investment of time and effort it represents on the part of its author. To satisfy this curiosity will, at the same time, enable me to express my gratitude to a number of individuals and institutions. For the book is the product of five years' work, during which period I have taken advantage of every opportunity to test its ideas by exposing them both to individual critics and to audiences. Accordingly I gladly accepted an invitation to deliver the Mead–Swing Lectures for 1962–3 at Oberlin College, Ohio, on the problem of evil, and the Mary Farnum Brown Lectures for 1964–5 at Haverford College, Pennsylvania, on the same subject. Both of these series of lectures are normally subsequently published, and my own contributions to them now appear, with grateful acknowledgements to these two distinguished colleges, within the present work. I should also like to thank the President and Trustees of Princeton Theological Seminary for a generous period of sabbatical leave; the John Simon Guggenheim Foundation for the award of a Guggenheim Fellowship; and the Master and Fellows of Gonville and Caius College, Cambridge, for their hospitality during my tenure of the S. A. Cook Bye-Fellowship. I cannot pretend that the high table and senior combination-room of Caius proved a specially favourable environment within which to investigate the problem of evil; they did, however, provide a very welcome reminder of the balancing problem of good!

I have, much to my profit, discussed the subject of this book with a number of friends and colleagues, some of whom have also read portions of the typescript and made comments and suggestions, as a result of which it has been very materially improved. I wish especially to thank my former teacher, H. H. Farmer, from whom I have learned so much, on this subject as on others; George Thomas and Arthur McGill of the Department of Religion, Princeton University; Dom Mark Pontifex of Benet House, Cambridge, and of Downside

Preface

Abbey; Svera Porat of the University of Tel Aviv — all of whom read and commented on either the manuscript as a whole or large parts of it; and Edward A. Dowey, George Hendry, Charles West, David Willis, and Daniel Migliore of Princeton Theological Seminary; Norman Pittenger of the General Theological Seminary, New York; Dom Illtyd Trethowan of Downside Abbey; F. G. Healey and John O'Neill of Westminster College, Cambridge; Terence Tice of the Presbyterian World Alliance in Geneva; Robert Adams; Angelos Phillips; and the students in my successive graduate seminars at Princeton on the problem of evil, especially Robert Wennberg, Conrad Hyers, Robert Cassidy, and Myron McClellan.

For advice on certain scientific questions I wish to thank Dr. Elston Ward, a consultant anaesthetist; and W. H. Thorpe, F.R.S., and Richard Bainbridge of the Department of Zoology, Cambridge University. These must not, however, be held responsible for any of the statements that I have made.

But my chief debt remains the wider and deeper one indicated in the dedication.

JOHN HICK

The Divinity School
St. John's Street
Cambridge
Easter 1965

xiv

LIST OF ABBREVIATIONS

A.H.	*Against Heresies*, by Irenaeus.
C.D.	*Church Dogmatics*, by Karl Barth.
C.F.	*The Christian Faith*, by Friedrich Schleiermacher.
C.G.	*The City of God*, by St. Augustine.
Conf.	*Confessions*, by St. Augustine.
En.	*Enneads*, by Plotinus.
Ench.	*Enchiridion*, by St. Augustine.
Essay	*An Essay on the Origin of Evil*, by William King.
F.F.F.	*For Faith and Freedom*, by Leonard Hodgson.
F.W.	*On Free Will*, by St. Augustine.
Inst.	*Institutes of the Christian Religion*, by John Calvin.
J.G.	*The Justification of God*, by P. T. Forsyth.
L.A.I.U.	*Love Almighty and Ills Unlimited*, by Austin Farrer.
M.E.	*The Meaning of Evil*, by Charles Journet.
N.E.	*New Essays in Philosophical Theology*, edited by A. Flew and A. MacIntyre.
N.G.	*The Nature of the Good*, by St. Augustine.
P.T.	*Philosophical Theology*, by F. R. Tennant.
S.c.G.	*Summa contra Gentiles*, by St. Thomas Aquinas.
S.T.	*Summa Theologica*, by St. Thomas Aquinas.

Note : Quotations from the Bible are taken from the American Revised Standard Version, 1952.

PART I

INTRODUCTORY

CHAPTER I

THE PROBLEM AND ITS TERMS

1. DEFINING THE PROBLEM

IF one were to construct a full descriptive title for this book, it might run as follows : A critical study of the two responses to the problem of evil that have been developed within Christian thought, and an attempt to formulate a theodicy for today.

The setting within which the subject is to be treated is, quite explicitly, that of Christian faith. No attempt will be made to establish the truth of the relevant Christian beliefs ; they constitute the starting-point of our inquiry, which is concerned with the challenge to them presented by the fact of evil. The problem dealt with in this book is thus a theological one : Can the presence of evil in the world be reconciled with the existence of a God who is unlimited both in goodness and in power? This is a problem equally for the believer and for the non-believer. In the mind of the latter it stands as a major obstacle of religious commitment, whilst for the former it sets up an acute internal tension to disturb his faith and to lay upon it a perpetual burden of doubt. Contrary to popular belief about the supposedly monolithic certitude of the Ages of Faith, the challenge of evil to religious conviction seems to have been felt in the early Christian centuries and in the medieval period as acutely as it is today. In the fifth century Augustine repeatedly discussed the problem and developed a Christian response that has proved so influential that we may speak of the Augustinian type of solution. And in the thirteenth century Thomas Aquinas listed as the two chief intellectual obstacles to Christian theism, first, that constituted by the reality of evil and, second, the difficulty of establishing the existence of God in view of the apparent explicability of the world without reference to a Creator.[1] The same two points would probably be offered by any thoughtful believer or disbeliever today, with the addition to the second topic of a new aspect, in the question of the *meaning* of the theistic assertion. One might thus be tempted to hail Aquinas's statement as very modern in its spirit, had he not himself already shown it to represent an authentically medieval insight. Indeed, so far as the problem of evil is concerned, reflection shows this to be equally challenging and unavoidable in all historical periods in virtue of the nature of the world and of the essential character of the Christian understanding of God.

It is important to bear in mind what this Christian understanding of God is. For the problem of evil does not attach itself as a threat to any and every concept of deity. It arises only for a religion which insists that the object of its worship is at once perfectly good and unlimitedly powerful. The challenge is thus inescapable for Christianity, which has always steadfastly adhered to the pure monotheism of its Judaic source in attributing both omnipotence and infinite goodness to God. We shall indeed have to take note of various deviations within the broad historical sweep of Christianity that have sought to limit its concept of God in such a way that it no longer provokes the problem of evil in any acute form. But if our terms are not to be uprooted from the ground of customary usage we must insist that these deviations — from the dualism of the Manichees and the Albingenses to the contemporary doctrine of a finite deity — do not constitute the normative or historic Christian faith. In spite of such

[1] The two first Objections to article 3 of part I, question ii, of the *Summa Theologica* are as follows : 'It seems that God does not exist; because if one of two contraries be infinite, the other would be altogether destroyed. But the name *God* means that He is infinite goodness. If, therefore, God existed, there would be no evil discoverable; but there is evil in the world. Therefore God does not exist. Further, it is superfluous to suppose that what can be accounted for by a few principles has been produced by many. But it seems that everything we see in the world can be accounted for by other principles, supposing God did not exist. For all natural things can be reduced to one principle, which is nature ; and all voluntary things can be reduced to one principle, which is human reason, or will. Therefore there is no need to suppose God's existence.' (Trans. by A. C. Pegis in *The Basic Writings of Saint Thomas Aquinas* (New York : Random House, Inc., 1945)).) These are the two primary topics of Christian apologetics; and, having attempted in the present book to write about the one, I hope in a subsequent volume to try to write about the other.

4

The Problem and its Terms

defections on its fringes the main stream of Christianity has
stood by its understanding of God as the most perfect conceivable Being, and has thus insisted upon acknowledging the
problem of evil as (in the Old Testament sense) the *Satan* that
perpetually accuses faith.

Given the traditional Christian belief in God as the unique
infinite, uncreated, eternal, personal Spirit, absolute in goodness and power, the accompanying problem of evil, in its
general form, is readily stated. It has, by an expository
custom going back to the Greeks, been formulated as a
dilemma.[1] If God is perfectly good, He must want to abolish
all evil; if He is unlimitedly powerful, He must be able to
abolish all evil: but evil exists; therefore either God is not
perfectly good or He is not unlimitedly powerful. No argument, it seems, could be simpler or clearer than this. And the
objection to theistic belief that is thus so quickly and easily
stated cannot be equally quickly or easily answered. To this
extent the sceptic has the advantage over the believer. But
it is worth noting that the superior simplicity and clarity of
problems over solutions is not a feature peculiar to theology.
It applies to all major intellectual issues, philosophical as well
as religious. One can raise in a few sentences puzzles about
the validity of sense-perception, about the possibility of
knowledge and the possession of truth, or about the authority
of conscience and the validity of moral judgements that have
exercised philosophers for more than two thousand years
and are still exercising them today. Under the customary
protocol of dialectic the critic is not obliged, in formulating his sceptical challenge, to enter deeply into the complexities of the problem. On the contrary, his question

[1] The dilemma was apparently first formulated by Epicurus (341–270 B.C.),
and is quoted as follows by Lactantius (c. A.D. 260–c. 340): 'God either wishes
to take away evils, and is unable; or He is able, and is unwilling; or He is
neither willing nor able, or He is both willing and able. If He is willing and is
unable, He is feeble, which is not in accordance with the character of God;
if He is able and unwilling, He is envious, which is equally at variance with
God; if He is neither willing nor able, He is both envious and feeble, and
therefore not God; if He is both willing and able, which alone is suitable to
God, from what source then are evils? or why does He not remove them?'
(*On the Anger of God*, chap. 13, trans. by William Fletcher in *The Writings of
the Ante-Nicene Fathers* (Grand Rapids, Michigan: Wm. B. Eerdman), vol. vii,
1951.)

stands out more clearly when these are left undeveloped. But, on the other hand, the issue, once opened up, may require for its further investigation and illumination far-reaching trains of reasoning and a very wide conspectus of thought. Thus fundamental questions are generally much more easily asked than answered; and there is no reason to be surprised or perturbed that this should be so in relation to so immense a problem as that of God and evil.

2. IS THEODICY PERMISSIBLE?

The accepted name for the whole subject comprising the problem of evil and its attempted resolution is theodicy, from the Greek θεός, God, and δίκη, justice. The word is thus a kind of technical shorthand for: the defence of the justice and righteousness of God in face of the fact of evil. The invention of the word (in its French form, théodicée) is commonly and credibly attributed to Leibniz.[1] I shall have frequent occasion to use the term — both to signify theodicy as a subject, and also in the sense of *a* theodicy, i.e. a particular proffered solution to the problem of evil — if only for the stylistic reason that in a lengthy book on a single topic it is helpful to have as many ways as can be found for referring to that topic.

However, as soon as we begin to speak of theodicy and of theodicies, of a systematic investigation of the problem of evil and of attempted solutions, we meet objections and protests.

We are told, for example, that the very notion of a theodicy is impious. It is said to represent a foolish pretension of the human creature, under the illusion that he can judge God's acts by human standards.[2] Instead of seeking to 'justify the

---

[1] *Encyclopedia of Religion and Ethics*, 'Theodicy': vol. xii, p. 289; *Die Religion in Geschichte und Gegenwart* (3rd ed.), 'Theodizee': vol. vi, p. 739; F. Billicsich, *Das Problem des Übels in der Philosophie des Abendlandes*, vol. ii, p. 111. John T. Merz (*Leibniz*, London: William Blackwood & Sons, 1884, p. 101) says of Leibniz that 'in 1697, in a letter to Magliabechi, he uses the word "Théodicée" as the title of an intended work'.

[2] See, for example, Anders Nygren, *Commentary on Romans*, trans. by Carl C. Rasmussen (London: S.C.M. Press Ltd., 1952), p. 365. Henry L. Mansel, in his Bampton Lectures, *The Limits of Religious Thought* (1858), likewise argued that we cannot apply moral categories to God: see his Preface to the 4th ed. (London: John Murray, 1859), p. 13.

ways of God to man' we should rather be trying to justify the
sinful ways of man to God, or, better still, we should like Job
be tremblingly silent before His incomprehensible majesty
and sovereignty.

Now certainly the problem of evil (like any other religious
question) can be approached in an impious spirit or with
irreligious presuppositions; but it can also be approached
with the utmost humility and sincerity of spirit and from a
standpoint of firm Christian commitment. The spirit of the
inquiry will naturally vary from individual to individual.
However, the subject does not demand an impious attitude
in the thinker who investigates it, and it should not on this
ground be singled out for deliberate neglect. Probably all
theological activity is in danger of impiety, but work on the
problem of evil is not in greater danger in this respect than
work on, say, the doctrine of the Person of Christ.

Nevertheless, some might say, the investigation is in itself
religiously improper, whatever may be the personal attitude
of the investigator. It is *ipso facto* irreligious for man to seek
to justify God! In this formulation of the difficulty the word
'justify' seems to cause the trouble. But suppose we use
instead the more neutral term 'understand'. Is it impious to
try to understand God's dealings with mankind? Surely, if
theology is permissible at all, it would be arbitrary to disallow
discussion of the topics that come under the rubric of theodicy: creation, the relation of human suffering to the will of
God, sin and the fall of man, redemption, heaven and hell.
Indeed, the objectors are usually themselves theologians, who
deal *inter alia* with these very topics. Their objection, then,
is not to the consideration of these themes as such but, presumably, to a consideration of them which results in a theodicy. That is to say, they object to the existence of sin and
suffering being thought about in a way that fails to conflict
with belief in the divine goodness and power. But what an
extraordinary restriction to place upon a theological investigation! We are debarred from finding God's ways with us
intelligible to human reason or acceptable to human morality. But why would that be so abhorrent, and what right
has anyone to exclude such an outcome *a priori*? By what

authority must we insist upon maintaining an unrelieved mystery and darkness concerning God's permission of evil? Surely this would be a dogmatism of the least defensible kind. It is, of course, permissible to hold, on the basis of an investigation of the issues, that there is in fact no theodicy, no legitimate way of thinking about the problem of evil that satisfies both mind and conscience; but in view of the fallibility of human reasoning it would be unwise to hold this with absolute confidence, and quite unjustifiable to forbid others from making their own attempts. It may be that what the theodicist is searching for does not exist. But, on the other hand, even if no complete theodicy is attainable, certain approaches to it may be less inadequate than others, and it may thus be possible to reach some modest degree of genuine illumination upon the subject and to discover helpful criteria by which to discriminate among speculations concerning it. If so, efforts in this direction need not be wasted.

The dogma of the impermissibility and undesirability of theodicy is sometimes supported by the following reasoning : Sin, which is basic to all other forms of evil, is essentially irrational and ndeed contra-rational. As such it is absolutely devoid of intel igible grounds or motives. It is an incomprehensible lapse from reason, as from adherence to the good, and cannot be rationalized or therefore theodicized in any way. A proposed understanding of it can only be a misunderstanding of it, treating the essentially irrational as though it were something else.[1]

I shall have occasion to refer to this argument again in another context.[2] But so far as the propriety of our subject is concerned, the argument from the irrationality of sin is of no effect. For even if we grant all that it claims, it is still proper, and indeed necessary, to ask how such deplorable irrationality could occur in a universe created out of nothing by infinite goodness and power. Even if we cannot hope to understand the motive or rationale of evil, we must still ask why God

[1] See, for example, Julius Müller, *The Christian Doctrine of Sin* (*Die Christliche Lehre von der Sünde*) (Breslau, 1839–44. 5th ed., trans. by William Urwick, Edinburgh : T. & T. Clark, 1868), ii. 173 ; Karl Barth, *Church Dogmatics*, iv/i. 410 ; and J. S. Whale, *Christian Doctrine* (Cambridge University Press, 1941), pp. 49–50. [2] See pp. 314–15 below.

permits it; and any answer to this question will be moving in the realm of theodicy.

One also meets as an initial objection the feeling that sin is so heinous and suffering so terrible that any attempt to think calmly and systematically about them must be lacking in either moral seriousness or human compassion. From this point of view, to make any other response to the world's evils than to agonize about them seems almost frivolous. One can sympathize with this feeling, which is provoked by all too many treatments of our subject. But to erect it into general embargo upon the reasoned consideration of sin and suffering would be to abandon the vocation of philosopher or theologian. The problem of evil is an intellectual problem about agonizing realities, and probably no one who has not first agonized in their presence is qualified to think realistically about them in their absence; but nevertheless the agonizing and the thinking are distinct, and no amount of the one can do duty for the other.

The partial validity and the ultimate invalidity of this feeling that the search for a theodicy is improper can be indicated by reference to Gabriel Marcel's distinction between a problem and a mystery, and his application of it to the subject of evil. After defining a mystery as 'a problem which encroaches upon its own data, invading them, as it were, and thereby transcending itself as a simple problem',[1] he says that 'there is no hope of establishing an exact frontier between problem and mystery. For in reflecting upon a mystery we tend inevitably to degrade it to the level of a problem. This is particularly clear in the case of the problem of evil':[2]

In reflecting upon evil, I tend, almost inevitably, to regard it as a disorder which I view from outside and of which I seek to discover the causes or the secret aims. Why is it that the 'mechanism' functions so defectively? Or is the defect merely apparent and due to a real defect of my vision? In this case the defect is in myself, yet it remains objective in relation to my thought, which

[1] Gabriel Marcel, *The Philosophy of Existence*, trans. by Manya Harari (London: The Harvill Press, 1948), p. 8. Cf. *Being and Having*, trans. by Katherine Farrer (London: Dacre Press, 1949), pp. 100–1.
[2] *The Philosophy of Existence*, p. 9.

discovers it and observes it. But evil which is only stated or observed is no longer evil which is suffered : in fact, it ceases to be evil. In reality, I can only grasp it as evil in the measure in which it *touches* me — that is to say, in the measure in which I am *involved*, as one is involved in a lawsuit. Being 'involved' is the fundamental fact ; I cannot leave it out of account except by an unjustifiable fiction, for in doing so, I proceed as though I were God, and a God who is an onlooker at that.[1]

Marcel is here expressing an important aspect of our situation in relation to evil. I would, however, suggest the following discrimination. As has often been observed,[2] in the case of human suffering the intellectual problem of evil usually arises in the mind of the spectator rather than in that of the sufferer. The sufferer's immediate and absorbing task is to face and cope with the evil that is pressing upon him and to maintain his spiritual existence against the threat of final despair. He does not want or need a theoretical theodicy, but practical grace and courage and hope. We can therefore say, in Marcel's terminology, that for him evil is not a problem to be solved, but a mystery to be encountered and lived through. But for the spectator, just because he is not personally undergoing this suffering but can reflect upon the fact that someone is undergoing it, the 'problem of evil' inevitably arises : namely, why should God allow such things to happen? It is true that the intellectual problem, which invites rational reflection, is distinct from the experienced mystery, which must be faced in the actual business of living, and that to a certain extent the one excludes the other from our attention. But it does not at all follow from this that the intellectual problem of evil is a false or an unreal problem, or that our obligation to grapple with it is in any degree lessened.

I shall also mention at this point an argument that does not, indeed, profess to rule out the subject of theodicy, but which has sometimes seemed to relax the tension of the problem. It is said that the problem of evil, although insoluble, is counterbalanced and, so to speak, cancelled out by the

---

[1] Marcel, *The Philosophy of Existence*. Cf. Marcel's *The Mystery of Being*, 1950 (Chicago : Henry Regnery Co., 1960), pp. 260–1.
[2] E.g. by Geddes MacGregor, *An Introduction to Religious Philosophy* (Boston : Houghton Mifflin Co., 1959), pp. 331–2.

mystery of good. In the words of the old Latin tag, 'Si deus est, unde malum? Si non est, unde bonum?'[1] 'The mystery of evil is very difficult when we believe in a good God,' said Harry Emerson Fosdick, 'but the problem of goodness seems to us impossible when we do not.'[2] He elaborates the point as follows:

Once I decided that I could not believe in the goodness of God in the presence of the world's evil, and then I discovered that I had run headlong into another and even more difficult problem: What to do about all the world's goodness on the basis of no God? Sunsets and symphonies, mothers, music, and the laughter of children at play, great books, great art, great science, great personalities, victories of goodness over evil, the long hard-won ascent from the Stone Age up, and all the friendly spirits that are to other souls a 'cup of strength in some great agony'— how can we, thinking of these on the basis of no God, explain them as the casual, accidental by-products of physical forces, going it blind? I think it cannot be done. The mystery of evil is very great upon the basis of a good God but the mystery of goodness is impossible upon the basis of no God.[3]

It does not appear to me that this is a sound piece of reasoning. For the atheist is not obliged to explain the universe at all. He can simply accept it at its face value as an enormously complex natural fact. It constitutes an environment that is for him partly pleasant, partly unpleasant, and partly neutral; but he need find no special intellectual problem either in its pleasantness or in its unpleasantness. There is no necessary question, Si [deus] non est, unde bonum? It is the Christian theist (among others) who claims that the situation is other than it appears, in that there is an invisible divine Being who is perfect in goodness and unlimited in power. And the problem of evil arises at this point as a genuine difficulty that he is bound to face. Si deus est, unde malum?

[1] Perhaps based on Boethius in *De Consolatione Philosophiae*, i. 105–6.
[2] H. E. Fosdick, *Living Under Tensions* (New York: Harper & Co., 1941), pp. 215–16. Others who make the same point include R. A. Tsanoff, *The Nature of Evil* (New York: The Macmillan Co., 1931), p. 5; Geddes MacGregor, op. cit., p. 253; J. S. Whale, *The Christian Answer to the Problem of Evil* (London: S.C.M. Press Ltd., 1939), pp. 20–21.
[3] Fosdick, op. cit., pp. 214–15.

3. 'Good' and 'Evil' as Non-theological Terms

'Evil' is one of a constellation of words that must be defined in relation to each other. Let us begin with the two pairs, 'right' and 'wrong', and 'good' and 'bad', and say that right and wrong are moral terms, describing human volitions and actions, whilst good and bad (although often used also as moral terms) refer in the context of theodicy not to what we do but to the experiences that we undergo; and that evil is used in theodicy as a wider generic term covering both the wrong and the bad — both wrong volitions and bad experiences.

As they occur in the language of non-theological ethics right and wrong have been analysed in many different ways: in terms of an action's motives or of its consequences, in terms of intuited duties and obligations, in terms of moral laws, and in various other ways. However, we are not concerned here to enter into these theories. Suffice to say that there is a non-theological concept, or rather that there are a number of non-theological concepts, of right and wrong, and that the existence of morally wrong action, however analysed, raises the theodicy problem.

The non-theological notion of bad as distinct from wrong — that is to say the notion of bad experiences, with its correlative notion of good experiences — may be approached by noting an aspect of human nature apart from which the word would, for us, have no function and accordingly no meaning. This is the fact that our experiences are not all equally acceptable to us, but that we have likes and dislikes, hopes and fears, desires and aversions. There are experiences that we welcome and others that we would shun. And the basic reference of good, as a notion required by the human mind in its interaction with its environment, is to that which we like, welcome, desire, seek to gain or to preserve, whilst bad refers to that which we dislike, fear, resist, shun and to which we are accordingly averse. As Hobbes said, 'Every man, for his own part, calleth that which pleaseth, and is delightful to himself, good; and that evil which displeaseth him.'[1]

[1] Thomas Hobbes, *Human Nature: or the Fundamental Elements of Policy* (1640), chap. 7, para. 3.

According to a very ancient tradition of human thought, constituting one of the longest-lived of coherent philosophical doctrines, that which we desire above all else, and which accordingly constitutes our highest good, is happiness. By happiness, however, is meant something more than pleasure and the absence of pain. The full meaning of human happiness, as that which satisfies man's deepest desires, must depend upon the character of reality as a whole, and the way to happiness is accordingly a secret bound up with that of the meaning and purpose of human life.

The formal account of the good as that which all men desire, together with the basic observation that what men desire above all else is happiness is, historically, associated with the *philosophia perennis* of Greece and of the classically inspired tradition of Catholic thought. However, it does not depend upon and entail other aspects of the Aristotelian–Catholic synthesis, but is a logically independent insight concerning the relation between the concepts of good, desire, and happiness, which we can test for ourselves and accept as valid. Aristotle long ago formulated the basic argument for this view. In answer to the question, 'What is the supreme good attainable in our actions?' he said :

Well, so far as the name goes there is pretty general agreement. 'It is happiness,' say both intellectuals and the unsophisticated, meaning by 'happiness' [εὐδαιμονία] living well or faring well. But when it comes to saying in what happiness consists, opinions differ and the account given by the generality of mankind is not at all like that given by the philosophers. The masses take it to be something plain and tangible, like pleasure or money or social standing. Some maintain that it is one of these, some that it is another, and the same man will change his opinion about it more than once . . .[1]

For man's supreme good, Aristotle points out, will be that which he seeks for its own sake and not as a means to anything else :

Now happiness more than anything else appears to be just such an end, for we always choose it for its own sake and never for the

[1] *Nicomachean Ethics*, trans. J. A. K. Thomson (London : George Allen & Unwin Ltd., 1953), bk. i, chap. 4.

sake of some other thing. It is different with honour, pleasure, intelligence and good qualities generally. We choose them indeed for their own sake in the sense that we should be glad to have them irrespective of any advantage which might accrue from them. But we also choose them for the sake of our happiness in the belief that they will be instrumental in promoting that. On the other hand nobody chooses happiness as a means of achieving them or anything else whatsoever than just happiness.[1]

Now the question as to what will and what will not make for our happiness is a question about how the world is. Our specific desires and more proximate goals are determined by our understanding of the character of our environment — determined, that is to say, by our judgement as to whether this or that course of action will lead to happiness. In so far as we believe the world to be such that wealth brings happiness, we will seek wealth; in so far as we believe that the good esteem of our peers is a source of happiness, we will seek that; and so with all our many other specific desires. Within the plurality of our appetites, of course, conflicts arise. For example, the desire for esteem sometimes operates as a curb upon, say, the avidity for wealth, by inhibiting us from pursuing wealth in ways which would forfeit esteem. Thus our many desires jostle one another until they fall into a more or less stable pattern that reflects our conception of the true character of this world as the scene of our search for happiness. An individual's or a group's distinctive way of life thus reveals, not an idiosyncratic conception of the human *summum bonum*, but distinctive convictions as to what life-procedures are effective as means to, or are elements in, the supreme end of happiness.

Often a man achieves the specific goals which he has set for himself, only to find that they do not provide the happiness that he had anticipated. This experience is impressive evidence for the contention that the basic aim which our nature has set for us is happiness and that our numerous specific goals are chosen because, too often mistakenly, we assume the world to be so constituted that these things will give us the happiness we seek.

[1] *Nicomachean Ethics*, bk. i, chap. 7.

14

Aristotle suggests that the happiness of any kind of creature consists in its fulfilment of its own *telos*, or the realization of its given nature and potentialities.[1] Everything, according to Aristotle, is constituted for some end, to achieve which is to fulfil itself. For example, the *telos* of a chrysanthemum seed is the full-grown flower ; and its happiness, if plant life were endowed with self-consciousness, would consist in its development into a perfectly formed chrysanthemum. Happiness is thus relative to structure, being the fulfilment of a thing's nature, whatever that nature and its fulfilment may be. And the happiness of a human being will accordingly consist in his fulfilment of the potentialities of specifically human nature.

Applying this viewpoint to 'bad' as the opposite of 'good', we find that its basic meaning is that which we dislike, do not welcome, and would shun. That which all men would shun is the opposite of happiness, namely the state of misery, reflecting the non-fulfilment and radical frustration of our nature. But once again, to know that man's supreme good is happiness, which is connected with the fulfilment of his nature, and that his supreme evil is misery, which is connected with the frustration of his nature, is not thereby to know how to attain good or to avoid misery. For this, a knowledge of the structure of reality, or of the character of our total environment, is required. It is at this point that the claims of religion concerning the true nature of the universe and the real purpose of life connect with the needs of our human nature in its universal search for happiness.

4. 'Good' and 'Evil' as Theological Terms

We have seen that from our human point of view, unaided by religious faith, the good is that which we welcome and the bad that which we would shun. The analogous theological definition will be in terms of the divine purpose for the created world. Whatever tends to promote the attainment of that purpose will be good and whatever tends to thwart it will be bad. The full and irreversible fulfilment of that plan

---

[1] Ibid., bk. i.

would be the complete good sought by God in His activity in relation to His creation, whilst any final and irrevocable frustration of that plan would constitute irredeemable and ultimate badness.

The Christian concept of God's purpose for man enables the two kinds of evil — sin and suffering — to be bracketed together under their common contrariety to the divine purpose. For according to Christianity the end for which human beings exist and which defines the *telos* of man's nature consists in a relationship to God. Men are created for fellowship with their Maker ; the destiny open to them is that described in alternative eschatological symbols as the vision of God and as the life of God's Kingdom. This is an eternal and limitless good, constituting the ultimate completion and happiness of our nature. For God has so made us for Himself that our highest possible fulfilment lies within a relationship to Himself in which the creature loves the Creator 'with all his heart, and with all his soul, and with all his mind' and is thereby set free to value his fellow creatures as himself. And conversely the deepest misery possible to our human nature would be to forfeit that happiness and to plunge instead into irrevocable and ever-increasing alienation from the source of our being and from the fulfilment of our own nature in community with our fellows.

Sin, which is the theological name for moral evil, will be discussed further in Chapter XIV ; but, according to the definitions there arrived at, sin, in the singular, consists in man's imperfect relationship to God whilst sins, in the plural, are men's wrong volitions and actions, occurring against God's will (or, more strictly, against His wishes) and arising within that distorted relationship. That sin tends to thwart the divine purpose, and is accordingly evil, is self-evident ; for sin is by definition the negation of that which the Creator is seeking in us and for us. And that human misery is likewise a negation of the perfect good that God is seeking for men is no less self-evident ; for that perfect good is the fulfilment and hence the happiness of our human nature. There is thus a sense in which Christianity is hedonistic in its conception of the good, declaring that God has made man for ultimate happiness.

However, it goes beyond both Aristotle and Hedonism, not
only (as we have already noted) in seeing man's highest good
in his relationship to God, but also in giving a cosmic backing
to man's quest for happiness. Happiness consists in the fulfilment of a conscious being's nature — provided, we must add,
that such fulfilment turns out to be in harmony with the
determining realities of its total environment. If the structure of human nature were fundamentally in conflict with the
wider structure of the universe in which our life is set, the
fulfilment of the human *telos* would not constitute man's ultimate well-being and happiness but, on the contrary, his
ultimate ill-being and frustration. If the whole nature of
things were either opposed or indifferent to those qualities
that constitute man's nature in its perfection, we should have
to conclude that full human happiness is an impossibility.
If, for example, the development and activity of love are
aspects of the perfection of our nature, and if it should prove
that the character of the universe is radically inhospitable to
love, then to achieve man's *telos* could not be to achieve a
secure and lasting happiness. But for Christianity, which
claims that the universe has been created and is ruled by
divine goodness and power, the eventual attainment of man's
highest good is guaranteed by God's sovereignty : He has
made His human creatures for fellowship with Himself and
will eventually bring them to this high end.

We must note finally that the eschatological character of
man's final good affects the present significance of the experiences which we undergo. We welcome these (i.e. regard
them as good) or shun them (i.e. regard them as bad) according as they cause us pleasure and happiness or pain and
misery. From the point of view, however, of the ultimate
good of our creaturely self-fulfilment in the divine Kingdom
they are to be welcomed or shunned according as they move us
nearer to or further from that ultimate and all-inclusive good.
In this eschatological perspective, as we shall see later, even
experiences that are unwelcome to us and that we would
shun may be good if they are so used, in our response to them,
as to bring us nearer to God's Kingdom. Conversely, even
experiences that we desire and welcome may in the same

ultimate perspective be bad if our response to them is such that we are thereby moved further from the divine Kingdom.

5. THE KINDS OF EVIL

The working vocabulary of theodicy, compared with that of some other branches of theology, is in a state of imprecision. In English 'evil' is usually, although not always, used in a comprehensive sense, and we then distinguish under it the moral evil of wickedness and such non-moral evils as disease and natural disaster. In German *Übel* is a general term, covering both moral and non-moral evil, though it can also be used specifically for the latter; whilst *Böse* refers more definitely to moral evil.[1] In French *le mal* can be used to refer to all types of evil.

What, then, are the various kinds of evil that have been identified in the literature of theodicy? There is, first, the important distinction just mentioned between moral and natural evil. Moral evil is evil that we human beings originate: cruel, unjust, vicious, and perverse thoughts and deeds. Natural evil is the evil that originates independently of human actions: in disease bacilli, earthquakes, storms, droughts, tornadoes, etc. In connection with these latter, it is a basic question whether events in nature which do not directly touch mankind, such as the carnage of animal life, in which one species preys upon another, or the death and decay of plants, or the extinction of a star, are to be accounted as evils. Should evil be defined exclusively in terms of human actions and experiences, with the result that events in the natural universe and in the sub-human world do not as such raise questions for theodicy? Or should the scope of the problem be extended to include the whole realm of sentient life, or perhaps only vertebrates, or perhaps only the higher mammals? St. Augustine, the greatest theodicist of all, felt it necessary to try to demonstrate the goodness of providence in respect of decay and death in the natural world, including in his purview the decay of plants as well as the death of

---

[1] On the distinction between *Böse* and *Übel* see Paul Häberlin, *Das Böse: Ursprung und Bedeutung* (Berne: Francke Verlag, 1960), pp. 5–6.

sentient animals.[1] A more moderate position, however, which would probably commend itself to most people, is that the organic cycle in non-sentient nature offers no problems to theodicy, but wherever there is pain, as there appears to be far down through the animal kingdom, there is a *prima facie* challenge to be met. On this view natural evil consists in unwelcome experiences brought upon sentient creatures, human or sub-human, by causes other than man himself.

It will be noticed that I have not thus far mentioned Satan and the satanic kingdom, in spite of the fact that they play such important roles in many of the main systems of theodicy. By this omission I do not intend to deny the existence of energies and structures of evil transcending individual human minds, but to indicate that any such forces are part of the general problem of evil, logically co-ordinate with human wickedness, and do not constitute a unique kind of evil that might provide a key to the solution of the problem as a whole. The puzzles attending human imperfection, free will, and sin are reiterated, but not further illumined, by transferring them to a superhuman plane. Indeed, the only effect of such a transference is to throw the discussion into metaphysical regions in relation to which the already sufficient difficulties of knowing whether we are talking sense or nonsense are compounded to a point that is, literally, beyond all reason.

Yet another kind of evil is discussed in the works of theodicy under the name of 'metaphysical evil'.[2] This phrase refers to the basic fact of finitude and limitation within the created universe. The Augustinian tradition of theodicy, on its more philosophical side, traces all other evils, moral and natural, back to this as their ultimate cause, or at least (in the case of sin) as their ultimate occasion. It is often denied by Augus-

---

[1] *The City of God*, trans. by Marcus Dods, George Wilson, and J. J. Smith (New York: Random House, Inc., 1950), bk. xii, chaps. 4–5; *The Nature of the Good*, chap. 8, trans. by John H. S. Burleigh in *Augustine: Earlier Writings* (London: S.C.M. Press Ltd., and Philadelphia: The Westminster Press, 1953).

[2] Leibniz seems to have been the first to use this term, although that which it signifies has been discussed almost from the beginning of the investigation of the theodicy problem.

tinian theodicists, however, that the unavoidable imperfection of created things is to be regarded as an evil; for finitude implying limitation, implying imperfection, is inevitable if there is to be a creaturely realm at all.[1] However, these are all matters that will be discussed at length in the course of the following chapters.

[1] See, for example, A. D. Sertillanges, O.P., *Le Problème du Mal*, vol. ii, 'La Solution' (Paris: Aubier, 1951), pp. 7 f.

CHAPTER II

THE TWO POLES OF THOUGHT —
MONISM AND DUALISM

1. MONISM AND DUALISM

CHRISTIAN thought concerning theodicy has always moved between the opposite poles set by the inherent logic of the problem — monism and dualism. These represent the only two wholly consistent solutions that are possible; and unfortunately neither of them is compatible with the basic claims of Christian theology. Monism, the philosophical view that the universe forms an ultimate harmonious unity, suggests the theodicy that evil is only apparent and would be recognized as good if we could but see it in its full cosmic context: 'All partial evil, universal good'.[1] Dualism as a theodicy, on the other hand, rejects this final harmony, insisting that good and evil are utterly and irreconcilably opposed to one another and that their duality can be overcome only by one destroying the other. Each of these polar positions has exerted a powerful pull upon Christian thought.

On the one hand, it appears that the absolute monotheism of the Christian (as of the Judaic) faith entails an ultimate monism. If God is God, and if God is good, there cannot be any co-equal contrary reality; and therefore evil must in the end be subject to God's sovereignty and must exist by a permission flowing from His purpose for His creation. There seems here to be an undeniable truth, to neglect which would be to forfeit the fundamental Christian belief in the reality of God as the sole Creator and ultimate ruler of all things. But this truth carries within it dangers for theodicy. Under its spell Christian thought may so strongly emphasize the divine sovereignty that evil is no longer recognized as being

[1] Alexander Pope, *Essay on Man*, i. 292.

21

genuinely *evil* and as utterly inimical to God's will and purpose. Evil can thus become domesticated within the divine household and seen as a servant instead of a deadly enemy; and then the theodicist finds himself calling evil good and preaching peace where there is no peace.

But, having reached this point, he is likely — or his readers are likely — to feel afresh the pull of the other polar truth, and to remember that the life of faith has always been an active warfare against evil. Through His prophets God uncompromisingly attacked greed, cruelty, and injustice within Israel; and in Christ He not only condemned these sins but also, by many acts of healing, relieved men's bodily diseases, thereby treating natural as well as moral evil as hostile to His purpose. Evil, then, is God's enemy. Here again we seem to be faced with undeniable truth. And yet here again there are dangers. We can so emphasize the wickedness of sin and the dread reality of suffering that we find ourselves according to evil an independent status over against the Creator Himself. Under the pressure of the problem of evil the theodicist may thus find that he has abandoned a fundamental religious insight of the prophets which stands as an essential foundation of the Christian faith : the universal Lordship of the sole and sovereign God.

Here, then, are two contrary truths, pulling in opposite directions. How are we to reconcile full Christian monotheism with a realistic view of both sin and suffering? This dilemma has haunted all attempts to arrive at a Christian understanding of evil. In subsequent chapters we shall see these conflicting motives at work and shall note their respective strengths and their attendant dangers as these affect the theodicies to which they have given rise. We shall see how different Christian thinkers have attempted to meet the problems provoked by their own insights. But in the end we have to acknowledge that both polar truths are valid and inescapable. Accordingly, in the constructive endeavour of Part IV an attempt will be made to present the Christian response to the problem of evil in eschatological terms, affirming a present interim dualism within the ultimate setting of an unqualifiedly monotheistic faith.

It will be illuminating, however, to observe first the monist and dualist insights in their pure and uncompromising forms. We shall see demonstrated there both the possibility of solving the problem of evil by a one-sided approach to it and also the high cost of any such solution through its neglect of other equally mandatory aspects of truth. The purest example of monism in Western thought is to be found in the philosophy of Spinoza; and as examples of the opposite and dualistic type of theodicy we shall take the rather different theories of John Stuart Mill and of Edgar S. Brightman.

2. THE PURE MONISM OF SPINOZA

In the philosophy of Spinoza the monistic vision of the universe, as it was felt by a deeply religious spirit and articulated by a powerfully logical mind, finds expression in a formal metaphysical system. Spinoza saw reality as forming an infinite and perfect whole — perfect in the sense that everything within it follows by logical necessity from the eternal divine nature — and saw each finite thing as making its own proper contribution to this infinite perfection. Thus every existing thing occupies a place within the system of universal perfection, and our human notion of evil as that which ought not to be is merely an illusion of our finite perspective.

Presenting this vision *more geometrico*, Spinoza deduces his monistic doctrine from the idea of substance, defined as 'that which is in itself, and is conceived through itself; in other words, that of which a conception can be formed independently of any other conception'.[1] It follows from this definition that there can be only one substance,[2] which necessarily exists,[3] which is infinite in an infinite number of ways, and of which all distinguishable things are modes or attributes.[4] More precisely, anything that can be named is either an infinite attribute of God — the two such attributes known to us being thought and extension — or a mode of one of

---

[1] Spinoza's *Ethics*, trans. by R. H. M. Elwes, 'Bohn's Philosophical Library' (London: George Bell & Sons, 1891), pt. i, def. 3.
[2] Ibid. i, props. 2–8 and 14.   [3] Ibid. i. 11.   [4] Ibid. i. 15.

these infinite attributes.[1] We call the one infinite all-inclusive reality God or nature (*deus sive natura*), which we may divide in thought into *natura naturans* ('nature naturing'), which is God in His creative activity,[2] and *natura naturata* ('nature natured'), which is the infinite world of being flowing continually from God.[3] For the infinite self-generating universe can be considered in these two alternative aspects, as creating (*naturans*) and as created (*naturata*).

It is an essential aspect of Spinoza's system, and one upon which he repeatedly insists,[4] that the created world, with its entire temporal history, follows throughout by strict logical necessity from the eternal divine essence :

from God's supreme power, or infinite nature, an infinite number of things — that is, all things have necessarily flowed forth in an infinite number of ways, or always follow from the same necessity ; in the same way as from the nature of a triangle it follows from eternity and for eternity, that its three interior angles are equal to two right angles.[5]

Nothing, then, is contingent, but all things are determined from the necessity of the divine nature to exist and to act in a certain manner.[6] Only God Himself is free, in the special sense that He is not determined by anything outside Himself.[7] Even the volitions of the human will are necessary elements within the all-embracing system of nature.[8] And since everything is thus determined by a perfect Determiner, everything must be perfect : 'It clearly follows from what we have said, that things have been brought into being by God in the highest perfection, inasmuch as they have necessarily followed from a most perfect nature.[9] Accordingly this is not merely, as Spinoza's contemporary Leibniz taught, the best practi-

---

[1] *Ethics*, ii. 1 and 2.

[2] By *natura naturans* we are to understand 'that which is in itself, and is conceived through itself, or those attributes of substance, which express eternal and infinite essence'. *Ethics*, i. 29, n.

[3] By *natura naturata* is to be understood 'all that which follows from the necessity of the nature of God, or of any of the attributes of God'. *Ethics*, i. 29, n.   [4] E.g. *Ethics*, i. 33 ; *Epistle* 23.

[5] Ibid. i. 17, n. Cf. ibid. i. 16.   [6] Ibid. i. 29.

[7] Ibid. i. 17. However, it would be incorrect to speak of God as having 'free will' — ibid. i. 32, corolls. 1 and 2.

[8] Ibid. i. 32 ; ii. 48.   [9] Ibid. i. 33, n. 2.

cable world, but it is in every respect perfect. For it is a necessary expression of the eternal and infinite perfection of God or Nature.

Against the background of this doctrine of universal determination by an infinite and perfect Determiner, Spinoza confronts the problem of evil. He asks why there should be what we regard as imperfect things and wicked men, and offers a twofold answer.

In the first place, good and evil are not objective realities (*entia realia*), but mental entities (*entia rationis*), formed by comparing things either in respect of their conformity to a general idea or merely in respect of their utility to ourselves. But each of these two types of comparison is fallacious. Behind the former lies the false assumption that there are general norms, like unchanging Platonic Ideas, to which individual things and animals and people are supposed to conform, and to diverge from which is *ipso facto* to be imperfect. Thus a shrivelled tree, a lame zebra, a diseased lion, or a sinful man are commonly regarded as defective specimens of their several species. But when we judge individuals by their congruence or noncongruence with generic norms we fall into a fundamental error. For these norms exist only in our own minds. God did not create eternal ideas of Tree, Zebra, Lion, and Man, but only the many particular trees, zebras, lions, and men, each of which is its own distinct and unique self. These individuals cannot be evaluated by reference to any general ideas to which they are supposed to conform; they exist in their own right as different expressions of the infinite divine fecundity.

And behind the other type of comparison, in terms of the usefulness of things to ourselves, there lies the deeply rooted delusion of the human mind that everything in nature obeys a purpose and works towards some end. But in reality, 'nature does not work with an end in view. For the eternal and infinite being, which we call God or nature, acts by the same necessity as that whereby it exists. . . . Therefore, as he does not exist for the sake of an end, so neither does he act for the sake of an end.'[1] It is this false assumption of the

[1] Ibid. iv, preface.

purposefulness of nature that has given rise to our evaluative terms :

> After men persuaded themselves, that everything which is created is created for their sake, they were bound to consider as the chief quality in everything that which is most useful to themselves, and to account those things the best of all which have the most beneficial effect on mankind. Further they were bound to form abstract notions for the explanation of the nature of things, such as *goodness, badness, order, confusion, warmth, cold, beauty, deformity*, and so on ; and from the belief that they are free agents arose the further notions of *praise* and *blame, sin* and *merit*. . . . We have now perceived, that all the explanations commonly given of nature are mere modes of imagining, and do not indicate the true nature of anything, but only the constitution of the imagination ; and, although they have names, as though they were entities, existing externally to the imagination, I call them entities imaginary rather than real.[1]

On this ground Spinoza denies any objective meaning to the statement that there is evil in the world. Everything in nature is, not indeed as it *ought* to be — for 'ought' presupposes a cosmic purpose or norm — but as it *must* be as a necessary part of the universal being that is God in his aspect of *natura naturata*. Everything is in order, because everything follows necessarily from the eternal divine essence ; or, more strictly, we must hold 'that all things are necessarily what they are, and that in Nature there is no good and no evil'.[2]

Spinoza supplements and completes this doctrine by an appeal to two ideas that we shall meet again in later chapters as elements of the more philosophical strand of the Augustinian tradition of theodicy, in which tradition Spinoza at these points participates. He regards what we call evil as negative rather than positive : 'Evil is in reality a lesser good.'[3] Sin, for example, is a state of self-imposed privation of virtue ; the sinful act is good in so far as it contains a certain degree of reality, but evil in so far as it lacks a greater degree ;[4] and error is likewise a privation, namely the lack of a more complete truth.[5] The question then is how these lacks can occur

---

[1] *Ethics*, i, appendix.
[2] *Short Treatise on God, Man and His Well-Being*, trans. A. Wolf (London : A. & C. Black, 1910), pt. ii, chap. iv, p. 75.
[3] *Ethics*, iv. 65.   [4] *Letter* 32.   [5] *Ethics*, ii. 35.

within the fabric of infinitely perfect reality. Spinoza's answer is derived from what Arthur Lovejoy has called 'the principle of plenitude', namely the view that a universe that contains as many different kinds of beings as possible, lower as well as higher, is a more perfect universe than would be one containing only the highest kind of being.[1] Spinoza asks why God did not make all men perfectly rational and good, so that they would never have sinned. And he answers, 'because matter was not lacking to him for the creation of every degree of perfection from the highest to the lowest; or, more strictly, because the laws of his nature are so vast, as to suffice for the production of everything conceivable by an infinite intelligence'.[2] Again in Letter 32, which summarizes much of his thought concerning our problem, Spinoza considers Adam's sin in the Garden of Eden. 'To ask of God, however, why he did not give Adam a more perfect will,' he says, 'were as absurd as to inquire why a circle had not been endowed with the properties of a sphere.'[3] For all the different possible kinds of being must exist to express the infinite creativity of God; and therefore the sinner exists as well as the saint.

Having thus shown in his metaphysic that there is no good or evil in the universe, but that everything that exists and everything that happens exists and happens alike by absolute necessity within the infinite Being which we call God or Nature, Spinoza nevertheless, in his doctrine of man, considers more closely the sins and sorrows that we commonly regard as evils. He has a lengthy discussion of the passions, including sorrow, hatred, envy, malice, fear, despair, remorse, contempt, pride, shame, regret, cruelty, cowardice, ambition, avarice, and lust, and treats these as real conditions of which one is obliged to take note, both in life and in thought, and indeed as conditions in the absence of which our state would be more perfect than it is. However, he insists that their reality is only a form of unreality, dependent upon a lack of

[1] See below, pp. 76 f. On Spinoza's use of the principle of plenitude see Arthur Lovejoy, *The Great Chain of Being* (Harvard University Press, 1936), chap. 5. [2] *Ethics*, i, appendix.

[3] Trans. R. Willis in *Benedict de Spinoza: His Life, Correspondence, and Ethics* (London: Trubner & Co., 1870), p. 299.

true knowledge, so that 'if we use our understanding and Reason aright, it should be impossible for us ever to fall a prey to one of these passions which we ought to reject'.[1] For 'the knowledge of evil is an inadequate knowledge', from which it follows as a corollary 'that, if the human mind possessed only adequate ideas, it would form no conception of evil'.[2]

However, men do in fact suffer from inadequate ideas, and consequently they do in fact have mistaken desires, which in turn lead to frustration, misery, and wrong-doing. And so the question returns: Why do we desire mistakenly? Does this represent a free and blameworthy act on our part, or are even our desires determined within the universal system of nature? In order to be consistent in his monism Spinoza must, and does, give the latter answer. Human desiring is no more free than human willing.[3] Our desires flow necessarily from our ideas of the things around us; and Spinoza asserts explicitly that the confused and inadequate ideas whereby we mistake lesser goods for the supreme good follow by the same necessity as our more adequate, or more clear and distinct, ideas.[4] For the relative clarity and adequacy or confusion and inadequacy of our ideas are necessary functions of our place within the whole. A universal perfection that includes (under the principle of plenitude) the more limited modes of existence cannot fail to contain their correspondingly limited awareness of the infinite whole within which they exist. Here again the unbroken determinism of nature is maintained. 'Thus we see now,' says Spinoza, 'that man, being *a part of the whole of Nature*, on which he depends, and by which also he is governed, cannot of himself do anything for his happiness and well-being.'[5] As a variation on the same theme Spinoza denies that men are free in the crucial matter of their loving or failing to love God: he rejects the notion that God might love those who love Him and hate those who hate Him, on the ground that in that case 'we should have to suppose that people do so of their own free

---

[1] *Short Treatise*, pt. II, chap. xiv, p. 99.
[2] *Ethics*, iv. 64.   [3] *Short Treatise*, II. xvii.
[4] *Ethics*, ii. 36.   [5] *Short Treatise*, II. xviii, p. 115.

will, and that they do not depend on a first cause ; which we have before proved to be false'.[1]

From the point of view of theodicy (which is the only point of view from which we are here concerned with Spinoza's system) the weakness of this way of thinking is not far to seek. In showing that the evils that we human beings experience are the illusory products of confused and inadequate ideas Spinoza has not made those evils any less dreadful and oppressive. For they are illusions only in a highly sophisticated sense. They are not unreal in the way in which a mirage, a dream, a hallucination, or an after-image is unreal. Pain, cruelty, and grief are still actual experiences, and they still hurt. If they are from an ultimate standpoint illusory, they are still very terrible illusions, liability to which must count as a great evil. And they are not less evil, but on the contrary more so, on account of their logical inevitability within an all-comprehensive system of absolute universal necessity.

Thus one might accept Spinoza's metaphysic as a true picture of the universe, and yet still find that God or Nature is, to us, largely evil — largely such that it would be better if it could have been otherwise. For there is still wickedness and injustice, disease and pain, grief and despair ; and Spinoza has said nothing to reconcile us to them except that they occur by an absolute logical necessity. In order to constitute a theodicy, a philosophical view of the evils that we experience must either show them to be justly deserved or to be means to a good end ; but Spinoza merely shows them to be parts of a totally determined universe whose character is uniformly necessary throughout.

3. A CONTEMPORARY VIEW OF EVIL AS ILLUSION — CHRISTIAN SCIENCE

The philosophy of Spinoza has never been adopted as a basis for living by any numerically significant community. As an appendix, then, to this study of the intellectually pure and

---

[1] Ibid. II. xxiv, p. 138.

lucid monism of Spinoza we should refer briefly to the contemporary institutionally embodied monism of Christian Science. The point at which Christian Science, on its own very different intellectual level, partially parallels the thought of Spinoza is in the doctrine that evil is an illusion, stemming in Spinoza's philosophy from inadequate ideas, and in Christian Science from 'mortal mind'.

What the Reformed Churches would call the subordinate standard of Christian Science, the document in which its understanding of the New Testament revelation is officially set forth, is Mrs. Mary Baker Eddy's *Science and Health with Key to the Scriptures* (1875). In this book, with its confused medley of half-digested philosophical themes, Mrs. Eddy taught 'the nothingness and unreality of evil',[1] and declared that 'Evil has no reality. It is neither person, place, nor thing, but is simply a belief, an illusion of material sense', [2] and that 'evil is but an illusion, and it has no real basis. Evil is a false belief.' Accordingly, 'If sin, sickness, and death were understood as nothingness, they would disappear.'[3] Hence the Christian Scientist proposes to overcome 'evil, disease, and death . . . by understanding their nothingness and the allness of God, or good'.[4] For example, 'When a sufferer is convinced that there is no reality in his belief in pain — because matter has no sensation, hence pain in matter is a false belief — how can he suffer longer?'[5]

I am not concerned here to expound or criticize the Christian Science scheme of thought as a whole — and indeed it is, from a philosophical point of view, so vague and undeveloped as to be incapable of clear exposition or therefore of philosophical criticism — but only to try to estimate the value for theodicy of its teaching concerning the unreality of all forms of evil. The fatal weakness of this doctrine, considered as a contribution to theodicy, is that the problem of evil is changed in form but unaffected in substance. The 'evils' of sin and pain, sorrow and death are not real; but nevertheless the illusion that they are real is real! As Mrs. Eddy wrote, 'the only reality of sin, sickness or death is the

---

[1] Authorized Edition, 1934, p. 205.   [2] Ibid., p. 71.
[3] Ibid., p. 480.   [4] Ibid., p. 450.   [5] Ibid., p. 346.

awful fact that unrealities seem real to human, erring belief, until God strips off their disguise'.[1] But in the meantime the illusions of sin, sickness and death really exist, and really require to be cured ; and thus all the evils that Christian Science has, by an act of definition, banished from the world continue to exist as modes of this evil state of illusion.[2]

There are, of course, philosophically interesting forms of the idea that evil is negative and privative, a kind of non-being or nothingness, and we shall meet these in our study of the Augustinian type of theodicy. But their proponents are always careful not to draw from them the absurd conclusion, drawn by Christian Science, that sin and pain and suffering are illusions.

4. PLATO'S DUALISM

From monism we turn to its polar opposite. The most extreme form of dualism and that which would, if it could be accepted, most fully solve the problem of evil, postulates two deities, one good and the other malevolent. Such a dualism was embodied in the ancient and now defunct Zoroastrian religion. Zoroaster (who lived in Persia about 1000 B.C.) proclaimed two rival gods, Ahura Mazdah (or Ormuzd), the source of good, the Angra Mainyu (or Ahriman), the source of evil, and summoned men to fight on the side of the good power against the evil. In the early Christian centuries a similarly dramatic dualism was taught by Mani (born about A.D. 215) and became the basis of the Manichaean religion, which so strongly attracted St. Augustine prior to his conversion to Christianity.[3] After nearly another millennium the Albigenses in the south of France in the twelfth and early thirteenth centuries revived a Manichaean-like dualism until they were persecuted out of existence by the Catholic Church.

But more recent Western thinkers who have solved the problem of evil by means of some form of dualism have preferred to the idea of two deities that of a single, benevolent,

---

[1] Ibid., p. 472.
[2] The general point that an illusion that evil exists would itself be an evil is elaborated by J. M. E. McTaggart in *Some Dogmas of Religion* (London : Edward Arnold, 1906), para. 171.
[3] On Manichaeanism see below, pp. 44–45.

divine Being, who is finite in power and accordingly unable to prevent evils from marring the universe over which he exercises his limited sovereignty. This notion has been applied in theodicy in two different ways. In its more straightforward and intelligible form a good but limited deity stands over against an independent realm of chaotic and intractable matter which he is only partially able to control to his own ends. This type of dualism traces its ancestry back to Plato (428/427–348/347 B.C.), the originator of so many of the elements of Western thought. Reacting against the sub-ethical Homeric conception of the gods, Plato repudiated the current assumption that the supreme power (Zeus, in the popular pantheon) is the source of evil as well as of good. 'He is responsible for a few things that happen to men, but for many he is not, for the good things we enjoy are much fewer than the evil. The former we must attribute to none else but God; but for the evil we must find some other causes, not God.'[1]

What Plato conceives these 'other causes' to be becomes clear, or at least as clear as Plato was to make it, in his late dialogue, *Timaeus*.[2] Here he describes our spatio-temporal world as having been formed by a divine power, the Demiurge, who made use of an existing chaotic material which he ordered within a likewise existing framework, the Receptacle (ὑποδοχή). It may well be that the Demiurge has a mythical status, and that Plato is not speaking of an actual beginning of the world but is rather indicating its intrinsic character as a compromise between Reason (νοῦς) and Necessity (ἀνάγκη) :[3] for, he says, 'the generation of this universe was a mixed result of the combination of Necessity and Reason'.[4] Reason,

[1] *Republic*, 379 C, trans. A. D. Lindsay. According to William Chase Greene, this passage is 'the first distinct statement in Greek literature of the problem of evil'. *Moira: Fate, Good and Evil in Greek Thought* (New York: Harper Torchbooks, 1944), p. 298. Aristotle refers to Plato's dualism in *Metaphysics* A, 988a.

[2] I leave aside here the myth in the *Statesman* (273–4) as Plato's more popular but philosophically less serious approach to his dualism; and likewise the almost Zoroastrian outcome in *Laws*, 896–7.

[3] Cf. Francis M. Cornford, *Plato's Cosmology: The Timaeus of Plato translated with a running commentary*, 1937 (New York: The Liberal Arts Library, 1957), p. 176.

[4] *Timaeus*, trans. Cornford, 48 A.

which always seeks the best, is represented by the Demiurge, who is not the unlimited Creator of Judaic-Christian faith but a divine artificer working upon a given material. The Demiurge desired 'that all things should be good and, so far as might be, nothing imperfect'.[1] Accordingly Reason 'overruled Necessity by persuading her to guide the greatest part of the things that become towards what is best; in that way and on that principle this universe was fashioned in the beginning by the victory of reasonable persuasion over Necessity'.[2] It should be noted that Necessity does not, in the *Timaeus*, signify rigid determination or an unbreakable chain of cause and effect, but something more like chaos and randomness. It is an 'errant cause' ($\pi\lambda\alpha\nu\omega\mu\acute{e}\nu\eta$ $\alpha\grave{i}\tau\acute{i}\alpha$) and represents, in Grote's words, 'the indeterminate, the inconstant, the anomalous, that which can neither be understood nor predicted'.[3] In other and earlier dialogues[4] this Necessity, which is the source of evil, is virtually identical with the matter that imprisons us as embodied beings, clogging and weighing down the soul and impeding it in its search for goodness and truth, so that the philosopher must aim so far as possible at a detachment from the body and its distractions.[5]

5. THE EXTERNAL DUALISM OF J. S. MILL

In the nineteenth century John Stuart Mill (1806–73) stated a form of external dualism with eloquence and force. He first excluded the traditional Christian belief in a perfectly good and infinitely powerful Creator:

It is not too much to say that every indication of Design in the Kosmos is so much evidence against the Omnipotence of the Designer. For what is meant by Design? Contrivance: the adaptation of means to an end. But the necessity for contrivance — the need of employing means — is a consequence of the limitation of power. Who would have recourse to means if to attain his end his mere word was sufficient? The very idea of means implies

[1] Ibid., trans. Cornford, 30 A.
[2] Ibid., trans. Cornford, 48 A.
[3] George Grote, *Plato* (London: John Murray, 1865), vol. iii, p. 249.
[4] Especially *Phaedo*, *Symposium*, and *Phaedrus*.
[5] On Plato's overlapping concepts of body, matter, and necessity, see William Chase Greene, *Moira*, pp. 301 f.

that the means have an efficacy which the direct action of the being who employs them has not. . . . But if the employment of contrivance is itself a sign of limited power, how much more so is the careful and skilful choice of contrivances? . . . Wisdom and contrivance are shown in overcoming difficulties, and there is no room for them in a Being for whom no difficulties exist. The evidences, therefore, of Natural Theology distinctly imply that the author of the Kosmos worked under limitations; that he was obliged to adapt himself to conditions independent of his will, and to attain his ends by such arrangements as those conditions admitted of.[1]

Mill then pointed out that nature's evils seem to arise from malfunctionings in a system that is basically designed for the preservation and enhancement of life, rather than from any deliberate cosmic aim to produce pain and suffering; and he concluded that

there is no ground in Natural Theology for attributing intelligence or personality to the obstacles which partially thwart what seem the purposes of the Creator. The limitation of his power more probably results either from the qualities of the material — the substance and forces of which the universe is composed not admitting of any arrangements by which his purposes could be more completely fulfilled; or else, the purposes might have been more fully attained, but the Creator did not know how to do it; creative skill, wonderful as it is, was not sufficiently perfect to accomplish his purposes more thoroughly.[2]

Mill does not say how this view is to be applied to moral evil. He is not willing to postulate a devil, as a personal malevolent power who could be regarded as the instigator of human sin; nor does he suggest any other explanation of its origin. Presumably he would have to hold that matter and energy, together with the laws of their operation, as the circumstances that God has not created and with which he has to contend, somehow necessitate man's moral frailty and failure. He would presumably argue that such a psycho-physical creature as man, organic to his material environment and subjected by it to a multitude of strains and stresses, must inevitably become self-centred, and that from this circumstance have developed the moral ills of human life. Nor

---

[1] J. S. Mill, *Three Essays on Religion* (London: Longmans, Green, Reader & Dyer, 1875), 'Theism: Attributes', pp. 176–7.   [2] Ibid., p. 186.

would this seem to be an unreasonable speculation. Thus this form of dualism is capable of being expanded into a comprehensive and consistent position, and one that has the great merit that it solves the problem of evil.[1]

From the point of view of Christian theology, however, a dualism of this kind is unacceptable for the simple but sufficient reason that it contradicts the Christian conception of God. Mill's type of dualism does not face, and therefore does not solve, the problem of evil as it arises for a religion that understands and worships God as that than which nothing more perfect can be conceived. Dualism avoids the problem — but only at the cost of rejecting one of the most fundamental items of Christian belief, belief in the reality of the infinite and eternal God, who is the sole creator of heaven and earth and of all things visible and invisible. This belief is so deeply rooted in the Bible, in Christian worship, and in Christian theology of all schools that it cannot be abandoned without vitally affecting the nature of Christianity itself. The absolute monotheism of the Judaic-Christian faith is not, so to say, negotiable; it can be accepted or rejected, but it cannot be amended into something radically different. This, then, is the basic and insuperable Christian objection to dualism : not that it is intrinsically impossible or unattractive, but simply that it is excluded by the Christian understanding of God and can therefore have no place in a Christian theodicy. However, the Christian rejection of an ultimate dualism is not only a corollary of revealed theology. Natural theology adds the related criticism that dualism is metaphysically unsatisfying. If neither of the two factors lying behind the creation of our world — matter and a finite deity — is ultimate and self-existent, we still have to ask, Who or what created *them*? The existence of a finite god raises the same metaphysical queries as the existence of our

[1] Some of the more recent forms of 'external dualism' are to be found in F. C. S. Schiller, *Riddles of the Sphinx* (London : Swan Sonnenschein & Co., 2nd ed., 1894), pp. 309–24 ; J. E. McTaggart, *Some Dogmas of Religion* (London : Edward Arnold, 1909), chap. 7; H. G. Wells, *God the Invisible King* (New York : The Macmillan Co., 1917) ; Christian Ehrenfels, *Cosmogony*, trans. by Mildred Focht (New York : Comet Press, Inc., 1948), pp. 217 f. ; and Edwin Lewis, *The Creator and the Adversary* (New York : Abingdon-Cokesbury Press, 1948).

Introductory

own finite selves. Accordingly dualism does not constitute a logical terminus as does the monotheistic conception of an eternal self-existent Being who is the creator of everything other than Himself. This latter idea brings its own problems with it — above all, the problem of evil — but as a primary theological premise it has a metaphysical self-sufficiency that dualism lacks. Further, as F. H. Bradley remarked concerning the doctrine of a finite god, 'it is an illusion to suppose that imperfection, once admitted into the Deity, can be stopped precisely at that convenient limit which happens to suit our ideas'.[1] Once our concept of God loses the firm shape provided by an inner backbone of metaphysical ultimacy it is liable to become fluid and elusive and to fail to satisfy the exigencies of religion.

6. THE INTERNAL DUALISM OF E. S. BRIGHTMAN

Various forms of what might be called internal dualism — that is to say, dualisms that locate the opposition to good within the divine nature itself — have been popular, and to some extent are still so, especially in the United States, partly as a result of the influence of the process philosophy of A. N. Whitehead.[2] I shall not take Whitehead himself as the basis for our study of this type of dualism, for the complexities of his thought would carry us too far afield, but shall turn instead to an independent thinker who has dealt more directly with the problem of evil itself, under the influence of the process philosophy. Edgar Sheffield Brightman (1884–1953) is usually and correctly classified as a member of the Boston Personalist school, even though his treatment of the particular topic with which we are concerned, namely the problem of evil, is not characteristic of that school of thought.[3] Brightman

[1] F. H. Bradley, *Essays on Truth and Reality* (Oxford : Clarendon Press, 1914), p. 430.
[2] See A. N. Whitehead, *Process and Reality* (Cambridge University Press, 1929), pt. v, chap. 2 ; *Religion in the Making* (Cambridge University Press, 1930), chap. iv, sect. 4. Cf. Charles Hartshorne, *Man's Vision of God* (Chicago : Willett, Clark & Co., 1941) ; *The Divine Relativity* (New Haven : Yale University Press, 1948).
[3] Cf. F. H. Ross, *Personalism and the Problem of Evil* (Yale University Press, 1940).

dealt with the subject in a succession of books,[1] his last and most thorough discussion, taking account of some of the published criticisms of his earlier formulations, being in *A Philosophy of Religion*, from which the quotations in this section will be taken.

'The problem of evil in its most acute form,' says Brightman, 'is the question whether there is surd evil and, if so, what its relation to value is.'[2] By 'surd evil' he means objects, events or experiences that are intrinsically and irredeemably evil and incapable of being turned to good. For example, the phenomenon of imbecility : 'Let us grant that imbecility may encourage psychiatry and arouse pity ; yet, if it be an incurable condition, there remains in it a surd evil embodied in the intrinsic worthlessness of the imbecile's existence and the suffering which his existence imposes on others.'[3] Although it might well be questioned, on the basis of Christian eschatology, whether this is a true example of surd evil in Brightman's sense — or even, if one adopts a doctrine of ultimate universal salvation, whether there can be any such thing as surd evil — the contrary is assumed without further argument in Brightman's discussion.

Brightman's distinctive solution to the problem of evil, or rather his distinctive manner of avoiding it in any acute form, consists in his 'theistic finitism'. This agrees with the orthodox Christian tradition in understanding God as 'an eternal, conscious spirit, whose will is unfailingly good'.[4] But it departs from this tradition by adding that 'there is something in the universe not created by God and not a result of voluntary divine self-limitation, which God finds as either obstacle or instrument to his will.'[5] In Brightman's doctrine this 'obstacle' lies within God's own nature.

In speaking of limitations upon the deity we have to distinguish between a real and a merely apparent circumscrib-

[1] *The Problem of God* (New York : The Abingdon Press, 1930) ; *The Finding of God* (New York : The Abingdon Press, 1931) ; *Is God a Person?* (New York : Association Press, 1932) ; *Personality and Religion* (New York : The Abingdon Press, 1934) ; *A Philosophy of Religion* (New York : Prentice-Hall, Inc., 1940) ; *Person and Reality*, ed. by Peter Bertocci (New York : The Ronald Press Co., 1958).   [2] E. S. Brightman, *A Philosophy of Religion*, p. 246.
[3] Ibid.   [4] Ibid., p. 281.   [5] Ibid., p. 314.

ing of His power. Orthodox Christian doctrine agrees with theistic finitism in holding that God cannot do the logically impossible (such as creating a square circle), but holds, as does Brightman,[1] that this is not an inability in God but rather an incoherence in the task proposed.[2] The Christian theological tradition likewise holds that the Creator voluntarily limits Himself in His dealings with the free finite persons whom He has created; and Brightman agrees that this also would not lead us to describe God as finite.[3] But to these merely apparent constraints upon the deity he adds his own peculiar conception of a real restriction imposed by the existence of what he calls 'The Given'.

The Given consists of the eternal uncreated laws of reason and also of equally eternal and uncreated processes of nonrational consciousness which exhibit all the ultimate qualities of sense objects (*qualia*), disorderly impulses and desires, such experiences as pain and suffering, the forms of space and time, and whatever in God is the source of surd evil.[4]

This definition is far from being a model of clarity; but so far as the problem of theodicy is concerned the essential clause is the last one: 'and whatever in God is the source of surd evil'. This source, whatever it may be, is something not created or willed by God but an eternal aspect of His nature. Brightman further explains his position by comparison with that of Plato. According to Brightman,

the creative factors in the universe for Plato are: God (the Demiurge or cosmic Artisan), the Pattern (the eternal ideal, corresponding to the Ideas in the earlier dialogues), and the Receptacle (the primordial chaos of space, discordant and disorderly motion). The actual world is caused by union of the forms (or Pattern) with the Receptacle.[5]

He now clarifies his own suggestion by saying:

It is easy to see that the Pattern corresponds to what we have called the formal aspect of the Given, while the Receptacle is the content aspect of the Given. But there is an essential difference between Plato's view and that which has been developed in this

---

[1] See ibid., p. 285.
[2] See below, pp. 301–2, for a discussion of this point.
[3] Brightman, *A Philosophy of Religion*, pp. 286–7.
[4] Ibid., p. 337.   [5] Ibid., p. 339. Cf. above, p. 32.

chapter ... The Receptacle (certainly) and the Pattern (probably) are external to God. The relations between them, and their ontological status, are therefore obscure. Much of this obscurity and unrelatedness is removed by our hypothesis which enlarges the idea of God so that Pattern and Receptacle are both included in God.[1]

It is no doubt true that Plato's form of dualism involves obscurities. But surely Brightman's form contains even more puzzling obscurities. For he insists that God is, on the one hand, unqualifiedly good and, on the other hand, the source of surd evil! One can understand, even if one does not accept, the view that outside God's nature and existing independently of Him there is a material that He did not create and with which He struggles with only partial success. But what is one to make of the claim that God is wholly good ('His will for goodness and love is unlimited'[2]), and yet that the Given with which He struggles, and which is the source of surd evil, is a part of His own nature? Surely, if the source of surd evil — of sheer, unqualified, unredeemable evil — is part of the Godhead, the Godhead can no longer be described as unreservedly good. It is, rather, partly good and partly evil. As one of Brightman's critics has said, 'He unites under the one label of "deity" two diametrically opposed realities, namely, the perfect and holy will of God and the evil nature which opposes that will.'[3]

[1] Ibid.   [2] Ibid., p. 337.

[3] Henry Nelson Wieman, in H. N. Wieman and W. M. Horton, *The Growth of Religion* (New York: Willett, Clark & Co., 1938), p. 356. Cf. James John McLarney, *The Theism of Edgar Sheffield Brightman* (Washington, D.C.: The Catholic University of America, 1936), p. 146.

PART II

THE AUGUSTINIAN TYPE
OF THEODICY

CHAPTER III

THE FOUNTAINHEAD: ST. AUGUSTINE —
EVIL AS PRIVATION OF GOOD STEMMING
FROM MISUSED FREEDOM

ST. AUGUSTINE (A.D. 354-430) has probably done more than any other writer after St. Paul to shape the structure of orthodox Christian belief — more even than such epoch-making figures as Thomas Aquinas or the Protestant Reformers; for Augustine's influence was exerted at an earlier and more plastic stage in the growth of the Christian mind and neither scholasticism nor Protestantism has significantly altered the grand design of his picture of God and the universe, or his conception of the drama of man's creation, fall, and redemption. Not that Augustine's teachings were novel. But it was his historic achievement to bring the diverse elements of Christian thought together to make an immensely powerful impact upon the intellect and imagination of the West. It is therefore with Augustine that we begin this study, though with glances back to Plotinus, through whom he absorbed so much from the surrounding thought-world of Neo-Platonism.

From his earliest to his latest writings Augustine was continually turning to the problem of evil. His characteristic teaching on the subject appears not only in the great works of his maturity, *The City of God*, the *Confessions*, and the *Enchiridion*, but also in a succession of earlier books going back to his controversies with the Manichaeans. In all these writings several distinct, though often overlapping, themes are developed, and I believe that justice can best be done to Augustine's many-sided treatment of the problem of evil by considering it from the following four angles: (1) Augustine first asks, What is evil? meaning, What is it metaphysically? i.e. not, Which empirical objects and events are to be accounted evil? but, Is evil an ultimate constituent of reality; and if not, what is its true nature and status? In his discussions under this head Augustine is reacting to Manichaean dualism, and his answer — that evil is not an entity in its own right but rather a privation of good — derives directly from Neo-Platonism. (2) Augustine next asks, Whence comes evil? His response to this question is the so-called 'free-will defence', which explains, for him, both the moral evil of sin and, derivatively, human suffering in its many forms—physical pain, fear and anxiety, etc. (3) Although this did not present itself to the men of Augustine's time as a problem, his thought offers a solution to what has later been called the problem of 'metaphysical evil', i.e. the fact of finitude, or of limited forms of existence, and the arbitrariness and imperfections thereof. Augustine's solution, which again represented a familiar Neo-Platonic theme, is found in the 'principle of plenitude', the idea that the most rich and valuable universe is one exemplifying every possible kind of existence, lower as well as higher, ugly as well as beautiful, imperfect as well as perfect. (4) Embracing these three aspects of the subject is the 'aesthetic' theme in Augustine's theodicy, his faith that in the sight of God all things, including even sin and its punishment, combine to form a wonderful harmony which is not only good but very good.

These divisions do not correspond to distinct sections of Augustine's teaching, clearly marked off as such by himself, but rather indicate four vantage-points from which to survey the complex streams of his thought. The first pair of topics will be discussed in the present chapter, and the second pair in the next.

I. Evil as Privatio Boni

1. Augustine and Manichaeism

For a period of about nine years Augustine was strongly attracted by Manichaeism and had the status of an *auditor*, or layman, in the Manichaean sect. This eclectic faith, which had been founded by Mani (*c.* A.D. 215–76) about a

century and a half earlier, dealt directly and explicitly with
the problem of evil by affirming an ultimate dualism of good
and evil, light and darkness. When Augustine renounced
Manichaeism and became a Christian the Manichaean solution now seemed to him utterly and dangerously mistaken,
above all in the conception of God that it entailed. It pictured the God whom men worship as less than absolute, and
as but one of two co-ordinate powers warring against each
other. Or, if one chooses to think of the higher, divine realm
as a single totality, Manichaeism saw this as divided against
itself and as including within it the principle and energy of
evil. To Augustine it was a 'shocking and detestable pro-
fanity' that 'the wedge of darkness sunders . . . the very
nature of God',[1] and against it he upheld vigorously the
integrity of God's goodness and the universality of His rule.
Indeed the reality and perfection of God both as the ultimate
of being and power and as infinite in goodness and beauty lay
close to the heart of Augustine's theodicy. There could be
no evil or possibility of evil in God Himself. Augustine
accordingly insisted, as a matter of religious life and death,
that whatever evil may be it neither comes from God nor
detracts in any way from His sole and majestic sovereignty.

What, then, is evil, and how does it come to be? Augustine
had no inclination to deny its presence and its virulent power.
The reality of sin was a constant theme, almost an obsession,
of his thinking; and he could also describe eloquently the
natural evils of our human lot.[2] His formulation of the philosophical *problem* of evil is as tough-minded as that of any
sceptic, and includes a rejection of the idea that evils do not
really exist and are therefore not to be feared; for, says
Augustine, 'either that is evil which we fear, or the act of
fearing is in itself evil'.[3]

How, then, may we think of evil so as to acknowledge its
oppressive power and yet at the same time preserve the full
Christian doctrine of God as the sovereign Lord of all?

[1] *Against the Fundamental Epistle of Manichaeus*, trans. Richard Stothert (Edinburgh: T. & T. Clark, 1872), xxiv. 26.   [2] E.g. *C.G.* xix. 4.
[3] *Confessions*, vii. 5, trans. Albert C. Outler in *Augustine: Confessions and
Enchiridion* (London: S.C.M. Press Ltd.; and Philadelphia: The Westminster
Press, 1955).

Augustine's answer was suggested to him by the Neo-Platonism to which he turned after renouncing Manichaeism and which, after his conversion to Christianity, still seemed to him of all human philosophies the closest to Christian truth. He presumably received the Neo-Platonic teaching concerning evil mainly from the writings of Plotinus (A.D. 204–70), which find numerous echoes in Augustine's own pages, and it will throw light on Augustine's own theodicy if we approach it through a brief review of Plotinus' way of dealing with the same set of problems.[1]

2. THE PLOTINIAN THEODICY

In Plotinus' metaphysical system the Supreme Being, the ultimate 'One', is in one of its aspects the Good; and evil has no place or part in it. For,

> If such be the Nature of Beings and of That which transcends all the realm of being, Evil cannot have any place among Beings or in the Beyond-Being; these are good. There remains, only, if Evil exists at all, that it be situate in the realm of non-Being, that it be some mode, as it were, of the Non-Being, that it have its seat in something in touch with Non-Being or to a certain degree communicate in Non-Being. By this Non-Being, of course, we are not to understand something that simply does not exist, but something of an utterly different order from Authentic-Being. . . . Some conception of it would be reached by thinking of measurelessness as opposed to measure, of the unbounded against bound, the unshaped against the principle of shape, the ever-needy against the self-sufficing : think of the ever-undefined, the never at rest, the all-accepting but never sated, utter dearth. . . .[2]

Before inquiring further into the nature of this chaotic evil, which is non-being and which yet has power to threaten and corrupt being, let us ask, How does evil, so conceived, come to be? The answer of the Plotinian philosophy (as distinct, as we shall see, from that of Augustine) is that evil represents the dead-end of the creative process in which the Supreme-

---

[1] For the development of Augustine's attitude to Plotinus, see Régis Jolivet, *Le Problème du Mal d'après saint Augustin* (Paris : Gabriel Beauchesne et fils, 2nd ed., 1936), appendix : 'Saint Augustin et Plotin'.

[2] *Enneads*, trans. Stephen MacKenna (London : Faber & Faber Ltd., 3rd ed., 1962), i. 8, 3.

Being has poured out its abundance into innumerable forms of existence, descending in the degrees of being and goodness until its creativity is exhausted and the vast realm of being borders upon the empty darkness of non-being.

Given that the Good is not the only existent thing, it is inevitable that, by the outgoing from it or, if the phrase be preferred, the continuous down-going or away-going from it, there should be produced a Last, something after which nothing more can be produced : this will be Evil. As necessarily as there is Something after the First, so necessarily there is a Last : this last is Matter, the thing that has no residue of good in it : here is the necessity of Evil.[1]

Plotinus' matter (ὕλη), however, is not matter as this has been conceived in the scientific thought of the seventeenth, eighteenth, and nineteenth centuries, and as we still tend for ordinary purposes to think of it today, namely as solid, extended, and enduring. It is rather the Platonic ὕλη : that, in itself formless and measureless, to which the eternal Ideas give form and so produce the sensible world. By itself, however, and apart from its condition as having been given form by the Ideas, matter cannot even be said to exist. 'We utterly eliminate every kind of Form ; and the object in which there is none whatever we call Matter.'[2] 'Matter has not even existence whereby to have some part in Good : Being is attributed to it by an accident of words : the truth would be that it has Non-Being.'[3]

Here we meet a very old and tenacious conception, which is to reappear again and again in the course of Christian reflection upon the problem of evil. Evil is non-being; it is a lack, a privation and a non-entity ; it comes from the *Ungrund*; as *das Nichtige* it opposes God Himself. We shall have to examine this idea in its Augustinian context and in the scholastic tradition, and then as it recurs in the present century in existentialist philosophy and in the thought of Karl Barth. It may be well to indicate already at this point a general conclusion concerning this perennial way of thinking. On the one hand, it bears witness to certain powerful psychological facts (above all, the fear of meaninglessness and

[1] *En.* i. 8, 7.   [2] *En.* i. 8, 9.   [3] *En.* i. 8, 5.

47

of death), and it is to that extent connected with reality. But, on the other hand, in its role as a mode of philosophical discourse it is both a product and a fertile source of conceptual confusion, and could profitably be abandoned; the truths which it has symbolized can more properly be expressed in other and less misleading ways. Some of the ambiguities involved in equating evil with non-being can already be seen at work in Plotinus. We should, however, begin with a distinction which is helpful and clarifying. This is the Platonic distinction between non-being in the absolute sense of sheer nothingness (οὐκ ὄν) and in the relative sense of the not yet realized potentiality to be some specific thing (μὴ ὄν).[1] It is the latter that has been regarded throughout a long philosophical tradition as the lurking-place of evil. The question immediately arises whether this non-being (τὸ μὴ ὄν) is a positive force with a malevolent nature of its own, liable actively to attack and infect all that it touches, or merely a passive and innocent occasion of evil. Like other participants in the tradition, Plotinus is obliged to treat it as both — as negative and passive when he is using it to support the thesis that the universe, as the emanation of Perfect Goodness, contains no independent power of evil; but as positive and active when he is using it to explain the felt potency of evil in human experience. This internal tension in Plotinus' thought is a symptom of a deep-seated weakness of the meontic theme in theodicy.

The conflicting tendencies lie open to view in the pages of Plotinus' *Enneads*. On the one hand, acknowledging the experienced power of iniquity, he speaks of the Absolute Evil[2] and refers to Matter (equated with evil) as 'a principle distinct from any of the particular forms in which, by the addition of certain elements, it becomes manifest'; and he draws the inference that 'we cannot be, ourselves, the source of Evil, we are not evil in ourselves; Evil was there before we came to be; the Evil which holds men down binds them against their will'.[3] And he accordingly identifies the fallen soul's entanglement in its material body as the cause of the

---

[1] See *Parmenides*, 160c–162a; and *Sophist*, 237a–254e, 255a–259d.
[2] *En.* i. 8, 3. [3] *En.* i. 8, 5.

evil which it both suffers and commits: 'This bodily Kind,' he says, 'in that it partakes of Matter, is an evil thing.'[1] On the basis of these and other passages Père Sertillanges claims that Plotinus affirms 'un sous-être' which is evil and which is a positive reality,[2] and says that Plotinus 'holds to his idea of the positivity of evil, which is consonant with his system as a whole'.[3] This matter, which is both non-being and non-good, is antagonistic to good.[4] Père Sertillanges is undoubtedly here responding to a real element within the Plotinian writings.

On the other hand, within the terms of Plotinus' metaphysical picture as a whole it would seem that evil can only be negative, an absence of good and of reality, a mere empty darkness outside the lucent realms of true being. And so Dean Inge, in his Gifford Lectures on *The Philosophy of Plotinus*, rejected in the name of the Plotinian system as a whole Plotinus' tendency at times to invest ὕλη with positive powers of resistance to the creative energy of the Good. 'Plotinus' Matter', he says, 'is the absence of order, which when isolated by abstract thought becomes the foe of order.'[5] And this too is a response to a real feature of the Plotinian system, perhaps even representing its dominant emphasis and tendency.

We are not concerned here, however, to try to resolve this apparent contradiction within Plotinus' thinking. Having noted these conflicting tendencies in his writings, we may turn to the first great Christian thinker to grapple persistently with the problem of evil, and see how Augustine uses and yet alters what he received from Plotinus. For we meet in Augustine (as it seems to me) a more intelligible, because more restricted, use of the non-being theme.

3. The Goodness of the Created Order

For Augustine, as for the Neo-Platonists, God is the ultimate of being and goodness. He is both the perfect good and the

---

[1] *En.* i. 8, 4.
[2] A. D. Sertillanges, O.P., *Le Problème du Mal*, vol. i, 'L'Histoire' (Paris: Aubier, 1948), p. 124. [3] Ibid. [4] Ibid., p. 125.
[5] W. R. Inge, *The Philosophy of Plotinus* (London: Longmans, Green & Co., 3rd ed., 1929), i. 134.

infinite beauty, and the eternal, immutable, and supremely real Being. He has created 'out of nothing' all that exists other than Himself. And as the work of omnipotent Goodness, unhindered by any recalcitrant material or opposing influence, the created world is wholly good. It is a richly varied good, one aspect of whose value is precisely its ordered variety. As such it contains innumerable different kinds of creature, some higher and some lower in the scale of being. 'To some He communicated a more ample, to others a more limited existence, and thus arranged the natures of beings in ranks.'[1] The main gradations of being are as follows :

> For, among those beings which exist, and which are not of God the Creator's essence, those which have life are ranked above those which have none ; those that have the power of generation, or even of desiring, above those which want this faculty. And, among things that have life, the sentient are higher than those which have no sensation, as animals are ranked above trees. And, among the sentient, the intelligent are above those that have not intelligence — men, e.g. above cattle. And among the intelligent, the immortal, such as the angels, above the mortal, such as men.[2]

Thus the created universe is an immensely abundant and variegated realm of forms of existence, each having its appropriate place in the hierarchy of being. The qualitatively fuller and richer a creature's nature, the higher it stands in the scale. But — and in this Augustine diverges significantly from Plotinus — there is here no Neo-Platonic descent from the goodness of pure being to the evil of matter. On the contrary, the whole creation, including the material world, is good ;[3] it is also, however, because it lacks the immutability of its Creator, capable of being corrupted. But there is no level of being, however humble, which is, as such, evil. To be an inferior creature, low down in the scale of being, is not to be evil, but only to be a lesser good. For it is good that

---

[1] *C.G.* xii. 2. Cf. xii. 2 ; ix. 16 ; *Against the Fundamental Epistle of Manichaeus*, xxv. 27 ; *N.G.* v ; *On Free Will*, iii. 13, 16, 24, and 32, trans. John H. S. Burleigh in *Augustine: Earlier Writings; Conf.* vii. 11, 12, and 16.
[2] *C.G.* xi. 16.
[3] See, for example, *Enchiridion*, iv. 12, trans. Albert C. Outler in *Augustine: Confessions and Enchiridion*, and *Conf.* vii. 12. Even formless ὕλη, if such there be, is good, because the capacity to receive form is good : *N.G.* xviii. Cf. F.W. 11. xx. 54.

there should be beings of all kinds, forming a universe of wonderful complexity that reflects the Creator's goodness from many angles and in every possible shape and colour. Therefore the simpler or (to us) the less beautiful or useful forms of existence should never be despised :

> The order of creatures proceeds from top to bottom by just grades, so that it is the remark of envy to say : That creature should not exist, and equally so to say : *That* one should be different. It is wrong to wish that anything should be like another thing higher in the scale, for it has its being, perfect in its degree, and nothing ought to be added to it. He who says that a thing ought to be different from what it is, either wants to add something to a higher creature already perfect, in which case he lacks moderation and justice ; or he wants to destroy the lower creature, and is thereby wicked and grudging. Whoever says that any creature ought not to be is no less wicked and grudging, for he wants an inferior creature not to exist, which he really ought to praise. For example the moon is certainly far inferior to the sun in the brightness of its light, but in its own way it is beautiful, adorns earthly darkness, and is suited to nocturnal uses. . . . If instead of saying that the moon should not exist he said that the moon ought to be like the sun, what he is really saying without knowing it is, not that there should be no moon, but that there should be two suns. In this there is a double error. He wants to add something to the perfection of the universe, seeing he desires another sun. But he also wants to take something from that perfection, seeing he wants to do away with the moon.[1]

Here, then, is a central theme of Augustine's thought: the whole creation is good ; the sun, moon, and stars are good ; angelic and human beings are good ; birds, reptiles, fish, and animals, trees, flowers, and plants, are all good ; light and darkness, heat and cold, sea and land and air : all are good, expressing as they do the creative fecundity of perfect goodness and beauty. So Augustine rejects the ancient Platonic, Neo-Platonic, Gnostic, and Manichaean prejudice against matter and lays the foundation for a Christian naturalism that rejoices in this world, and instead of fleeing from it as a snare to the soul, seeks to use it and share it in gratitude to God for His bountiful goodness.

Where, then, in a creation that is all good, consisting in a

[1] *F.W.* III. ix. 24. Cf. III. ix. 25, and III. v. 13.

multitudinous host of greater and lesser, higher and lower goods, is the place of evil; and whence does it arise? Having rejected, in the name of the absolute sovereignty of God, the idea that there can be any independent force of evil or of resistance to good in the universe, coeternal with the Almighty, what account was Augustine to give of the undoubted presence and power of evil? His answer — adapted rather than adopted from Plotinus — is that evil is not any kind of positive substance or force, but consists rather in the going wrong of God's creation in some of its parts. Evil is essentially the malfunctioning of something that in itself is good. For 'omnis natura bonum est';[1] and yet everything, other than God Himself, is made out of nothing and is accordingly mutable and capable of being corrupted; and evil is precisely this corruption of a mutable good.

4. Man Mutable because 'Made out of Nothing'

Augustine repeats a number of times that man is mutable *because* he is made out of nothing. 'Nature', he says, 'could not have been depraved by vice [vitium] had it not been made out of nothing. Consequently, that it is a nature, this is because it is made by God; but that it falls away from Him, this is because it is made out of nothing.'[2] He seems to speak in such passages as though 'nothing' were a material of inferior quality which has gone into the making of men and which, being an inherently weak material, introduces an element of instability into us.[3] If he *is* thinking of 'nothing' in this way, Augustine is a victim of the deceitfulness of words. The familiar habits of our language would lead us to assume that because we have the noun 'nothing' there must be something of which it is the name, and hence that 'nothing' is a peculiar kind of 'something'. However, in another place Augustine sees very clearly the fallacy of any such inference. In his exegesis of John i. 3, 'and without him nothing [nihil]

---

[1] *Ench.* iv. 13.
[2] *C.G.* xiv. 13. Cf. xii. 1 and 8; *N.G.* i, x, and xlii.
[3] Cf. J. H. S. Burleigh, *The City of God* (London: Nisbet & Co., 1949), p. 77.

was made', Augustine asserts against those who would hypo-statize this 'nihil', that 'nothing' means simply 'not any-thing'.[1] In other words, the world was not fashioned out of any pre-existing material. The nothing out of which the creation came is the οὐκ ὄν of the Platonic distinction; hence, simply, 'not out of anything'. But because it is thus a secondary, dependent, and contingent realm, there is the possibility of its falling into disorder and so becoming evil.

5. Evil Privative and Parasitic

Augustine's most frequent phrase to define evil is *privatio boni*, 'privation of good'; but more or less synonymously with *privatio* he uses also *deprivatio, corruptio, amissio, vitium, defectus, indigentia*, and *negatio*.[2] Augustine never means by privation of good a simple lack of goodness, in the sense in which a tree, for example, lacks the spiritual qualities of an angel. It is not an evil to have been created as a lesser rather than a greater good — as a worm, for example, instead of a dog, or a dog instead of a man. For, according to the principle of plenitude,[3] there is positive value in the existence of less exalted as well as more exalted forms of creaturely being in a well-ordered scale. Evil enters in only when some member of the universal Kingdom, whether high or low in the hier-archy, renounces its proper role in the divine scheme and ceases to be what it is meant to be.

When such malfunctioning occurs it cannot be said to exist as a separate entity; it is, on the contrary, the absence of proper being in a creature. Thus 'Evil has no positive nature; but the loss of good [amissio boni] has received the name "evil"'.[4] Evil is negative, a lack, a loss, a privation.

What, after all, is anything we call evil except the privation of good? In animal bodies, for instance, sickness and wounds are nothing but the privation of health. When a cure is effected, the

---

[1] *N.G.* xxv.
[2] Prior to Augustine a privative view of evil appears in Christian literature in Origen, *De Principiis*, ii. 9, 2, and *Commentary on St. John*, ii. 13; Athanasius, *Contra Gentes*, chap. 7; Basil the Great, *Hexaemeron*, homily 2, para. 4; and Gregory of Nyssa, *The Great Catechism*, chap. 7.
[3] See pp. 76 f.   [4] *C.G.* xi. 9.

evils which were present (i.e. the sickness and the wounds) do not retreat and go elsewhere. Rather, they simply do not exist any more. For such evil is not a substance; the wound or the disease is a defect of the bodily substance which, as a substance, is good.[1]

Speaking of 'measure, form and order' (*modus, species, ordo*) as constituting a creature's goodness, Augustine says that evil is

nothing but the corruption of natural measure, form or order. What is called an evil nature is a corrupt nature. If it were not corrupt it would be good. But when it is corrupted, so far as it remains a natural thing, it is good. It is bad only so far as it is corrupted.[2]

Everything that exists, then, is good. But many things are now less good than they were when they first came forth from the Creator's hand. They have fallen away from their original state and have forfeited a proportion of the worth with which they were endowed by God. This decrease in the goodness or corruption of the nature of some entity means that the thing in question has to that extent become evil. For evil is *privatio boni*; it is the absence-of-goodness that prevails when anything has defected from the mode of being that is proper to it in God's creative intention. Augustine accordingly stresses the secondary and dependent as well as the negative and privative character of evil : 'there can be no evil where there is no good. . . . Nothing evil exists *in itself*, but only as an evil aspect of some actual entity. . . . Evils, therefore, have their source in the good, and unless they are parasitic on something good, they are not anything at all.'[3] Evil is thus fundamentally self-defeating and absurd ; for to the extent that it succeeds it can only destroy that upon which it lives. Accordingly a totally evil entity could not possibly exist ; so far as anything has being it is good, and if it had no goodness it could not be at all. 'If the good is so far diminished as to be utterly consumed, just as there is no good left so there is no existence left.'[4]

---

[1] *Ench.* iii. 11. Cf. *F.W.* iii. xiii. 36 ; *N.G.* iv ; *Conf.* vii. 12, and iii. 7 ; *C.G.* xi. 9 and 22.
[2] *N.G.* iv. Cf. vi.   [3] *Ench.* iv. 13–14.
[4] *N.G.* xvii. Cf. vi, ix, and xiii ; *Ench.* xii ; *Conf.* vii. 12 ; *C.G.* xii. 3.

6. THE IDENTITY OF BEING AND GOODNESS

This thought is continuous in Augustine's theodicy with another : the identification of being with goodness. The highest good has (or is) the most intensely real being, and the diminishing degrees of goodness are at the same time diminishing degrees of being. This principle is not stated as such, nor is its meaning explicitly developed, by Augustine ; it is rather a pervasive presupposition which he breathed in from the philosophical atmosphere of his time, dominated as it was by Neo-Platonism. From the premise that 'every entity [natura], even if it is a defective one, in so far as it is an entity, is good'[1] Augustine deduces that fallen angels 'became, not indeed no nature at all, but a nature of less ample existence, and therefore wretched'[2] and that if things 'are deprived of all good, they will cease to exist'.[3]

The identity of being and goodness is one of the great philosophic conceptions that have inspired Western thought through the passage of several centuries. But, like most ideas of comparable historical influence, it is somewhat general — a theme rather than a precise doctrine — and is capable of different uses and interpretations. Seeking to unfold its meaning we may begin by noting that in Neo-Platonism, which is the true source and natural habitat of this idea, it bears a literal meaning. The ultimate One radiates its nature to create the universe, which is thus an extension of what at its source is both Being itself and the Good itself. The descending emanations of the One are thus increasingly attenuated forms of that Being which is also the Good. This continuity of the universe with its divine source down the emanating stream of being gives literal significance to the equation of the degrees of goodness with degrees of being. But Augustine had renounced the emanationist picture in favour of the Christian doctrine of creation *ex nihilo*. Instead of degrees of creaturely participation in the divine reality there is for him an absolute gap between God and His creation. In terms of this radically theistic understanding of creation

[1] *Ench.* iv. 13.   [2] *C.G.* vii. 6.   [3] *Conf.* vii. 12, 18.

the identity of being and goodness cannot, or at any rate ought not to, have the same meaning as for Plotinus; and in so far as Augustine does seem to intend it in this way we can only say that he was still being unconsciously a Neo-Platonist at a point at which Neo-Platonism is incompatible with Christian teaching.

But although there is undoubtedly a Neo-Platonist cast to much that Augustine says when he is assuming the identity of being and goodness, this identity is also capable of being developed independently of Neo-Platonism. In order to appreciate Augustine's more Christian use of the idea, and the way in which he might have explicated it in response to the questionings of a twentieth-century reader, let us formulate what is for us today a clear and valid distinction, and then note why this distinction would have had no point or use for Augustine. It is natural to us, when we read Augustine, to distinguish between two different concepts of 'being', and to suspect that he was guilty of confusing them together. There is, first, 'being' as bare existence, mere being there as one of the occupants of space-time. In this sense anything is properly said to exist or to have being if it would be listed in a complete inventory of the contents of the universe. 'Being', so defined, is not susceptible of degrees; something either exists or does not exist, and it makes no sense to say that one thing exists more, or in a higher degree, than another. However much the various items that exist may differ in moral or aesthetic value or in metaphysical importance, considered simply as existent things each counts only for one. There is, however, at least an approach to the concept of 'being' or 'existence' as capable of degrees when, for example, we speak of the poet or the genius existing, or living, more intensely than other people. There is here an evaluative use of 'existing'; the more intense the existence the higher its worth. And such a use could be extended and elevated into the Neo-Platonic definition of 'being' and 'good' as synonyms.

Reading Augustine with this distinction in mind, it might well seem that he illicitly contravenes it by treating the two concepts as one, for he repeatedly says that if an entity lost all its goodness it would thereby cease to exist. (For example,

'si omni bono privabuntur, omnino non erunt'.[1]) And yet the
degrees in which created entities approximate to the Good
are clearly not degrees of bare existence. The baboon, for
example, is lower in the scale of goods than a man, but it is
not correspondingly less existent. It is not, in comparison
with a man, a thin, shadowy, barely real creature, but full
of solid flesh and bone, coursing with blood and physical
energy. A baboon exists just as much as a man — or rather,
since existence does not admit of degrees, we must simply say
that both exist. On the other hand, a man is more valuable
than a baboon and accordingly embodies more of being in
the sense in which 'being' connotes value. But if one neg-
lected this distinction one might conclude, erroneously, that
if anything came to lose its quota of (axiological) being it
would thereby cease to exist. And one is tempted to see
Augustine as committing precisely this error.

However, this would, I think, be an injustice. For Augustine there was no such thing as bare existence. He held that
existence always and necessarily exhibits certain categorial
attributes that are intrinsically valuable, the most fundamental of these being 'measure, form and order' (*modus,
species, ordo*).[2] To exist is to possess some degree of measure,
form, and order, and this degree indicates the creature's
place in the scale of goods :

> These three things, measure, form and order, not to mention
> innumerable other things which demonstrably belong to them,
> are as it were generic good things to be found in all that God has
> created, whether spirit or body. . . . Where these three things are
> present in a high degree there are great goods. Where they are
> present in a low degree there are small goods. And where they
> are absent there is no goodness. Moreover, where these three
> things are present in a high degree there are things great by
> nature. Where they are present in a low degree there are things
> small by nature. Where they are absent there is no natural good
> at all. Therefore, every natural existent [omnis natura] is good.[3]

Thus Augustine might grant, if the point could have had
any significance for him, that bare existence in contrast to

[1] *Conf.* vii. 12.
[2] *N.G.* vi. Or *mensura, numerum, ordo*, 'measure, number and order', in
F.W. ii. 54. Cf. Plotinus, *En.* i. 7, 2.   [3] *N.G.* iii. Cf. *C.G.* xi. 15.

non-existence is not a matter of degree. But he would at the same time insist that all existing entities necessarily share certain basic categorial characteristics, without which they could not be; and he would assert that these characteristics, which are exhibited in higher or lower degree, are good. Accordingly to exist is to have these good qualities of 'measure, form and order'; and everything has them in its own appropriate manner and is accordingly located at a certain point in the rising scale of goods.

However, we can still properly ask why, or in what sense, the possession of measure, form, and order is good. Good from whose point of view — the existing creature's, or God's, or both? It would be substantially true to say that from the point of view of existing entities, or at any rate such of them as possess consciousness, it is preferable to go on existing than to cease to exist; otherwise living creatures would presumably seek to destroy themselves. Generally speaking, existence seems good to those who have it; and this generalization covers mankind as well as the lower animals. What, however, of the divine point of view? Is it in God's eyes always good that existing things should continue to exist? If Augustine were to pose the question to himself in this form he would (as we can see by bringing together different parts of his teaching) give an unqualified yes as regards spiritual creatures, and a qualified yes, with a time condition attached to it, as regards creatures below the level of spirit. Of these latter, such as animals and plants, Augustine would say that it is good in God's sight that they exist when they do exist, but also good that they are instances of measure and form whose proper order prescribes for them a transient role within the on-going life of nature. Of spiritual beings, angels and men, Augustine would say that it is always good in God's sight that they should (having once been created) continue to exist. For He retains in being even the devil and his angels, as well as those of the human race who are destined for damnation; and even when these have finally been consigned to hell He will continue to hold them in existence throughout eternity. And if we ask Augustine *why* it is good that spiritual creatures, even though wicked, should exist for ever, his answer will be that this particular *species*, exhibiting its own *modus* of being and conforming to an *ordo* which is either that of righteousness leading to happiness or of wickedness leading to punishment, is good, and makes its own contribution to the composite goodness of the created universe.

Commenting upon this aspect of Augustine's thought, it seems to me correct to characterize its basic standpoint as aesthetic rather than ethical. The kind of analogy that is appropriate to his doctrine of creation is that of the Artist enjoying the products of his creative activity, rather than the Person seeking to bring about personal relationships with created persons. For in spite of the fact that Augustine has been called the first psychologist, his metaphysical reasoning proceeds unhesitatingly in terms of essences and substances rather than in terms of persons and personal relationships. In our own time, however, when the theological significance of the personal and of the I–Thou relationship has been rediscovered, we may well think differently about the divine motive in creation and may thereby find ourselves led towards a different theodicy. We shall still hold, with Augustine, that the goodness of each existing thing consists in the fact that God has willed it to be, both spiritual and non-spiritual creatures having been summoned into existence to serve a divine purpose ; but we need not commit ourselves, as Augustine does, to the theory that 'measure, form and order' are intrinsically valuable in God's sight and that the goodness of creaturely existence consists in its embodiment of these characteristics.

7. THE LOGICAL CHARACTER OF AUGUSTINE'S DOCTRINE

Seeking now to evaluate Augustine's doctrine, we must ask a standard question of contemporary philosophical investigations : Of what logical type is the statement that 'evil is a privation of good'? Taken by itself and apart from its theological context it might be no more than the expression of a semantic preference — like that of one who, on being told that the glass is half empty, replies, 'There is no such state as "being half empty"; what you call "being half empty" is really being half full.' Such a retort, although having the form of a factual assertion, would express a linguistic recommendation. That is to say, it does not allege that the observable state of affairs is other than as described in the rejected sentence, but urges that a situation of this kind should be reflected in language in a different way. The statement that evil consists in *privatio boni* could be construed as a recommendation of this kind. Indeed in the case of any pair of opposite terms, such as large-small, hot-cold, fast-slow, good-evil, it is possible to eliminate one by defining it in terms of the other. One could, for example, abolish the word 'small', and instead of describing objects as larger and smaller, speak of them instead as being more or less large; or one could do without 'cold' by referring only to degrees of heat. Similarly, one could eliminate the word 'evil' and speak of greater and lesser degrees of goodness. On such an interpretation Augustine would not be offering an analysis of the actual nature of evil, but would rather be recommending an optimistic vocabulary, and with it an optimistic way of thinking about the world.

In a sense it is not entirely incorrect to say that this is what Augustine is doing. But if we choose to put it in this way, we must give great weight to the new way of thinking that Augustine advocates. For his privative doctrine of evil is part of a total metaphysical picture of the universe, and if we choose to regard it as a linguistic recommendation we must recognize that what it recommends is nothing less than the whole Christian interpretation of life. For the *privatio boni* conception receives its meaning and validity from a context of distinctively Christian (and Jewish) presuppositions. It connects most directly with the idea of creation *ex nihilo*. Given that the universe has been created by an omnipotent and good God, what is evil? It cannot be anything substantial, a positive constituent of the universe, but can only be a loss of natural 'measure, form and order', a malfunctioning of something that is in itself good. This privative definition makes it clear, within the Christian theological picture, how there can be evil in a good creation: evil was not created, but consists in the voluntary turning away of free beings from the good.

The valid point, then, of the Augustinian doctrine of evil as *privatio boni* is the metaphysical claim that evil has not been created by God and that its status within His universe is secondary and parasitic rather than primary and essential. But it does not follow from this metaphysical analysis that evil is empirically — that is to say, as a fact of experience — accurately describable as a loss or lack of goodness. As an element in human experience, evil is positive and powerful. Empirically, it is not merely the absence of something else but a reality with its own distinctive and often terrifying quality and power.

The error that would result from construing the *privatio boni* doctrine as empirical rather than metaphysical in character — as a fact of observation rather than an inference from theological premises — is more serious in relation to 'moral' than to 'natural' evil. Even as applied to the latter, however, the privative account of evil could function only very inadequately as an empirical description. What we call evil in nature can, it is true, often be regarded as consisting in the corruption or perversion or disintegration of something which, apart from such disruption, is good. When a living organism, such as a tomato, corrupts or (and here colloquial speech supports Augustine) goes bad, it does tend towards disintegration and dissolution. The tomato tends to cease to exist — not indeed absolutely, but in the sense that the matter comprising it enters into other combinations and becomes part of the earth or of another plant or animal body. This process can be brought without strain within the description of evil as a loss of 'measure, form and order'. Disease, too, is a loss of normal bodily order and functioning. Volcanic eruptions, droughts, tornadoes, hurricanes, and planetary collisions can perhaps likewise be regarded as breakdowns in some imagined ideal ordering of nature. In all such cases the evil state of affairs can plausibly be seen as the collapse of a good state of affairs, and as tending towards non-existence, at least in the relative sense of the dissolution of a previously established arrangement of life or matter. But does such an account really lay bare that aspect of the event or of the situation that makes us call it evil? Do we regard a volcanic eruption, for example, as evil considered simply as a loss of a previous 'measure, form and order'? Do we not, on the contrary, regard it as evil only if it causes harm to human, or at least to sentient, life? Is the eruption of a volcano on an uninhabited island, or (assuming it to be uninhabited) on Venus, an evil? Or again, is the natural decay of vegetation in virgin jungle to be accounted evil? Or the burning up of a star or the fragmentation of a meteor a million million light years distant from us in space? If not, the quality of evil is not attributed to physical disintegration as such, but only in so far as it impinges deleteriously upon the realm of the personal, or at least upon the sphere of animal life. It is in fact not loss of 'measure, form and order' *per se* that is evil, but only this considered as a cause of pain and suffering. But the resulting pain and suffering, which make us stigmatize their cause as evil, are positive. They are at least as emphatic and intrusive realities of experience as are pleasure and happiness. Pain and pleasure are experientially co-ordinate, and it would be as arbitrary and as inadequate to describe pain as a privation of pleasure as to describe pleasure as a privation of pain.

When we turn from the realm of nature to that of moral personality, an interpretation of the *privatio boni* doctrine as a description of experience would be even less adequate. For a corrupted will does not always tend to disintegrate and to cease to exist as a will or a personality. On the contrary it may retain its degree of mental integration, stability, coherence, intelligence, lucidity, and effectiveness. And when this happens — one thinks, for example, of Milton's Satan or of Iago in fiction, and of such men as Goebbels in recent history — the combination of personal evil with mental power constitutes a very positive and terrible moral evil. To describe, for example, the dynamic malevolence behind the Nazi attempt to exterminate the European Jews as merely the absence of some good, is utterly insufficient. As Dean Inge remarks in the course of his discussions of Plotinus' theodicy, 'In the scale of existence there are no *minus* signs. . . .

But in the scale of values, as in our thermometer, we have to register temperatures far below freezing-point.'[1] The evil will as an experienced and experiencing reality is not negative. It can be a terrifyingly positive force in the world. Cruelty is not merely an extreme absence of kindness, but is something with a demonic power of its own. Hatred is not merely lack of love, or malevolence merely a minimum degree of goodwill. It would be an arbitrary and unfruitful amendment of the dictionary, rather than an illuminating way of looking at the facts, to describe moral evils as merely privations of their corresponding moral goods.[2]

This has to be maintained against those writers who fail to distinguish sufficiently clearly between a metaphysical and an empirical understanding of the *privatio boni* doctrine. Père Sertillanges, for example, in the second volume of his work on *Le Problème du Mal*, defends a strictly privative account of moral evil and seems to regard this as experientially as well as metaphysically valid. It may therefore serve to test the matter further if we take account of his discussion. Sertillanges distinguishes two aspects of an immoral action : there is the physical action, or series of actions, as such, and there is the neglect of the moral law in the mind of the agent. The physical action as such is not evil. On the contrary, 'Everything positive in it relates to good. A savant who analyses it will only find manifestations of the laws which govern nature and life, laws which are properly to be admired. The agent himself is intent upon a good which his action seeks and without which it would accomplish nothing. The satisfaction which he pursues would be legitimate in other cases. If it is not at all legitimate here, this is solely because of the moral rule which applies to the situation and which brands as a sinner one who does not obey it.'[3]

Sin [says Sertillanges] is never in what one does but in what one neglects. Evil is not in the pursuing of the values of life which the

---

[1] *The Philosophy of Plotinus*, i. 133.
[2] Cf. Julius Müller, *The Christian Doctrine of Sin*, i, pp. 277–9 and 293–4. For a psychologist's critique of the *privatio boni* conception of evil see Carl Jung, *Psychology and Religion: West and East* (New York : Pantheon Books, 1958), pp. 304–5 and p. 168.
[3] A. D. Sertillanges, O.P., *Le Problème du Mal*, vol. ii, p. 17.

sinner covets, but in the abandoning of the higher values to which his action blocks the way. Speaking generally, everything positive in the criminal action is good. What is bad is the negation entailed in such and such an act, and this does not prove the positive character of evil.[1]

Père Sertillanges concedes that the deficient willing (as he calls it, adopting Augustine's phrase) in which evil consists, 'has a positive character in so far as it is a phenomenon of conscience. The sin is indeed an act, even in its negativity, because it is voluntarily negative.'[2] But this concession does not lead to the recognition of the positive reality in moral evil; for the passage continues, 'But even considered as a phenomenon of conscience, it is good; it is a manifestation of freedom. Evil results, once more, from the present non-consideration of the rule of reason in favour of the law of the flesh or the pride of the spirit. And there is here, in all truth, only deficiency. No *réalité mauvaise* appears as an element in the analysis of moral evil, any more than in that of physical evil.'[3]

This would be valid as a metaphysical statement based upon the theological conception that man was not created as a sinner but has voluntarily fallen into sin; for this conception entails that no 'bad reality', that is to say no reality that is bad naturally and essentially, enters into the picture. But it would not be valid if it were construed instead as an observational report upon the phenomenon of moral evil. On the empirical level we must say that, just as earthquakes and the like are authentically and positively evil in virtue of the pain and suffering which they cause, so morally evil actions are authentically and positively evil, not only as causes of pain and suffering but also as expressions of the evil willing that produces them. For as suffering is an experienced reality, no less positive than pleasure, so morally evil motives, dispositions, and volitions are realities as positive as good motives, dispositions, and volitions.

[1] *Le Problème du Mal.*   [2] Ibid., pp. 17–18.   [3] Ibid., p. 18.

II. The 'Free-Will Defence' in St. Augustine

8. SIN AS THE BASIC EVIL

According to St. Augustine, the correct order of inquiry concerning evil is to ask first about its nature and second about its origin.[1] We have seen that his answer to the first question is that evil does not, as the Manichaeans maintained, exist in its own right as one of the original constituents of the universe. On the contrary, the whole creation is good, and evil can consist only in the corrupting of a good substance; it is the privation of some good which is proper to the world as God made it.

But how does such privation of goodness come about? Where does the corruption come from? Rejecting the Neo-Platonic view that evil is a metaphysical necessity, inevitably appearing where being runs out into non-being, Augustine attributes all evil, both moral and natural, directly or indirectly to the wrong choices of free rational beings. 'An evil will [improba voluntas], therefore, is the cause of all evils.'[2] Again, 'The cause of evil is the defection of the will of a being who is mutably good from the Good which is immutable. This happened first in the case of angels and, afterwards, that of man.'[3] For when Augustine forsook Manichaeism for Christianity he was taught, and came wholeheartedly to believe, 'that free will is the cause of our doing evil and that thy just judgment is the cause of our having to suffer from its consequences'.[4] Here, then, is the heart of Augustine's theodicy. As he succinctly states it in his commentary on Genesis, 'omne quod dicitur malum, aut peccatum esse, aut poenam peccati'.[5] Again, in the treatise *Of True Religion* he says, 'This covers the whole range of evil, i.e. sin and its penalty.'[6]

The primary sin, which makes angels and men evil and

---

[1] *N.G.* iv. Cf. Plotinus, *En.* i. 8, 1.   [2] *F.W.* iii. xvii. 48.
[3] *Ench.* viii. 23.   [4] *Conf.* vii. 3, 5. Cf. iv. 15, 26.
[5] *De Genesi Ad Litteram*, Imperfectus liber, i. 3.
[6] *Of True Religion*, xii. 23. Cf. xx. 39; and *Against the Fundamental Epistle of Manichaeus*, chap. 26.

brings upon them the further punitive evils of pain and sorrow is an 'aversio a Deo, conversio ad creaturas', a wilful turning of the self in desire from the highest good, which is God Himself, to some lesser good. 'For when the will abandons what is above itself, and turns to what is lower, it becomes evil — not because that is evil to which it turns, but because the turning itself is wicked.'[1] It is this that occurred both in the premundane fall of the angels and in the primeval fall of man, and this is the continuing nature of man's sinfulness today:

The will which turns from the unchangeable and common good and turns to its own private good or to anything exterior or inferior, sins. It turns to its private good, when it wills to be governed by its own authority; to what is exterior, when it is eager to know what belongs to others and not to itself; to inferior things, when it loves bodily pleasures. In these ways a man becomes proud, inquisitive, licentious, and is taken captive by another kind of life which, when compared with the righteous life we have just described, is really death.[2]

When we ask, What makes free beings thus perversely turn from the higher to the lower? we meet again the privative conception of evil. According to Augustine's doctrine of 'deficient causation' the evil will has no positive or 'efficient' cause but only a 'deficient' cause. 'Now, to seek to discover the causes of these defections — causes, as I have said, not efficient, but deficient — is as if someone sought to see darkness, or hear silence. Yet both of these are known by us, and the former by means only of the eye, the latter only by the ear; but not by their positive actuality, but by their want of it. Let no one, then, seek to know from me what I know that I do not know; unless he perhaps wishes to learn to be ignorant of that of which all we know is, that it cannot be known.'[3] One could dwell upon the oddity of such a way of speaking, in which there is alleged to be a cause of evil-willing, and yet only a negative cause, a mysterious kind of non-cause. But what Augustine's doctrine really amounts to is, I think, clear enough : evil willing is a self-originating act,

---

[1] *C.G.* xii. 6. Cf. xii. 7, 8, and 9; *N.G.* xxxiv, xxxvi, and xx; *F.W.* i. 34 and 35; ii. 53; *Conf.* vii. 16.   [2] *F.W.* ii. ix. 53.
[3] *C.G.* xii. 7. Cf. xii. 6 and 9; *F.W.* ii. 54.

and is as such not explicable in terms of causes that are distinguishable from the agent himself. Thus the origin of evil lies for ever hidden within the mystery of finite freedom ; for 'what cause of willing can there be which is prior to willing?'[1]

But although there is no external force pushing or pulling the soul when it 'abandons Him to whom it ought to cleave as its end, and becomes a kind of end to itself',[2] yet there is a motive within the sinner which leads him astray. There is 'pride, which is the beginning of sin'.[3] While some angels 'steadfastly continued in that which was the common good of all . . . others, being enamoured rather of their own power, as if they could be their own good, lapsed to this private good of their own . . . and, bartering the lofty dignity of eternity for the inflation of pride, the most assured verity for the slyness of vanity, uniting love for factious partisanship, they became proud, deceived, envious'.[4]

The dramatic picture that Augustine bequeathed to later ages is accordingly as follows. God creates a universe out of nothing. The universe consists of the richest possible abundance of kinds of creatures, each good but together spanning many different grades of goodness ; and high in this scale of being are the angels.[5] Since they are created out of nothing they are mutable[6] and capable of turning away from God, in adherence of whom their happiness consists.[7] By a wilful misuse of their freedom some of the angels revolt from their blessed dependence upon the divine source of their good and seek instead to be lords of their own being. They thereby lose their true relation to the good and are thrust down in misery from heaven. The majority of the angels, however, remain in steadfast allegiance to God and enjoy eternal felicity. And so even before the creation of mankind the two

---

[1] *F.W.* iii. xvii. 49. Cf. iii. xxii. 63 and ii. xx. 54. Augustine, despite his high predestinarian doctrine, is emphatic that in order to be morally good or ill an action must be free : 'sin is so much a voluntary act that it is not sin at all unless it is voluntary'. (*Of True Religion*, xiv, 27. Cf. *F.W.*, ii. i. 3.)

[2] *C.G.* xiv. 13.

[3] *Ecclesiasticus* x. 13.

[4] *C.G.* xii. 1. Cf. xii. 6 ; xi. 15.

[5] For the creation of the angels with, rather than before, the world, see *C.G.* xi. 32 and xii. 15.

[6] *C.G.* xii. 1.   [7] *C.G.* xii. 1 and 6.

cities are formed whose strife is to shape our human history, peopled at first by good and evil angels :

the one dwelling in the heaven of heavens, the other cast thence, and raging through the lower regions of the air ; the one tranquil in the brightness of piety, the other tempest-tossed with beclouding desires ; the one, at God's pleasure, tenderly succouring, justly avenging — the other, set on by its own pride, boiling with the lust of subduing and hurting ; the one the minister of God's goodness to the utmost of their good pleasure, the other held in by God's power from doing the harm it would ; the former laughing at the latter when it does good unwillingly by its persecutions, the latter envying the former when it gathers in its pilgrims.[1]

These two angelic communities, then, are 'dissimilar and contrary to one another, the one both by nature good and by will upright, the other also good by nature but by will depraved'.[2]

9. The Self-creation of Evil 'ex Nihilo'

The theodicy that follows from belief in the fall of the angels, and its repetition in mankind, is built upon two central pillars of doctrine : first, that God created all things good ; and second, that free creatures, by an inexplicably perverse misuse of their God-given freedom, fell from grace, and that from this fall have proceeded all the other evils that we know. Of these two pillars, the first maintains the innocence of God, the second the guilt of the creature. This great traditional picture, together with the theodicy implicit within it, has persisted through the centuries — not, however, because it is an inherently satisfying response to the mystery of evil but because the Christian mind was for so long content to refrain from examining it critically. But whenever it has been freely probed by Christian thinkers in its relation to the problem of evil — as it was, for example, by Schleiermacher in the nineteenth century[3]— its radical incoherence has become all too evident. The basic and inevitable criticism is that the idea of an unqualifiedly good creature committing

[1] *C.G.* xi. 33.　　　　　　　　　　[2] *C.G.* xi. 33.
[3] See *The Christian Faith* (E.T., Edinburgh : T. & T. Clark, 1928), pp. 293–304.

sin is self-contradictory and unintelligible. If the angels are finitely perfect, then even though they are in some important sense free to sin they will never in fact do so. If they do sin we can only infer that they were not flawless — in which case their Maker must share the responsibility for their fall, and the intended theodicy fails. As Schleiermacher says, 'the more perfect these good angels are supposed to have been, the less possible it is to find any motive but those presupposing a fall already, e.g. arrogance and envy'.[1]

That the notion that a flawless being who turns to sin amounts to the absurdity of the self-creation of evil *ex nihilo*, is implicitly acknowledged in Augustine's theology — as, later, in that of Calvin — by a doctrine of absolute divine predestination. This in effect brings the origin of evil within the all-encompassing purpose of God, and lays upon Him who is alone able to bear it the ultimate responsibility for the existence of evil. Thus Augustine asks himself whether the angels who fell were previously in the same state of perfection as those which remained steadfast, and is obliged to answer 'No'. For the blessedness of an intelligent being contains two elements : 'that it uninterruptedly enjoy the unchangeable good, which is God ; and that it be delivered from all dubiety, and know certainly that it shall eternally abide in the same enjoyment'.[2] The blessedness of the loyal angels was always complete, but that of those who were to fall lacked the second element : 'the [heavenly] life of these angels was not blessed, for it was doomed to end'.[3] For 'How shall we say that they participated in [complete blessedness] equally with those who through it are truly and fully blessed, resting in a true certainty of eternal felicity? For if they had equally participated in this true knowledge, then the evil angels would have remained eternally blessed equally with the good, because they were equally expectant of it.'[4]

Discussing the same problem again in book xii, Augustine takes a more explicitly predestinarian view of the first appearance of sin in the angelic world. Speaking of the angels who fell, he says, 'These angels, therefore, either

---

[1] Ibid., p. 161.   [2] *C.G.* xi. 13. Cf. xi. 4.
[3] *C.G.* xi. 11.   [4] *C.G.* xi. 11. Cf. xi. 13, and *Ench.* ix. 29.

received less of the grace of the divine love than those who
persevered in the same ; or if both were created equally good,
then, while the one fell by their evil will, the others were
more abundantly assisted, and attained to that pitch of
blessedness at which they became certain that they should
never fall from it — as we have already shown in the preceding book.'¹ This can be interpreted only as implying an
ultimate divine preordination of Satan's fall. For Augustine
says of the fallen angels that 'their blessedness was destined
to come to an end',² and that 'the angelic darkness *though it
had been ordained*, was yet not approved'.³ This seems to be
a stronger doctrine of predestination than Augustine usually
taught with regard to men. Men, he held, fall freely and
responsibly, but God foresees, even before He created them,
that they will do this.⁴ But in the case of the angels, God
withheld His assisting grace from some, and thus selected them
for a different destiny from those whom He enabled to remain
steadfast. Presumably they were themselves aware of this
by inference from the fact that, in contrast to their more fortunate fellows, they lacked the blessed assurance of eternal
bliss. One is tempted to a certain sympathy with them as
they appear at this early stage of their careers ; for they seem
to have been in the same kind of discouraging situation as
someone endowed by nature with irresistible criminal tendencies who discovers that a prison cell was reserved for him
before he was born !⁵

10. Sin and Predestination

The same pattern prevails when Augustine turns from the
fall of the angels to that of man. Adam and Eve, as they
dwelt in the Garden of Eden prior to their fatal sin, are
described in the most idealized terms. They were (but for
their subsequent fall) immortal ;⁶ they did not grow old ;⁷
they had complete control over their bodily passions ;⁸ they

¹ *C.G.* xii. 9. ² *C.G.* xi. 13.
³ *C.G.* xi. 20 (my italics). ⁴ See below, pp. 83 f.
⁵ Further on the fall of the angels see *C.G.* xi. 11, 13, 15, 20, 32–33 ; xii. 1, 9;
Ench. ix. 28. ⁶ *C.G.* xiv. 1, 10; xii. 21.
⁷ *C.G.* xiv. 26. ⁸ *C.G.* xiv. 26.

were endowed with infallible moral insight;[1] they lived in the enjoyment of God and were 'good by His goodness' ('ex quo bono erat bonus') ;[2] their love for Him was unclouded and tranquil ('imperturbatus') ;[3] and they had no desire for the forbidden fruit.[4] Much of this is summed up in the following idyllic picture :

> In Paradise, then, man . . . lived in the enjoyment of God, and was good by God's goodness ; he lived without any want, and had it in his power so to live eternally. He had food that he might not hunger, drink that he might not thirst, the tree of life that old age might not waste him. There was in his body no corruption, nor seed of corruption, which could produce in him any unpleasant sensation. He feared no inward disease, no outward accident. Soundest health blessed his body, absolute tranquillity his soul. As in Paradise there was no excessive heat or cold, so its inhabitants were exempt from the vicissitudes of fear and desire. No sadness of any kind was there, nor any foolish joy ; true gladness ceaselessly flowed from the presence of God, who was loved 'out of a pure heart, and a good conscience, and faith unfeigned'. The honest love of husband and wife made a sure harmony between them. Body and spirit worked harmoniously together, and the commandment was kept without labour. No languor made their leisure wearisome ; no sleepiness interrupted their desire to labour.[5]

How, then, we may well wonder, did sin enter into this paradisal state? If it occurs to us at this point to remember the devil, who had already fallen and who was present to tempt the primal pair to their destruction, Augustine will not allow any such extenuating suggestion. He insists that 'the evil act had never been done had not an evil will preceded it' ('Non enim ad malum opus perveniretur, nisi praecessisset mala voluntas'),[6] and that Adam had already in his heart

---

[1] *F.W.* iii. xviii. 52.   [2] *C.G.* xiv. 26.
[3] *C.G.* xiv. 10.   [4] *C.G.* xiv. 10.
[5] *C.G.* xiv. 26. The idealizing of Adam's pre-fallen state has a long history in Judaism prior to some of the early Church Fathers and then Augustine. In the rabbinical literature there are passages expressing the belief that 'the first man was endowed with extraordinary stature (he is frequently said to have filled the world), with physical beauty, with surpassing wisdom, with a brilliancy which eclipsed that of the sun, with a heavenly light which enabled him to see the whole world, with immortality, and with a ministration of angels'. F. R. Tennant, *The Sources of the Doctrines of the Fall and Original Sin* (Cambridge : The University Press, 1903), pp. 149–50.
[6] *C.G.* xiv. 13.

turned away from God in order for the devil's solicitations to be able to make any appeal to him. 'The wicked deed, then — that is to say, the transgression of eating the forbidden fruit — was committed by persons who were already wicked. . . . The devil . . . would not have ensnared man in the open and manifest sin of doing what God had forbidden, had man not already begun to live for himself . . . this wicked desire [to be self-sufficient] already secretly existed in him, and the open sin was but its consequence.'[1]

As in the case of the fallen angels, then, we have presented to us the incomprehensible conception of the self-creation of evil *ex nihilo*. And once again we find this mystery wrapped in a further mystery, which, without rendering the first any less mysterious, largely nullifies it as a refuge for theodicy by casting upon God the more ultimate responsibility for the creation of beings who He knew would, if created, freely sin. This latter mystery is provided by Augustine's doctrine of an arbitrarily selective predestination to salvation. The sin of Adam was at the same time the sin of all his descendants, who were 'seminally present' in Adam's loins.[2] Thus all mankind is from birth in a state of guilt and condemnation, and there would be perfect justice in the consignment of the entire human race to the eternal torments of hell. But in the (to us) mysterious workings of His sovereign grace, God has chosen some to be saved out of this *massa perditionis*, leaving the rest to undergo their just punishment:

Within that number of the elect and the predestinated, even those who have led the worst lives are by the goodness of God led to repentance. . . . Of these our Lord spoke when He said, 'This is the Father's will which hath sent me, that of all He hath given me I should lose nothing' (John vi. 39). But the rest of mankind who are not of this number, but who, out of the same lump of which they are, are made vessels of wrath, are brought into the world for the advantage of the elect. God does not create any of them without a purpose [ac fortuito]. He knows what good to work out of them: He works good in the very fact of creating them human beings, and carrying on by means of them this visible system of things. But none of them does He lead to a wholesome and spiritual repentance. . . . All indeed do, as far as themselves

---

[1] *C.G.* xiv. 13.   [2] *C.G.* xiii. 14.

are concerned, out of the same original mass of perdition treasure up to themselves after their hardness and impenitent heart, wrath against the day of wrath ; but out of that mass God leads some in mercy and repentance, and others in just judgment does not lead.[1]

In this and other passages Augustine is on the verge of, or even sometimes over the verge into, teaching that God creates some with the express intention of damning them and others with the opposite intention of saving them.[2] However, the balance of his many statements perhaps favours the slightly milder view that men fall freely and culpably and that out of the fallen race God saves some, leaving others to perish ; although God knows from the beginning which He intends to save and which to abandon. In the case of the saved He then gives a grace to respond which they would not otherwise be able to have. Commenting upon the text, 'As He chose us in Him before the foundation of the world',[3] Augustine says,

And assuredly, if this were said because God foreknew that they would believe, not because He Himself would make them believers, the Son is speaking against such a foreknowledge as that when He says, 'Ye have not chosen me, but I have chosen you' (John xv. 16); when God should rather have foreknown this very thing, that they themselves would have chosen Him, so that they might deserve to be chosen by Him. Therefore they were elected before the foundation of the world with that predestination in which God foreknew what He Himself would do; but they were elected out of the world with that calling whereby God fulfilled that which He predestinated.[4]

None, then, can be saved unless God Himself, by His own irresistible grace, brings them to repentance and then grants them pardon. God's choice is His own free and inscrutable act, expressing what must be for us a hidden justice : 'As to the reason why He wills to convert some, and to punish others for turning away . . . the purpose of his more hidden

[1] *Contra Julianum Pelagianum*, bk. v, chap. 14.
[2] E.g. *On the Soul and its Origins*, bk. iv, chap. 16 ; *On the Merits and Remission of Sins, and on the Baptism of Infants (De Peccatorum Meritis et Remissione, et de Baptismo Parvulorum)*, bk. ii, chap. 26, trans. Peter Holmes in *Nicene and Post-Nicene Fathers*, first series, vol. v.
[3] Ephesians i. 4.
[4] *On the Predestination of the Saints*, chap. 34, trans. Peter Holmes in *Nicene and Post-Nicene Fathers*, first series, vol. v.

judgment is in His own power [consilium tamen occultioris justitiae penes ipsum est].'[1]

We must now see what part man's free will plays in all this. Augustine's conception of our human freedom is identical with that of a number of contemporary philosophers who define a free act as one that is not externally compelled but that flows from the nature and will of the agent.[2] Augustine too holds that man has free will and responsibility in the sense that his acts are his own personal deeds, expressing his own nature in its response to the various situations in which he finds himself. Thus to say that man has free will is, essentially, simply to say that he wills. It follows, as Augustine points out, that the fact that an action is willed and is thus a free act is compatible with its being an object of divine foreknowledge. Let us not be afraid, he says, lest 'we do not do by will that which we do by will, because He whose foreknowledge is infallible, foreknew that we would do it'.[3] Again, 'it does not follow that, though there is for God a certain order of all causes, there must therefore be nothing depending on the free exercise of our own free wills, for our wills themselves are included in that order of causes which is certain to God, and is embraced by His foreknowledge, for human wills are also causes of human actions; and He who foreknew all the causes of things would certainly among those causes not have been ignorant of our wills'.[4]

Man's freedom, so conceived, is not equivalent to randomness, indeterminacy, and unpredictability. On the contrary, a man's actions are determined by his own inner nature. Our deeds proceed from our character; we do what we do because we are what we are. Thus Augustine rejects the idea that 'our will could remain in a certain neutrality,—so as to be neither good nor bad; for we either love righteousness, and then our will is a good one . . . or if we do not love it at all, our will is not good. . . . Since therefore the will is

---

[1] *Merits and Remission of Sins*, bk. ii, chap. 32.
[2] See, for example, Antony Flew, 'Divine Omnipotence and Human Freedom' in *New Essays in Philosophical Theology*, ed. Flew and MacIntyre (London : S.C.M. Press Ltd., 1955).
[3] *C.G.* v. 9. Cf. xii. 1, 22, 27; xiv. 11, 27.
[4] *C.G.* v. 9. Cf. *F.W.* iii. 6–10.

either good or bad, and since we have not the bad will from God, it remains that we have a good will from God. . . . Therefore . . . to whoever He gives it, He gives in His mercy, not their merits, and from whoever He withholds, He withholds in His truth.'¹ Thus whilst fallen men have the power of willing, and are responsible for their volitions, their wills are wholly in bondage to sin. Their state is one of 'non posse non peccare' (not able not to sin), and accordingly, 'Mortals cannot live righteously and piously unless the will itself is liberated by the grace of God from the servitude to sin into which it has fallen.'²

The situation is thus (1) that man was created with a freedom for either good or evil; (2) that he willed wrongly, and so lost his freedom for good; and (3) that God foresaw man's fall 'from before the foundation of the world' and planned its compatibility with the balanced perfection of His universe. We have already seen that, considered as the basis for a Christian theodicy, this scheme of thought is open to radical questioning at two levels. The notion that man was at first spiritually and morally good, orientated in love towards his Maker, and free to express his flawless nature without even the hindrance of contrary temptations, and yet that he preferred to be evil and miserable, cannot be saved from the charge of self-contradiction and absurdity. And even if this strange hypothesis be allowed, it would still be hard to clear God from ultimate responsibility for the existence of sin, in view of the fact that He chose to create a being whom He foresaw would, if He created him, freely sin.

¹ *Merits and Remission of Sins*, bk. ii, chaps. 30–31.
² *Retractiones*, i. 9. The freedom to refrain from sin ('posse non peccare') belonged to man in the beginning, but was lost by the fall. 'When we speak of the freedom of the will to do right we are speaking of the freedom wherein man was created.' On the complexities of Augustine's thought concerning human free will, see J. B. Mozley's *Treatise on the Augustinian Doctrine of Predestination* (London: John Murray, 2nd ed., 1878), chap. 8.

CHAPTER IV

THE FOUNTAINHEAD: ST. AUGUSTINE—
THE PRINCIPLE OF PLENITUDE AND THE
AESTHETIC THEME

III. The Principle of Plenitude

1. THE PROBLEM

THE universe as we know it is complex and variegated. It is not a mere empty space, or an undifferentiated continuum of, say, light or mass; nor, on the other hand, is it a sheer structureless chaos. It is instead a cosmos in which complex conjunctions and interrelations form a virtually infinite and ever-changing variety and arrangement of objects. There are the great wheeling nebulae of suns and planets, and on our own earth the endlessly different shapes and sounds, colours and odours, movements, structures, and patterns of nature. There are, for example, hundreds of millions of species of living organisms on the surface of our globe. Within this vast realm of terrestrial life, to confine our attention to this, it is obvious that the wider the variety of kinds the more unavoidable are distinctions and inequalities among the creatures — variations in size, intelligence, strength, complexity, beauty, longevity, and in adaptation to the different circumstances and climates. Given the principle of variety, some must needs be stronger, some more beautiful, some more intelligent, some more hardy than others; for variety entails inequality. And if one is disposed to criticize the state of nature from a moral point of view this inequality will have the appearance of arbitrariness; and arbitrariness in the distribution of life's fundamental endowments will, in turn, be equated with injustice. Thus it may seem unfair that there should be creatures that are stronger than ourselves, that some can fly whilst we cannot, and that others are at home under the waters that would drown us. Or it may seem unjust that a beautiful flower should be so fragile, or the fleet horse without a greater degree of intelligence, or the friendly dog unable to talk. Indeed, if one cares to think along these lines, one can regard every creature as 'underprivileged' in some respect and as the victim of arbitrary and unjust treatment. And so there seems to be a question, within the wide field of theodicy, as to why God should have created a world of complexity and variety, with its inevitable inequalities of creaturely equipment and capacities. The question does not properly take the form, Why did God not make butterflies as long-lived as elephants, or elephants as beautiful as butterflies? (for then they would not be butterflies and elephants), but simply, Why did God make such variously imperfect creatures as these? And again, not, Why did He fail to give men the wisdom and intelligence of angels? (for then they would not be men), but simply, Why did He make so imperfect a creature as man? Given that God intended to summon into existence out of nothing a reality other than and dependent upon Himself, why did He not restrict His creation to the most perfect kind of contingent beings? The universe would then presumably consist exclusively of angelic creatures; at any rate it would certainly not contain such inferior natures as slugs or snakes or baboons or even human beings. Why, then, is there a *world* rather than only the highest heaven of heavens?

Instead of considering the matter from our own point of view as creatures, and in terms of the perhaps eccentric question of fairness or unfairness to the different species of created beings, one can think, as Augustine did, in terms of an absolute cosmic scale of being and goodness. Thus because God has life, living creatures are higher in the scale of being than inanimate objects; and because God is Spirit, spiritual beings (men and angels) are higher than the animals, and so on. Everything has its place in the great chain of being, which either culminates in or points towards God Himself. In terms of this metaphysical picture, which for many centuries seemed self-evidently true, one can ask why God should have gone beyond the highest level of creation.

Having formed the archangels, why did He go on to make lower and lower forms of life, down through all the grades of the universe as we find it?

This is not the question to which the doctrine of the fall of man is the traditional answer, but another and prior question. We are not concerned at the moment with the malfunctioning of the world, whether in terms of human sin or of disease and natural conflict, but with the existence of a world at all in the sense of a realm composed of various kinds of creatures which cannot in the nature of the case all belong to the highest category.

Two different answers have been given to this question. One, which was first suggested by Plato, and which descended through Neo-Platonism to Augustine and so into the common stream of Western thought, in which it continued as a pervasive assumption until about the end of the eighteenth century, is based upon what Arthur Lovejoy, in his classic of intellectual history, *The Great Chain of Being*, has called the principle of plenitude. The other answer, of less famous philosophical lineage, though not necessarily on that account less adapted to the exigencies of Christian truth, sees the intelligibility of the world primarily in its relation to man : the world has been created as a relatively autonomous sphere in which men might exist as free moral beings, responding to the tasks and challenges of their common environment and being summoned to serve God as He reveals Himself to human faith in the midst of life's mingled meanings and mysteries. We shall develop and examine this second answer at a later stage.[1] At present we are concerned with the solution that Augustine accepted : that a universe containing every possible variety of creatures, from the highest to the lowest, is a richer and better universe than would be one consisting solely of the highest kind of created being.

2. AUGUSTINE'S NEO-PLATONIST ANSWER

The origin of this conception is in Plato's *Timaeus*, where Timaeus says, 'Let us state then for what reason becoming

[1] See Chaps. XIII, XIV, and XVI.

and this universe were framed by him who framed them. He was good; and in the good no jealousy in any matter can ever arise. So, being without jealousy, he desired that all things should come as near as possible to being like himself. That this is the supremely valid principle of becoming and of the order of the world, we shall most surely be right to accept from men of understanding.'[1] And then the Demiurge, addressing the lower gods whom he had created and urging them in turn to create a further realm of existence below themselves, says, 'There are yet left mortal creatures of three kinds that have not been brought into being. If these be not born, the Heaven will be imperfect; for it will not contain all the kinds of living being, as it must if it is to be perfect and complete. . . .'[2] That 'all things' and 'every kind of creature' should have the boon of existence could only mean in the context of Plato's philosophy that the whole range of eternal Ideas should become incarnated in the world of sense.

It is 'this strange and pregnant theorem of the "fullness" of the realization of conceptual possibility in actuality' that Lovejoy has called the principle of plenitude. More fully stated, it is 'the thesis that the universe is a *plenum formarum* in which the range of conceivable diversity of *kinds* of living things is exhaustively exemplified', and 'that no genuine potentiality of being can remain unfulfilled, that the extent and abundance of the creation must be as great as the possibility of existence and commensurate with the productive capacity of a "perfect" and inexhaustible Source, and that the world is the better, the more things it contains'.[3]

Deploying this theme, which has proved to be one of the recurrent ideas entertained by the Western mind, Plotinus depicted the ultimate One as flowing out in the creation of the reality next below itself in the order of possible beings, and this in turn as producing that next below itself, and so on in a series of emanations filling the whole realm of the possible down to matter, which is so far removed from the One that it

[1] *Timaeus*, trans. Cornford, 29 E.
[2] Ibid. 41 B–C.
[3] Lovejoy, *The Great Chain of Being*, p. 52.

represents the vanishing point of being, or the very border of non-being :

> Seeking nothing, possessing nothing, lacking nothing, the One is perfect and, in our metaphor, has overflowed, and its exuberance has produced the new.[1]

If the First is perfect, utterly perfect above all, and is the beginning of all power, it must be the most powerful of all that is, and all other powers must act in some partial imitation of it. Now other beings, coming to perfection, are observed to generate; they are unable to remain self-closed ; they produce : and this is true not merely of beings endowed with will, but of growing things where there is no will ; even lifeless objects impart something of themselves, so far as they may; fire warms, snow chills, drugs have their own outgoing efficacy ; all things to the uttermost of their power imitate the Source. . . . How then could the most perfect remain self-set — the First-Good, the Power towards all, how could it grudge or be powerless to give of itself . . . ?[2] To this power we cannot impute any halt, any limit of jealous grudging ; it must move for ever outward until the universe stands accomplished to the ultimate possibility. All, thus, is produced by an inexhaustible power giving its gift to the universe, no part of which it can endure to see without some share in its being.[3]

Thus the divine emanations proceed outwards and downwards, unfolding and realizing all the possibilities of existence, until the full range of creatures has been produced.

Augustine shared with Plotinus before him and with a long medieval and modern tradition after him a reliance upon the principle of plenitude as rendering intelligible the general character of the world. We have already encountered this aspect of Augustine's thought in his conception of the universal scale of beings, consisting of higher and lower goods each with its proper place and dignity in the scheme of creation ; and various illustrative passages have been quoted.[4] At another point Augustine states very concisely the inner logic of the principle of plenitude : 'from things earthly to things heavenly, from the visible to the invisible, there are some things better than others ; and for this purpose are they unequal, in order that they might all exist'.[5] And Lovejoy

---

[1] *En.* v. 2, 1.   [2] *En.* v. 4, 1.   [3] *En.* iv. 8, 6.
[4] See pp. 50–51 above.   [5] *C.G.* xi. 22.

quotes a further passage in which Augustine, finding in the principle of plenitude 'his answer to the old question, "Why when God made all things, did He not make them all equal?" reduces the Plotinian argument to the epigram, *non essent omnia, si essent aequalia* : "if all things were equal, all things would not be; for the multiplicity of kinds of things of which the universe is constituted — first and second and so on, down to the creatures of the lowest grades — would not exist".[1]

3. THE PRINCIPLE OF PLENITUDE IN PLOTINUS

As in the case of other themes that Augustine took over from Plotinus, there is a real question as to how far the principle of plenitude, whose natural home is within the setting of Neo-Platonism, can usefully serve the exigencies of Christian truth. For the principle of plenitude is part of a family of ideas and images that cohere well enough within the framework of an emanationist philosophy, but which cannot fit without serious disruption into the very different framework of Christian theism — and this in spite of the historical fact that they entered, through Augustine, into the substance of medieval Catholic thought.

Two main features of the Plotinian universe are the spontaneously emanating, ultimate, divine One, and (by implication at least) a given structure of possibilities determining the forms to be taken by the emanations of being. Plotinus saw the ultimate reality as so superabundantly full that it 'gives off' being as the sun radiates light. The divine plenitude overflows, pouring itself outwards and downwards in a teeming cascade of ever-new forms of life until all the possibilities of existence have been actualized and the shores are reached of the unlimited ocean of non-being. This conception leads to the thought of a hierarchy of higher and lower forms of existence. For the picture of waves of creativity falling out

[1] *The Great Chain of Being*, p. 67. Another form of the principle of plenitude was developed and applied to the theodicy problem by George Tyrrell in 'The Divine Fecundity', *Essays on Faith and Immortality* (London: Edward Arnold, 1914).

into the void and congealing into an ordered world naturally results in a vertically structured universe, like a frozen waterfall. Thus the two notions of emanation and of the creation of all the possible kinds of being in a descending series, together constitute the Neo-Platonic vision of reality.

In this emanationist theology, the radiating of being to form a dependent order does not follow from any divine decision. It is as unconscious and inevitable as the streaming forth of light from the sun. And as the beams of light must gradually exhaust themselves and fade out into darkness, so the cascading streams of being must finally end, setting a lower limit to the scale of created nature.

In addition to the emanating energy of being, there is, either obscurely within the One or standing somehow outside it, a fully determinate pattern of possible forms of existence, the eternal programme of a universe waiting to be. Responding to this thought, the visualizing imagination may perhaps see a wide terraced hillside filled with variously shaped holes and crevices, down which there pours a stream of molten lava from its living source at the summit, until it has flowed over every step of the terrace and filled every hole and been moulded into their many shapes. Note that in this picture there is a finite *quantum* of creativity — for the lava is only sufficient to reach to the lowest terrace — as well as a given pattern of possibilities within which the divine creativity operates.

4. EMANATION AND CREATION

If we are to understand Augustine and the medieval tradition whose main lines he so largely determined, we must see this Neo-Platonist picture as a natural point of departure for Christian thought. The great metaphors of overflowing reality and the great chain of being were already there, providing well-worn paths for the imagination, and the Christian theologian thought only of adapting them so as to make them more adequate as vehicles of Christian truth.

Thus in Augustine the unconscious emanating of the Ultimate is replaced by conscious divine volition. God acts deliberately to form a universe, and He acts in terms of the principle of plenitude, considering it better to produce all possible forms of being, lower as well as higher, poorer as well as richer, all contributing to a wonderful harmony and beauty in His sight, than to produce only a society of blessed archangels. And no doubt, if one starts from a Neo-Platonist combination of emanationism and the principle of plenitude, this is the minimum alteration required to purge the system of its most manifestly non-Christian element. But is there any good reason why the twentieth-century Christian should embrace even this modified version of Neo-Platonism? How do we know, or on what grounds may we suppose, that God adopted the principle of plenitude as His policy for creation? Today, when the personal character of God and of His dealings with mankind have become increasingly central and normative for theological thinking, we are more inclined to say that God willed to create finite beings who should be capable of personal relationship with Himself, and that He created our enigmatic world as an environment whose apparently arbitrary character provides the concrete occasions and opportunities for free and faithful obedience to Him. But this is to move in a different direction from that of the medieval tradition, and towards an alternative type of theodicy. The answer of the medieval thinkers, true to the original Plotinian influence which had come to them through Augustine, was that the goodness and love of God consist in His creativity, His bestowal of the boon of existence as widely as possible. Indeed, throughout medieval theology the love of God tends to be thought of in metaphysical rather than personal terms. It is not so much the love of the personal Infinite for finite persons, as the inexhaustible creative divine fecundity, expressed in the granting of being to a dependent universe with its innumerable grades of creatures.[1] And so

---

[1] Cf. Lovejoy, op. cit. 67–68. It might be urged in qualification of this picture that, for the medievals, rational and personal beings, namely the angels, formed much the greater part of the created universe. However, the angels do not seem to have been thought of primarily as individuals, each with his own distinctive personal characteristics. Perhaps it was their disembodied nature that precluded full personal individualization. At any rate they were generally seen as mirrors of God's glory and instruments of His will, rather than as genuinely individual persons.

while for Plotinus there is the self-emanating One, overflowing into all the possible forms of being to constitute a universe of infinite variety, there is in the parallel Christian doctrine of Augustine the out-going divine love and goodness deliberately bestowing existence upon every conceivable kind of being and so creating the same universe of infinite variety. But the divine love, conceived in this way as an impetus to the creation of finite being as widely as possible (rather than as a mode of personal relationship), is still closely akin to the Plotinian principle of emanation. In both systems creation means being multiplying itself in a range of lower and dependent forms. But such a conception is bound to conflict with the doctrine of Christianity that creation is an entirely unnecessitated act of grace on God's part, and that the universe is the outcome of a free divine choice, not of an automatic self-emanation. It may be granted that the more traditionally orthodox teaching on this point is not free from problems; and it may indeed in the end prove difficult to maintain the contrast between the two positions. Even so strong an exponent of the absolute divine 'otherness' and freedom as Karl Barth has come to see revealed in the Incarnation 'the humanity of God', which is God's 'free affirmation of man, His free concern for him, His free substitution for him'.[1] 'It is when we look at Jesus Christ', says Barth, 'that we know decisively that God's deity does not exclude, but includes His *humanity*. . . . In Him the fact is once for all established that God does not exist without man. . . . In the mirror of this humanity of Jesus Christ the humanity of God enclosed in His deity reveals itself.'[2] Without attempting to follow further the hints contained in Barth's remarkable essay (written in 1956), we can see in the suggestion that humanity has, in Christ, a place on the divine side of the gulf of creation, an indication of the difficulty of maintaining in Christian thought the absolute transcendence and aloofness of a God who has no need for or essential interest in His creation.

[1] Karl Barth, *The Humanity of God*, trans. Thomas Wieser and John Newton Thomas (Richmond : John Knox Press, 1960), p. 51.
[2] Ibid., pp. 49, 50, and 51.

5. THE PRE-EXISTING PATTERN

The other aspect of the Plotinian metaphysic, the fixed pattern of forms, is translated into Augustinian and medieval theology as a realm of essences or archetypes existing in the mind of God. These are Plato's ideas baptized to Christian use.[1] We find this theme in Augustine,[2] in Anselm,[3] in Aquinas,[4] and again as a set of alternative creatable worlds, in Leibniz.[5] There is here a conceptuality that can perhaps be used to solve some of theology's problems. We should remember, however, that the Platonic realism, according to which universals have an existence of their own, transcending their exemplification in concrete particulars, is one philosophical vision or opinion among others, and not one that is widely held today. There is therefore no necessity for Christian theology to commit itself permanently to this Platonist tenet or for this to be regarded as an essential element in the Christian doctrine of God.

Quite apart, however, from the Platonist and Neo-Platonist roots of this aspect of the principle of plenitude, the principle itself is open to objection on another ground. For whilst it might explain why the world contains a variety of kinds, it cannot, without a modification that would disqualify it for theistic use, explain why it contains *only* the particular selection of kinds that it in fact contains out of the infinite realm of the possible. The logic of the principle of plenitude demands that there be creatures of every conceivable kind above the lowest. But while the world contains an immense abundance of species, so that we are not at all inclined to complain of nature's poverty, yet it still does not by any means contain every possible form of life. It does not, so far as we know, contain mermaids or unicorns or griffins or centaurs or winged horses or talking cats or elephant-like animals with camel-like humps or trees that grow a fruit

[1] See C. C. J. Webb, *Studies in the History of Natural Theology* (Oxford: The Clarendon Press, 1915), p. 247.   [2] E.g. *C.G.* xi. 10.
[3] *Monologium*, chaps. 9–10.
[4] *Summa Theologica*, pt. 1, Q. xv, arts. 1 and 3, and Q. xliv, art. 3.
[5] *Theodicy*, para. 225.

yielding a full and balanced human diet or an indefinitely large number of other conceivable animals, vegetables, and minerals. There are in the world, for example, many different species of birds. But however numerous these species may be, one can still ask why there are not twice as many. And if, in order to produce and sustain twice as many kinds of birds, the world would have to be bigger, one can equally well ask why it is not bigger.

To save the hypothesis of plenitude one might be tempted to speculate that all the possible forms of life *are* realized somewhere in the universe, though not all on our own earth.[1] Perhaps the centaur and other beasts of human imagination, which *in intellectu* inhabit our mythologies, exist *in re* on the planets of other stars. Perhaps there are also to be found there other forms of life and organizations of matter that are beyond the scope of man's imagination. And outside of space and time there are perhaps yet further unimaginable realms of being, forming a Spinozistic universe which is infinite in an infinite number of ways.

Now certainly no one can disprove such speculations. And if the universe should be as thus imagined, the principle of plenitude would be thoroughly exemplified in it. Every conceivable form of existence would be realized. Whether the principle requires, not only that every possible species exists, but also every possible different individual of each species, is perhaps an optional matter of definition. But even the latter and stronger version of the principle could in theory be realized. By making unlimited drafts upon the unlimited account of the logically possible this hypothesis (like almost any other) can be saved. However, in this process the hypothesis would become a sheer *a priori* speculation, entirely unrelated to human experience. It is *conceivable* that there are centaurs and mermaids on Mars or elsewhere ; but to affirm this merely because it is conceivable, and because it must be the case if every possible kind of being is to exist,

[1] Thus Leibniz, who upholds the completeness of the chain of being, suggests in passing that the species may be 'not always in one and the same globe or system'. (*New Essays Concerning Human Understanding*, bk. III, chap. vi, para. 12).

would be to have turned the corner from rational speculation into uncontrolled fantasy.

If we refuse to proceed in this way we are obliged to modify our picture in one or other of two ways that are equally incompatible with Christian theism. Starting from the principle of plenitude, and acknowledging that not every conceivable form of being has in fact been realized, we must either allow the divine creative energy to be limited, or else postulate a fixed realm of possibilities to which even God cannot add and which accordingly limits the exercise of His creative power. There would then be in the nature of things just so many possible forms of life : a definite scheme of species such that when all of them have come to be, God's creative work is perforce completed. This latter would perhaps not be so unacceptable to theism but for the fact that the range of possibilities that has in fact been realized does not seem to have any logically or factually ultimate character, such that the human mind is unable to visualize anything beyond it. For we can imagine all manner of possible extensions of nature as it is. Thus, if we follow the logic of the principle of plenitude, we are driven to think either that God lacked the creative power to realize those further possibilities, or that He is restricted by the limitations of the set of Platonic Ideas or archetypes. That the latter exist within God Himself, as an aspect of the divine mind, rather than being a factor external and alien to His nature, would not lessen, but on the contrary increase, the weight of a limitation that even our finite human minds are able to transcend in thought.

A further difficulty will reveal itself in the use of the principle of plenitude in the interests of theodicy when we see that principle carried to a logical conclusion by Leibniz in the eighteenth century. For Leibniz this is the best of all possible worlds, even though it contains, under the principle of plenitude, all kinds and levels of beings and their inevitable conflicts. But if this world, with its evils, is the best that is possible, there is no scope or hope for improvement. If all things are now perfect, we have no recourse left but to despair !

The principle of plenitude is one of the theodicy-themes that seek a solution to the problem of evil by looking to the past, whether to the character of the original creative act or, in the case of the fall of man, to a primeval creaturely act. This backward-looking stance is a main characteristic of the Augustinian type of theodicy, and needs (I shall argue) to be either supplemented or supplanted by a more forward-looking theology.

IV. The Aesthetic Theme

6. THE AESTHETIC THEME IN AUGUSTINE

What I am calling Augustine's aesthetic theme is his affirmation of faith that, seen in its totality from the ultimate standpoint of the Creator, the universe is wholly good; for even the evil within it is made to contribute to the complex perfection of the whole.[1] As Harnack says, 'Augustine never tires of realizing the beauty (pulchrum) and fitness (aptum) of creation, of regarding the universe as an ordered work of art, in which the gradations are as admirable as the contrasts. The individual and evil are lost to view in the notion of beauty. . . . Even hell, the damnation of sinners, is, as an act in the ordination of evils (ordinatio malorum), an indispensable part of the work of art.'[2] This is a fourth theme in Augustine's treatment of the problem of evil, in addition to his arguments concerning evil's privative character and its origin in misused free will, and his use of the principle of plenitude to account for the world's inclusion of lower as well as higher forms of being. The aesthetic idea embraces the other three, but is most closely related to the world's graded diversity, as appears in the following passage from Plotinus, from whom Augustine received this as well as some of the other main aspects of his theodicy:

[1] Billicsich aptly calls this kind of view 'aesthetic optimism'. (*Das Problem des Übels*, ii. 182.)
[2] Adolf von Harnack, *History of Dogma*, 3rd ed., trans. James Millar (London : Williams & Norgate, 1898, and New York : Russell & Russell, 1958), vol. v, p. 114.

The Reason-Principle is sovereign, making all : it wills things as they are and, in its reasonable act, it produces even what we know as evil : it cannot desire all to be good : an artist would not make an animal all eyes; and in the same way, the Reason-Principle would not make all divine; it makes Gods but also celestial spirits, the intermediate order, then men, then the animals ; all is graded succession, and this in no spirit of grudging but in the expression of a Reason teeming with intellectual variety.

We are like people ignorant of painting who complain that the colours are not beautiful everywhere in the picture : but the Artist has laid on the appropriate tint to every spot. Note also that cities, however well governed, are not composed of citizens who are all equal. Again, we are censuring a drama because the persons are not all heroes but include a servant and a rustic and some scurrilous clown ; yet take away the low characters and the power of the drama is gone; these are part and parcel of it.[1]

In similar vein we find Augustine writing, 'All have their offices and limits laid down so as to ensure the beauty of the universe. That which we abhor in any part of it gives us the greatest pleasure when we consider the universe as a whole. . . . The very reason why some things are inferior is that though the parts may be imperfect the whole is perfect, whether its beauty is seen stationary or in movement. . . . The black colour in a picture may very well be beautiful if you take the picture as a whole.'[2] And in one of his anti-Manichaean writings Augustine points out that such 'evils' as the substance that poisons, the fire that burns, and the water that drowns are evil only in a relative sense. Substances are not poisonous or harmful in themselves (if they were, the scorpion, for instance, would be the first to die of its own poison) but harmful only when brought into conjunction with other substances with which they disagree.[3]

However, Augustine's main concern, in the numerous

---

[1] *En.* iii. 2, 11. Cf. ii. 9, 13 ; iii. 3, 7.
[2] *True Religion*, xl. 76. Cf. ibid., 77 ; *N.G.* xvi ; *F.W.* iii. v. 13 ; ix. 24–27; *C.G.* xi. 16–18, 22, 23 ; xii. 4. For the Platonic starting-point of this strand of thought, see Plato's Laws, x. 903. Between Plato and Augustine the 'aesthetic' conception of evil occurs in Epictetus (born *c.* A.D. 60), in the *Discourses*, i. 12, and in Marcus Aurelius (A.D. 121–80) in the *Meditations*, vi. 42.
[3] *On the Morals of the Manichaeans*, i. Cf. *Conf.* vii. 22 ; *C.G.* xi. 22.

pages in which he expresses the 'aesthetic' view, is to show
that 'the universe even with its sinister aspects is perfect'[1]

To thee there is no such thing as evil, and even in thy whole
creation taken as a whole, there is not; because there is nothing
from beyond it that can burst in and destroy the order which thou
hast appointed for it. But in the parts of creation, some things,
because they do not harmonize with others, are considered evil.
Yet those same things harmonize with others and are good, and
in themselves are good. ... I no longer desired a better world, because my thoughts ranged over all, and with a sounder judgment
I reflected that the things above were better than those below, yet
that all creation together was better than the higher things alone.[2]

From the divine point of view, then, each being realizes
a creative possibility; each is in its own way good; and each
in its proper place in the system contributes to the perfection
of the whole. A human being, however, regarding the creation from a limited perspective within it instead of judging
the creatures according to their value in the sight of God, is
liable to think only of their practical usefulness or otherwise
to himself:

according to the utility each man finds in a thing, there are various
standards of value, so that it comes to pass that we prefer some
things that have no sensation to some sentient beings. And so
strong is this preference, that, had we the power, we would abolish
the latter from nature altogether, whether in ignorance of the
place they hold in nature, or, though we know it, sacrificing them
to our own convenience. Who, for example, would not rather
have bread in his house than mice, gold than fleas?[3]

As an element within the aesthetic view Augustine singles
out the transiency of so many natural phenomena — the
decay of vegetation that provides the beauty of nature's
face and forms, and the incessant devouring of life by life in
the animal kingdom. All this, he says, is part of the perfection of the creation when it is seen, not as a static structure,
but as an organic process:

they are so ordered that the weaker yield to the stronger, and the
feebler to those that have greater might, and the less powerful

---

[1] *Soliloquies*, i. i. 2.   [2] *Conf.* vii. 13.   [3] *C.G.* xi. 16.

to the more powerful. . . . When things pass away and others succeed them there is a specific beauty in the temporal order, so that those things which die or cease to be what they were, do not defile or disturb the measure, form or order of the created universe.[1]

7. ANIMAL PAIN IN A PERFECT WORLD

Augustine applies this view to both beasts and plants; but it is more readily applicable to the latter than the former. The leaves whose decay makes the gorgeous reds and yellows and purples of the autumn trees feel no pain or fear. But the deer that is torn to pieces by the lion, or the mouse by the cat, belongs to another dimension of nature which cannot so easily be fitted into the aesthetic view. Augustine deals with the problem of animal pain in *On Free Will*, but in so forced a way as to suggest that he was insensitive towards the animal kingdom. He accuses those who see a problem in animal suffering of demanding 'that the bodies of animals shall not suffer death or any corruption', as though there were no difference between a natural death in old age and a violent and painful death in the midst of the natural life span. He then proceeds to argue that the pain that an animal suffers bears witness to the desire for bodily unity inherent in the animal soul. 'We should never know what eagerness there is for unity in the inferior animal creation, were it not for the pain suffered by animals. And if we did not know that, we should not be made sufficiently aware that all things are framed by the supreme, sublime and ineffable unity of the Creator.'[2] If taken seriously, this implies that the vast mass of animal suffering is divinely arranged for our benefit; and this in turn implies a strikingly different conception of God from that of the heavenly Father who cares for the birds and flowers, and without whose knowledge not a sparrow falls.

Elsewhere, however, Augustine has another view of animal pain which, even if not wholly satisfying, is at least free from offensive moralizing. It was suggested by Plotinus, who in

---

[1] *N.G.* viii.   [2] *F.W.* iii. 69.

a striking passage treats of the apparent evil of life preying upon life in the animal kingdom :

> This devouring of Kind by Kind is necessary as the means to the transmutation of living things which could not keep form for ever even though no other killed them : what grievance is it that when they must go their dispatch is so planned as to be serviceable to others? Still more, what does it matter when they are devoured only to return in some new form? It comes to no more than the murder of one of the personages in a play ; the actor alters his make-up and enters in a new role. The actor, of course, was not really killed ; but if dying is but changing a body as the actor changes a costume, or even an exit from the body like the exit of the actor from the boards when he has no more to say or do — though he will still return to act on another occasion — what is there so very dreadful in this transformation of living things into one another?[1]

This thought is repeated, less poetically, by Augustine when he says,

> Since, then, in those situations where such things are appropriate, some perish to make way for others that are born in their room, and the less succumb to the greater, and the things that are overcome are transformed into the quality of those that have the mastery, this is the appointed order of things transitory. Of this order the beauty does not strike us, because by our mortal frailty we are so involved in a part of it, that we cannot perceive the whole, in which these fragments that offend us are harmonized with the most accurate fitness and beauty.[2]

There is a certain degree of merit in Augustine's application of the aesthetic principle to nature, considered as a great organism whose life involves ceaseless process and change. Such change necessarily consists in the continual emergence and elimination of the individual vegetable and animal units. The on-going life of nature flows, so to speak, through these units, and each has its own brief period of individual existence before the elements composing it are reclaimed, only to be organized again into new forms within the larger whole. This way of viewing the fleeting transiency of the individual items of nature is, however, more satisfying

---

[1] *En.* iii. 2. 15.
[2] *C.G.* xii. 4. Cf. xii. 5 ; *F.W.* iii. xv. 42–43 ; *N.G.* viii.

with regard to the mineral and vegetable kingdoms than to the realms of sentient life. It leaves unillumined a large part of the mystery of animal pain, constituting a problem that apparently did not deeply move or concern Augustine.[1]

8. HELL AND THE PRINCIPLE OF MORAL BALANCE

There is, however, an even larger question : Can Augustine's aesthetic theme include *moral* evil? For if it is a serious contention that 'the universe is perfect'[2] and that 'To thee there is no such thing as evil',[3] because in the sight of God all things harmonize, this principle must apply to human sin as well as to physical transiency and animal pain. And accordingly Augustine does boldly embrace the fall and damnation of sinners within his aesthetic theodicy. He does so by invoking the principle of moral balance already referred to. Sin, which is a culpable misuse of freedom, is not allowed to mar the perfection of God's universe, because the balance of the moral order is preserved by the infliction of appropriate punishment. A universe in which sin exists but is precisely cancelled out by retribution is no less good than a universe in which there is neither sin nor punishment : 'since there is happiness for those who do not sin, the universe is perfect ; and it is no less perfect because there is misery for sinners . . . the penalty of sin corrects the dishonour of sin'.[4] Again, 'There is no interval of time between failure to do what ought to be done and suffering what ought to be suffered, lest for a single moment the beauty of the universe should be defiled by having the uncomeliness of sin without the comeliness of penalty.'[5] For 'no multitude of spiritual creatures condemned for their demerits could throw into confusion an order which has a proper and suitable place for as many as are damned'.[6] And so the moral-aesthetic perfection of the universe is maintained : 'For as the beauty of a picture is increased by well-managed shadows, so, to the eye that has skill to discern it,

[1] For further discussions of the problem of animal pain see below pp. 108 f. and 345 f.   [2] *F.W.* iii. ix. 26.   [3] *Conf.* vii. xiii. 19.
[4] *F.W.* iii. ix. 26. Cf. ibid. 25 ; iii. xii. 35 ; xv. 43–44 ; *True Religion*, xxiii. 44 ; *N.G.* vii, ix, xx, and xxxvii ; *C.G.* xi. 23 ; xii. 3.
[5] *F.W.* iii. xv. 44.   [6] *F.W.* iii. xii. 35.

the universe is beautified even by sinners, though, considered by themselves, their deformity is a sad blemish.'¹

Augustine does not go so far as to teach that sin is positively necessary to the perfection of the universe, though he does at one point incautiously allow himself to say: 'The fact that there are souls which ought to be miserable because they willed to be sinful contributes to the perfection of the universe.' However, he goes on at once to explain his meaning more fully :

Neither the sins nor the misery are necessary to the perfection of the universe, but souls as such are necessary which have power to sin if they so will, and become miserable if they sin. If misery persisted after their sins had been abolished, or if there were misery before there were sins, then it might be right to say that the order and government of the universe were at fault. Again, if there were sins and no consequent misery, that order is equally dishonoured by lack of equity. But since there is happiness for those who do not sin, the universe is perfect ; and it is no less perfect because there is misery for sinners. . . . Hence the penal state is imposed to bring [the universe] into order, and is therefore in itself not dishonourable. Indeed it compels the dishonourable state to become harmonized with the honour of the universe, so that the penalty of sin corrects the dishonour of sin.²

As well as repairing by proportionate punishment the breach produced by sin in the moral fabric of the universe, God so overrules human affairs as to use sinners, including even the devil himself, for the furtherance of His own good purposes. In what has become one of the key sentences in the whole literature of theodicy, Augustine says, 'God judged it better to bring good out of evil, than to suffer no evil to exist.'³ Again, 'For God would never have created any, I do not say angel, but even man, whose future wickedness He foreknew, unless He had equally known to what uses in behalf of the good He could turn him, thus embellishing the course of the ages, as it were an exquisite poem set off with antitheses.'⁴

This last phrase introduces the remaining and very minor

¹ *C.G.* xi. 23.　　²　*F.W.* III. ix. 26.　　³ *Ench.* xxvii.
⁴ *C.G.* xi. 18. Cf. xi. 17 and 23; xii. 1 and 2; xiv. 27; *Ench.* vii. 27; *N.G.* xxxii and xxxvii.

of the several Augustinian themes. This is the 'contrast' theory of evil, which Augustine invokes when he says that 'In this universe, even what is called evil, when it is rightly ordered and kept in its place, commends the good more eminently, since good things yield greater pleasure and praise when compared to the bad things.'[1] This represents, however, little more than a passing thought in Augustine and should not be made central or even near-central to his theodicy.

That God mysteriously overrules the malicious deeds of the wicked (and, when necessary, the well-intentioned but ill-judged efforts of the virtuous) and eventually brings good out of evil, and indeed brings an eternal and therefore infinite good out of a temporal and therefore finite evil, is a thought of great promise for Christian theodicy. But it is completely vitiated in its Augustinian context by the proviso that this divine activity of drawing good out of evil is to operate only in a minority of cases and that the great majority of mankind is eternally to sin and eternally to suffer torment. For a combination of sin and suffering that is endless is, by definition, an evil that is never turned to good but remains for ever a blot upon God's creation.

Thus hell, understood as Augustine so explicitly understood it, must be accounted a major part of the problem of evil! Whatever gain there may be from other aspects of Augustine's discussion is outweighed again and again by the endlessly accumulating burden of sin and pain contained in the hell that he regards as a permanent feature of the universe. This is a matter that we shall have later (in Chapter XVII) to discuss more fully. At the moment, however, we simply note that those exingencies of Christian theology that have led to the doctrine of eternal punishment are directly in conflict with those other Christian impulses that underlie the search for a theodicy.

[1] *Ench.* iii. 11. Cf. *C.G.* xi. 18 and 23. The first appearance of the contrast theme is in Plato's *Theaetetus*, 176A, and it reappears in the ancient world in, for example, Plutarch's *De communibus notitiis*, 13–15.

CHAPTER V

CATHOLIC THOUGHT FROM AUGUSTINE
TO THE PRESENT DAY

1. AUGUSTINE'S THEODICY WRIT LARGE:
HUGH OF ST. VICTOR

THERE is no need to trace Catholic teaching concerning the mystery of evil step by step through the centuries by referring to all the medieval and modern theologians of the Roman communion who have treated the subject. Such a catalogue would be almost wholly repetitious, for the theodicy developed by Augustine and later restated by Aquinas has remained essentially unchanged and is still being presented today by Catholic writers. They generally cite St. Thomas as their principal source. But the student of the theodicy-problem who is not himself a Thomist is likely to give more attention to Augustine than to the angelic doctor. For the latter added nothing essential to what his great predecessor had said on the subject, although he did at certain points refine and strengthen the presentation of the tradition. But the topic occupies a smaller and less central place in Aquinas' thought than in that of Augustine, who must be acknowledged as the true fountainhead of the Catholic theodicy. The streams of thought that were fused together in the fires of Augustine's genius — the aesthetic view of the world, with its allied notion of evil as non-being, and the Judaic-Christian sense of sin as a personal rebellion against God — were already in existence, flowing in their separate courses, long before Augustine. But his combination of them within the framework of a massive and coherent Christian vision of the universe was a major event, of which the Thomist theodicy which is still in use today is a continuing echo.

There is one figure in the period between Augustine and Aquinas at whom it may be instructive to glance briefly.

This is Hugh of St. Victor (1096–1141), a theologian who in
general followed Augustine so closely that he has been called
Alter Augustinus. He carries some of the Augustinian ideas
further, or at least makes them more starkly explicit, than did
Augustine himself; and he also at one point introduces a new
note that is at variance with one of the original Augustinian
themes. Hugh of St. Victor assumes, without argument, the
privative nature of evil.[1] But it is the aesthetic conception of
a complex and balanced cosmic perfection that is most prominent in his response to the mystery of evil. Where Augustine
is content to leave this aesthetic vision enveloped in a certain
aura of mystery, Hugh of St. Victor pushes it straight to its
logically inevitable conclusion. He was perhaps therein less
discreet than Augustine; but we may be grateful for his
indiscretion, since it facilitates critical discrimination concerning this type of theodicy. According to Hugh, God made
all that is good, and permitted all that is evil; and in this permission, 'He permitted well and it was good that He permitted,
even if that was not good which He permitted'.[2] Thus, 'He
wills evil to be, and in this He wills nothing except good, because it is good that there be evil; He does not will evil itself,
because evil itself is not good.'[3] In the realm of moral evil:

We have said that the greater good is to be good from evil and
from good rather than from good alone. So evils had to be permitted to be, since it was good that they be from which good was
to be. So God commanded what was good for each and permitted
what was good for the whole world, since the good of the whole
world, which he ought not to hinder, was greater. . . . If we do
evil He wills that we do not do good, and He approves this because it is good. . . . His omnipotent will is established certain
and firm in His good pleasure, approving goods and permitting
evils for the sake of the good . . .[4]

And in the eternity beyond this life it is still good that there
should be perpetual evil to set off the good. The tormented
souls in hell will be visible to the redeemed in glory, but not

[1] *On the Sacraments of the Christian Faith*, trans. Roy J. Deferrari (Cambridge,
Mass.: the Medieval Academy of America, 1951), bk. 1, pt. v, chap. 29. Cf.
ibid. 1. vii. 16.   [2] Ibid. 1. iv. 4.
[3] Ibid. 1. iv. 13. Cf. 1. iv. 5, 'it was good that there be both good and evil'.
[4] Ibid. 1. iv. 23. Cf. ibid. 4.

vice versa, and (as Augustine had taught before him) the heavenly spectators' view of hell will have the effect of enhancing, by contrast, their own bliss :

> The unjust will surely burn to some extent so that all the just in the Lord may see the joys that they receive and in those may look upon the punishments which they have evaded, in order that they may realize the more that they are richer in divine grace unto eternity, the more openly they see that those evils are punished unto eternity which they have overcome by His help.[1]

In the writings of Hugh of St. Victor we thus have presented with a total lack of reticence the monstrous moral paradox that God intends His creatures to sin (since it is better that there should be sin and redemption in the universe than that there should not), but that He nevertheless inflicts eternal punishment upon them when they do sin, though arbitrarily reprieving some and granting to them the joy of His presence — a joy that is supposed not to be eclipsed, but on the contrary heightened, by their contemplation of the torments of others who were no more wicked than themselves but who were left behind when the arbitrary divine decree lifted some out of the *massa perditionis*. This is a morally repugnant set of ideas ; and a theology cannot go unchallenged when it is repugnant to the moral sense that has been formed by the religious realities upon which this theology itself professes to be based. It is without biblical warrant, and can be connected with the teaching of our Lord only by long speculative and theoretical extensions and constructions. Not only are these theoretical constructions the work of the human mind, rather than any kind of divinely revealed truths, but they are not among the more beautiful or morally elevated creations of human speculation. There is no reason why they should not be subjected to uninhibited Christian scrutiny ; and in the light of this they may well have to be rejected as products of a sinful imagination. For sin — the spiritual sins of pride, lovelessness, and cruelty — can infect theological work as well as other forms of human activity. And any attribution to the Supreme-Being of qualities that are incompatible with the demanding and yet always

[1] *On the Sacraments of the Christian Faith*, II. xviii. 2.

gracious agape revealed in the Incarnation must come under suspicion of being projections of our own darkened hatreds, resentments, and fears.[1]

In all this side of his thought Hugh of St. Victor is Augustinian; but at another point he diverges from his tradition. Instead of invoking the Christianized Neo-Platonic doctrine that the 'motive' behind creation was the impulse of the infinite divine goodness to manifest itself as widely as possible in a dependent order, so that the universe contains as many different kinds of creature as are compossible, Hugh puts man (or rational creatures in general) at the centre of creation. He speaks of 'Adam and Eve, for the sake of whom all other things were made',[2] and says,

> For man was made that he might serve God for whose sake he was made, and the world was made that it might serve man for whose sake it was made. . . . For God wished man to serve Him, in such a way, however, that by this service, not God, but man himself in serving should be helped, and He wished that the world serve man, and that from it man likewise should be helped, and that all good should belong to man, because for man's sake all this was made. . . . To such an extent then is the foundation of the rational creature proven to be superior to all other things which were made for its sake, because it itself is the cause of these.[3]

This represents a significant departure from the Neo-Platonic side of the Augustinian tradition, and points in its further implications away from the principle of plenitude, and the aesthetic type of theodicy that tends to accompany it, towards a different approach to the mystery of evil. This alternative was not developed by Hugh of St. Victor, but it has been developed by later thinkers in a different tradition, and we shall examine it in subsequent chapters.

2. Thomas Aquinas

St. Thomas Aquinas (1226–74) writes about the problem of evil in an abstract and detached manner that seems well

---

[1] It has been suggested that the theology implied by the title of Jonathan Edwards' famous sermon, 'Sinners in the Hands of an Angry God', reflects God in the hands of angry sinners!

[2] *On the Sacraments of the Christian Faith*, i. i. 29.

[3] Ibid. i. ii. 1. Cf. i, Prologue, chap. 3; i. v. 3.

suited to the somewhat impersonal aesthetic principle that governs his theodicy. Following Augustine, he defines evil in general in negative terms. But he renders the traditional definition more precise by giving priority, among the several terms used by Augustine, to 'deprivation' and 'defect'. Evil is 'the absence of the good which is natural and due to a thing'[1]— as, for example, blindness is the deprivation of a good that is proper to a man but not proper to a stone. Evil, so conceived, exists only in the sense in which blindness exists. As blindness is not a reality additional to the eyes, but a defect of them, so evil in general is not a substance or entity on its own, but consists in the lack of some positive power or quality that a thing ought by its nature to possess.[2] Since there cannot be a defect except within something good, it follows that there cannot be a purely evil being.[3]

Being thus wholly negative, evil has no causal efficacy of its own.[4] It can, however, operate as a cause *per accidens* through the activity of the good thing of which it is a defect.[5] Again evil as such can never be desired or intended, for all desire is, by definition, directed to the good; when evil is produced this must be beside the intention of the agent.[6] Paradoxically, then, the cause of evil can only be something good, since evil as such cannot act as a cause. Good is accordingly the cause of evil — but only accidentally and in virtue of some defective power of the agent.[7]

Given this metaphysical analysis, how does Aquinas construct his theodicy — his defence of the righteousness of God in face of the fact of evil?

He divides evils into those affecting 'voluntary things' (i.e. angels and men) and those affecting the rest of creation. So far as the latter is concerned, the variety and inequalities of nature are accounted for on the basis that God 'produced things into being in order that His goodness might be com-

---

[1] *Summa Theologica*, trans. Fathers of the English Dominican Province (London: R. & T. Washbourne Ltd., 1912), pt. I, Q. xlix, art. 1 : 'Malum enim est defectus boni quod natum est et debet haberi.'

[2] *S.T.* pt. I, Q. xlviii, art. 2.

[3] *Summa contra Gentiles*, bk. iii, chap. 15.

[4] *S.T.* pt. I, Q. xlviii, art. 1.

[5] *S.c.G.* iii. 14.   [6] *S.c.G.* iii. 3–4.   [7] *S.c.G.* iii. 10.

municated to creatures [propter suam bonitatem communicandam creaturis], and be represented by them ; and because His goodness could not be adequately represented by one creature alone, He produced many and diverse creatures, that what was wanting to one in the representation of the Divine goodness might be supplied by another'.[1] Again, 'as the Divine wisdom is the cause of the distinction of things for the sake of the perfection of the universe, so is it the cause of inequality. For the universe would not be perfect if only one grade of goodness were found in things. [2]

Nevertheless, although God has operated in creation *propter perfectionem universi*, it is not to be inferred that the universe is perfect in the sense that God could not, had He wished, have made a better one. In connection with the power of God, Aquinas raises the question, Could God have created a different universe from ours? This being answered in the affirmative, he then asks, Could He have created a *better* universe than ours? Aquinas' reply is that God could indeed have created a different and better universe, but that, as regards the divine act of creating, He could not have better created our present universe than He did, nor, as regards the thing created, is this universe (whilst remaining *this* universe) capable of being improved :

Given the things which actually exist, the universe cannot be better, for the order which God has established in things, and in which the good of the universe consists, most befits things. For if any one thing were bettered, the proportion of order would be destroyed ; just as if one string were stretched more than it ought to be, the melody of the harp would be destroyed. Yet God could make other things, or add something to the present creation ; and then there would be another and a better universe.[3]

In considering why, within the universe that he has chosen to create, God has made intellectual beings as well as subrational natures, Aquinas goes back to the Platonic and Neo-Platonic principle of plenitude. Because the perfection of the universe requires that there be every kind of creature, it requires, *inter alia*, that there be rational creatures. This

[1] *S.T.*, pt. I, Q. xlvii, art. I. Cf. *S.c.G.* ii. 45.
[2] *S.T.*, pt. I, Q. xlvii, art. 2.
[3] *S.T.*, pt. I, Q. xxv, art. 6.

thought is reinforced by a further argument, employing a principle which is at once aesthetic and mathematical :

> For then is an effect most perfect when it returns to its source ; wherefore of all figures the circle, and of all movements the circular, are the most perfect, because in them a return is made to the beginning. Hence, in order that the universe of creatures may attain its ultimate perfection, creatures must return to their principle. Now each and every creature returns to its principle, in so far as it bears a likeness to its principle, in keeping with its being and nature, wherein it has a certain perfection. . . . Since then God's intellect is the principle of the creatures' production, as we proved above, it was necessary for the universe's perfection that there should be some intelligent creatures.

Within the realm of spiritual beings all evil consists either in sin or its punishment,[1] and its occurrence is covered by the same general principle as that applied by Aquinas to nature as a whole :

> As, therefore, the perfection of the universe requires that there should be not only beings incorruptible, but also corruptible beings ; so the perfection of the universe requires that there should be some which can fail in goodness, and thence it follows that they do fail [ad quod sequitur ea interdum deficere].[2]

One might well inquire into the grounds of the view that what can fail sometimes will fail. Is this an empirical generalization reflecting our observation that things that are capable of failing sooner or later do so? Or does it, more narrowly, simply restate the circumstance that man, created free, has in fact fallen? Or again, does it postulate some necessity that a fallible creature must fail? Or that out of a collection of free beings a certain proportion must fail? On neither type of interpretation can Aquinas' statement do more than gloss over a major difficulty. For if a free creature *necessarily* fails, the Being who created him must at least share the responsibility for his failure. If, on the other hand, it is just an observable fact that free beings have fallen, to report this does not serve to *explain* anything, and we are left with the undiminished

---

[1] *S.T.*, pt. i, Q. xlviii, art. 5.
[2] Ibid. art. 2.

paradox of a being who is created good but who spontaneously becomes evil.

At any rate, we now have the following overall picture:

God [makes] what is best in the whole, but not what is best in every single part, except in relation to the whole, as was said above. And the whole itself, which is the universe of creatures, is all the better and more perfect if some things in it can fail in goodness, and do sometimes fail, God not preventing this. . . . Hence many good things would be taken away if God permitted no evil to exist; for fire would not be generated if air was not corrupted, nor would the life of the lion be preserved unless the ass was killed. Neither would avenging justice nor the patience of a sufferer be praised if there were no injustice.[1]

Here, as Maritain says, 'St. Thomas considers reality from a particular point of view, from the point of view of the order of nature, of the universe as a work of art made by God. . . .'[2] The same aesthetic standpoint is evident in the following passage :

The order of the universe requires, as was said above, that there should be some things that can, and do sometimes, fail. And thus God, by causing in things the good of the order of the universe, consequently and as it were by accident [quasi per accidens], causes the corruption of things. . . . Nevertheless the order of justice belongs to the order of the universe; and this requires that penalty should be dealt out to sinners. And so God is the author of the evil which is penalty, but not of the evil which is fault.[3]

One must, however (as Maritain points out),[4] relate this essentially aesthetic vision of the universe, with its orderly balancing of good and evil, to Aquinas' teaching concerning grace and redemption. For in his Treatise on the Incarnation, he says (though without elaboration) that 'God allows evils to happen in order to bring a greater good therefrom', and quotes the startling and pregnant sentence of the 'O felix culpa';[5] and in his Treatise on Grace he exalts as the supreme

[1] Ibid.
[2] Jacques Maritain, *St. Thomas and the Problem of Evil* (Milwaukee : Marquette University Press, 1942), p. 10.
[3] *S.T.*, pt. I, Q. xlix, art. 2.
[4] Maritain, op. cit., pp. 9–10.
[5] *S.T.*, pt. III, Q. i, art. 3. ('O felix culpa, quae talem ac tantum meruit habere redemptorem' — 'O fortunate crime which merited such and so great a redeemer'.)

activity of divine grace the saving of sinners : 'the justification of the ungodly, which terminates at the eternal good of a share in the Godhead, is greater than the creation of heaven and earth, which terminates at the good of a mutable nature'.[1]

There are many points in St. Thomas' treatment of the problem of evil that call for comment or questioning — not least the cutting of a whole complex of Gordian knots by the principle that 'what can fail, sometimes will'— but perhaps it will be better to defer criticism until we have drawn into the discussion a presentation of the Thomist solution that is addressed directly to the contemporary world.

3. A CONTEMPORARY THOMIST PRESENTATION : CHARLES JOURNET

Roman-Catholic writings on the problem of evil that have any kind of official or semi-official standing usually follow closely the Thomist teaching, which in turn derives in its main lines from Augustine. In addition to these more traditional treatises there are a few writings by contemporary or recent Catholic authors that probe in new directions and that express, by implication, dissatisfaticon with certain aspects of the traditional teaching. Some of these will be referred to in Chapter XII. At present, however, we are still concerned with the 'main-line tradition' of Catholic theodicy, and I have selected as a contemporary representative of this the Abbé Charles Journet's *Le Mal*, which is available in English as *The Meaning of Evil*.[2]

Journet's discussion centres primarily upon evil as it is directly experienced by mankind, such evils consisting (following Augustine and Aquinas) either in sin or in punishment for sin. 'All the trials in our human lives', says Journet, 'are due to sin, though not all in the same way : some are attributable to original sin, others to our personal sins'.[3] However, he also discusses secondarily what he calls evil in nature, in

---

[1] *S.T.*, pt. II/I, Q. cxiii, art. 9.

[2] Charles Journet, *Le Mal* (Paris: Desclée de Brouwer, 1961). The quotations in this chapter are from the Eng. trans. by Michael Barry, *The Meaning of Evil* (London: Geoffrey Chapman, 1963).

[3] *M.E.* p. 218.

the forms of 'antagonism, destruction, deformation, disease, accident, death, affliction, suffering, etc.,'[1] the two circles intersecting where natural evil impinges upon mankind as part of the appointed punishment for man's sin.[2]

Following Aquinas, Journet defines evil as a privation — a privation, not in the sense of a mere absence of good, but in the sense of the loss of some good that should be present.[3] Journet regards this 'apophatic' or negative definition of evil as a distinctively Christian contribution.

It represents [he continues] the most delicate and penetrating intellectual handling of evil which the mind can attain to, either on the metaphysical or theological plane. Evil is accorded a great deal of room, so that it can be seen to its full extent, and at the same time the metaphysical poverty of evil is laid bare. But with the affirmation that evil exists, yet lacks any substance, comes the triumph over the dilemma to which those succumb who either deny the reality of evil because of God's goodness and infinite power, or deny God's goodness and infinite power because of the reality of evil.[4]

Pointing out that, although metaphysically negative, evil is neither non-existent nor powerless, Journet uses a striking simile : 'Let us therefore not talk of pure non-existence, but of an existence which, like letters hollowed out of stone, can be a terrible reality.'[5] He can accordingly say: 'The paradox of evil is the terrible reality of its privative existence.'[6] In all this Journet reflects faithfully the Augustinian–Thomist tradition.

Following Aquinas, Journet insists that God was under no kind of obligation or need to create, and that ours is not the sole universe that He could have created, but only one out of a limitless realm of possibilities. Further, he repudiates the

[1] *M.E.*, p. 50.
[2] Journet has a more complex theoretical division (pp. 50–54) based upon the 'three worlds' in which evil appears : 'the natural or physical world', 'the world of freedom', and 'the world of grace'. However, this schema proves more confusing than illuminating as a principle of division, for it leads to overlapping. It turns out that the first world, described as 'natural and physical', includes spiritual and incorruptible beings, namely angels and human souls ; and that the second and third worlds have the same content so far as evil is concerned, namely guilt and punishment. [3] *M.E.*, pp. 27–28.
[4] *M.E.*, p. 35. [5] *M.E.*, p. 43. [6] *M.E.*, p. 47.

Leibnizian thesis that if He elects to create, God must produce the best possible universe :

> The mistake is to think that by reason of his infinite goodness God is bound to create rather than not to create such and such a better world, rather than some other one which is simply good ; or to create such and such a world from which evil and sin would be banished rather than that one in which, with a view to some great good, evil and sin would be admitted ; to create the 'best of all possible worlds'. . . .
> An unbridgeable gap separates the infinite, uncreated good [i.e. God] from the whole universe of created and creatable things. He is completely disproportionate to them. . . . Even for God it will never be possible to pour out the fullness of uncreated being into the vessels of created beings, or to enclose the infinite within the finite. Whatever world he decides to make, what will be manifested of his infinite fullness will never be equivalent to what remains to be manifested. There will always be an infinite margin in which other worlds could occur.[1]

Acknowledging that God could have so made His world, or so intervened in its development, as to have obviated its worst horrors, Journet continues,

> Would the world as a whole be better? Perhaps it would. But if God creates, what is he bound to do in virtue of His justice, wisdom and infinite goodness? He is bound to make a good world in which evil cannot ultimately prevail over good. Is there a level or degree of goodness in the world that is not infinitely insufficient to express infinite Goodness? Is there a world, better than such and such another one, that God would have to choose if He did not want to be blameworthy? Here again is the illusion which makes us think that God's justice and wisdom, and especially His goodness, would be better to the extent that the world he created was better — in which case he would have to create an infinitely good world — and that, if he decides to create, he is constrained to make the best of all possible worlds.[2]

There is here a direct repudiation of a principle that seemed self-evident to Leibniz, and the issue between them will be explored further when we come to Leibniz and his *Theodicy*.[3]

We may now see how Journet deals, against the background of his teaching, with the various kinds of evil in the world.

---

[1] *M.E.*, pp. 104, 110–11.　　[2] *M.E.*, p. 115.　　[3] See pp. 166 f, below.

Concerning evils other than sin itself — that is to say, pain
and suffering, whether borne punitively by mankind or non-
punitively by the lower animals — Journet confronts the
question whether they are willed by God. He replies that
they are not willed by God directly and as such, but are
indirectly willed by Him, or permitted, as unavoidable or
accidental side-effects of the promotion of some great good.
'The evil in nature and evil of punishment, which are not
willed for their own sakes or intended, and which are only
willed by reason of the good which they include, can be said
to be willed *accidentally*, or more simply, *permitted*.'[1] Journet
is here using an analogue of a principle of justification that is
familiar in Catholic moral theology under the name of the
rule, or principle, of double effect. This lays down that
when someone performs an action for the sake of a sufficiently
important good that he seeks as its result, but in a situation
in which bad secondary effects are unavoidably entailed, the
agent is not morally blameworthy in respect of these latter.
As Aquinas, the originator of this principle, says, 'Nothing
hinders one act from having two effects, only one of which is
intended, while the other is beside the intention. Now moral
acts take their species [i.e. as good or bad] according to what
is intended, and not according to what is beside the intention,
since this is accidental . . .'[2] For example, a man lethally
attacked by another may defend his own life in the only way
available in the circumstance even though this should involve
killing his assailant. In such a case the principal effect of his
action is good, namely the saving of his own life, and the fact
that there is an undesired and unavoidable second effect in
the death of the aggressor does not prevent the action from
being morally permissible. Or again, in a 'just war' a certain
city may be bombarded as a licit military target, and as an
unsought concomitant a number of civilians may be killed.
This unfortunate secondary effect does not nullify the justice
of the action as a whole, which aims at a morally righteous
outcome. Or again, in a pregnancy a cancerous uterus may

[1] *M.E.*, p. 77. Instead of 'the good which they include', Journet's discussion as a whole suggests rather 'the good which includes them'.
[2] *S.T.*, pt. II/II, Q. lxiv, art. 7.

be excised in order to save the mother's life even though the known secondary effect of the operation will be the death of the unborn child.[1]

On essentially the same basic principle, Journet (1) justifies the element of suffering in nature as an unavoidable secondary effect of the creation of a rich and varied universe; and (2) likewise justifies the evil of human suffering (which is always, for Journet, punishment for sin) as the secondary effect of God's preservation of the moral order.

(2) Concerning the animal pain that occurs so incessantly within the economy of nature, Journet says that the 'evil of nature is *permitted* in the sense that it is in itself inseparably connected with a good which is intended and directly willed by God';[2] for 'the evolution of the universe does not go on without destruction, new forms of life appear to take the place of the old'.[3]

> The evil in nature [he says] is not intended by God. What he intends is the good in all its countless proliferations. . . . But this good intended by God inevitably holds an element of nothingness and privation which must be accepted, and which in this sense is willed indirectly. The forms of evil in nature, with their proliferations, are willed indirectly by God in so far as they accompany the forms and proliferations of good which are directly willed by him.[4]

Journet elaborates this position as it concerns the sufferings of animals. He points out that 'what accounts for the presence of suffering in animals is precisely the perfection and delicacy of their organic constitution' and continues, 'To do away with the vulnerability of an animal's organs would be to do away with the animal itself; and that would be to deprive the universe of this immense reach of life, incomparably rich and variegated, which stretches from the lower world of minerals and plants to that of man'.[5] Journet then raises the inevitable question, 'granted the existence of our

---

[1] On the other hand, in the case of other operations in which the death of the baby would be *directly* intended, even though in order to save the mother's life or health, the rule of double effect cannot (according to Catholic teaching) be applied to justify the operation.

[2] *M.E.*, p. 147.   [3] *M.E.*, p. 74.   [4] *M.E.*, p. 181.
[5] *M.E.*, p. 139. Cf. p. 140.

animal world, and hence of suffering, could not the sum of suffering be less? The only answer is, of course, that it could be less. And also it could be greater. Here again we must avoid the perpetual trap, in which Leibniz was caught, of a God bound by his infinite goodness to create one particular world rather than any other one, and bound ultimately to create the best of all possible worlds.'[1]

Now the sufferings of animals constitute one of the most baffling aspects of the problem of evil. Although this is perhaps not the gravest and most oppressive of evil's many forms, it may nevertheless be the hardest for us to understand; for as well as being part of the general problem of evil it is wrapped in an impenetrable additional mystery of inaccessibility. We have no knowledge of the quality of experience below the level of self-consciousness, and can only make guesses in which there is every likelihood that we are projecting our human form of experience into creatures whose life is lived at a very different level. This psychic distance between ourselves and the lower animals aggravates the already profound mystery constituted for us by their existence and their pains. It may be that in the end Christian thought can only take refuge in this ignorance, without professing a positive solution of any kind. At all events, any positive solutions that are proposed must be subjected to the most careful and candid examination. What, then, are we to say about the solution here propounded by Journet?

It weaves together two strands of theological tradition, each of which has weighty authority behind it; but the resulting teaching, when viewed as a whole, must be frankly declared to be self-contradictory and lacking in rational feasibility.

The appeal to the distinction between divine 'willing' and 'permitting', which is the theological analogue of the rule of double effect in moral casuistry, may well be appropriate and illuminating in relation to God's dealings with free personal beings whom He has created. When such beings proceed in ways contrary to their Maker's desire, it may well be proper to say that their actions are permitted but not willed by Him.

[1] *M.E.*, p. 139.

But the distinction between permitting and willing can have no validity when applied to the planning and creating of the physical universe by an omnipotent Being.[1] The moral rule of double effect presupposes the limited power of the agent on whose behalf it is invoked. It is because the individual attacked by an aggressor, or the soldier waging a just war, or the physician attending a dangerous pregnancy is not able to effect his good primary purpose without at the same time permitting a secondary evil that he is held blameless in respect of that secondary effect. But if, on the contrary, it was within his power to avoid this, and he failed to do so, he would no longer be blameless. For this reason there can be no foundation for the plea of double effect in the case of the omnipotent creator. His power is, by definition, not limited and not insufficient. It cannot be said that He *must* or that He *can only* create a world in which life devours life and in which creatures are wounded, maimed, starved, frozen, diseased, and hunted to their death. Nor does Journet in fact say this. On the contrary, we have already noted that he affirms against Leibniz that 'God could, according to his absolute power, make better worlds than ours'.[2] Applying this doctrine to animal pain in a passage quoted above, he asks, 'granted the existence of our animal world, and hence of suffering, could not the sum of suffering be less? The only answer, of course, is that it could be less.'[3] Again, 'Could the level of the world's internal order be raised, thereby reflecting better the goodness of its supreme End? To this question there must be no hesitation about answering that all this would indeed be possible. God could, and still may, make this present world better.'[4]

What, in view of this, has become of the distinction with which Journet began, between divine willing and divine permitting? Journet is not after all claiming (as did Leibniz) that God willed the best world that He could and permits only such evil as is unavoidably attached to it. Instead he is saying that God willed this specific world, with this rather than

---

[1] Cf. Schleiermacher, *The Christian Faith*, p. 338.   [2] *M.E.*, p. 113.
[3] *M.E.*, p. 139. Quoted more fully above on pp. 108–9.
[4] *M.E.*, p. 113. Cf. pp. 114–15.

a lesser volume of natural evil in it.   In the context of a divine
decision to create this universe rather than a better one, the
suggestion that God permits but does not intend its evils loses
all force.   Indeed, the two positions that Journet espouses —
that God merely permits natural evil as the unavoidable
concomitant of His achieving some great good, and that He
could had He wished have created instead a better universe
free from these evils — are mutually destructive.   One cannot say both that God is blameless in respect of the natural
evil in our world because He only allows it as something
inseparable from the world's good, *and* that He could had He
wished have created a better world in which there would have
been less natural evil.   The Leibnizian teaching that this is
the best of all possible universes can sustain the claim that
God merely permits such evil as is unavoidable in the best
practicable world.   But Journet rejects the foundation for
such a defence while yet wanting to retain the defence itself;
and his resulting teaching is accordingly involved in hopeless
contradiction.

(2) In the case of human suffering, which is always ultimately punishment for sin, the same basic principle leads to
the claim that God does not will directly, and as such, the
pains that men suffer, but indirectly intends them as the side-
effects of His willing something else, which is good.   'It is the
order of the universe that God wills ; punishment, which He
does not will or intend — and God should not be thought of
as a torturer — is the injury self-inflicted by those who rebel
against an order which, being divine, could never be upset
by a creature.'[1]   This is an appeal to what was called, when
discussing Augustine, the principle of moral balance, a balance which is not impaired by sin so long as that sin is fully
cancelled by just punishment.   Thus,

the order of freedom and morality is a particular order made
to return, by one path or another, into the universal order.   In
those who refuse the inflowing of creative Love, the privation
which they choose for their share is in its own way a perpetual
admission that fullness is in God, and a proof of the intimate
superabundance proper to Love.   To see this is to see how the

[1] *M.E.*, p. 75.

evil of sin and hell itself return into the universal order, certainly not as a component, a structural part, but privatively, as a hollow witness, in an inadequate and finite way, to the infinite fullness of God. Such a perception of evil will disturb no one's inner peace.[1]

I have already commented critically on the principle of moral balance as it was evoked by Augustine, and am likewise unable to accept Journet's suggestion that evil, in the form of punitively inflicted pain, 'will disturb no one's inner peace' so long as it exists in an orderly polarity of sin and punishment. The principle involved arises out of an impersonal and legalistic way of thinking which is foreign in its spirit to that of our Lord Himself. In the teachings of Jesus the heavenly Father is seen as loving His human children individually and as seeking their salvation as a human father might seek his lost son, or a thrifty housekeeper her lost coin, or a good shepherd his lost sheep. Such teaching surely suggests that if the prodigal son had not returned home but had instead been finally lost to his father, the fact that this was the son's own fault and that the moral balance of the universe was accordingly unimpaired would by no means assuage the father's wounded love. On the contrary, he would be left with a deep and permanent sorrow. And likewise we feel, as we see in the gospels the divine love going to every length to seek and to save those who are lost, that if in the end of the ages any are finally self-excluded from the eternal life and joy of His Kingdom, God will not merely note with satisfaction that the moral accounts are in balance, but that He 'without whom not a sparrow falls' must be eternally sorrowful at the loss of beings who 'are of more value than many sparrows'. The principle of moral balance, according to which a universe containing sin and its appropriate punishment is as valuable in the sight of God as a universe without sin, because it is no less morally orderly, excludes the great central reality of the Christian faith, which is God's utterly free and miraculously transforming love for His human creatures. But a principle that would rule out the heart of the gospel is irreparably sub-Christian, and can only be

[1] *M.E.*, p. 279.

expected to lead to the development of a sub-Christian theology.

4. JOURNET ON SIN AND HELL

We turn now from what are for Journet the theologically secondary topics of human suffering and the evil in nature to what is for him the crucial matter of sin. This is treated on quite different principles. Whereas the evil of nature is 'tolerated and accepted by God . . . willed indirectly and by accident',

> The evil of sin, on the other hand, in itself is inseparably connected to nothing good and acts only to destroy the work of God ; so the question as to whether it might, even indirectly or by accident, be willed by God, does not arise. It is *permitted*, tolerated and suffered in a completely different sense from the evil of nature ; it is permitted as a rebellion, an offence, *which God cannot will in any way*, which he cannot acquiesce in or *consent* to without denying his own being, which he could certainly suppress by force and eradicate completely, but which also, he can, if he decides to respect even the resistance of our wills, allow to happen and bear fruit indirectly in other things.[1]

Developing the 'free-will defence', Journet appeals to a necessary connection between finite freedom and the possibility of sin. 'God is able not to create free beings, but if he does create them, they will be able to fall away.'[2] If we ask why God has created dependent and fallible creatures, the answer is that 'To be able to sin and actually not to sin presupposes an act of free preference and voluntary love. And such acts are so dear to God that in his eyes they justify the whole world of creation, especially that of free beings.'[3]

Journet raises the question whether God could, if He had wished, have created men so that they would be 'impeccable', unable to sin. It should be noted that this is not quite the same question as that raised by those contemporary philosophers who ask whether God could have so made men that they would always in fact freely act rightly. Their question is,

---

in Journet's terminology, whether God could have so made a free and peccable creature that he would nevertheless always choose to be good. If the question were posed in these terms, I believe that Journet's answer would be that God could not do this because what is here proposed is self-contradictory. But Journet would also say that God knew with regard to each possible free being whose creation He contemplated that he would or would not voluntarily cleave to the good; with the implication that God could, if He had wished, have given existence only to those whom He knew would, if He created them, persist in good.[1] In this more complicated sense God could have created a race of free beings who would freely cleave to Him.

Although this is implicit in Journet's discussion, he chooses to reach his conclusion by a circuitous route. Asking whether God could have made men impeccable, he answers:

> God cannot make free beings impeccable by nature, but he can make them impeccable by the supernatural dispositions of his providence. . . . Thus the angels and the elect, overwhelmed by the immediate, uninterrupted vision of the divinity, enjoy the happy impossibility of sinning. They adhere unchangingly to God by a consent which is, as it were, beyond our division of acts into necessary and free. . . . For all that, they are not deprived of free will, for over and above this adhesion to God who fundamentally rectifies their will, 'there exists a multitude of things which they can either do or not do'.[2]

This apparently means that a free finite creature cannot be by nature impeccable, but that God can make him impeccable by an inward operation of divine grace that does not destroy his creaturely freedom.

Journet next asks, in a passage which directly continues that just quoted,

> If God does this in the next world for his faithful creatures, after their testing time, why could he not have done it from the outset, creating angels and men in heaven and immersing them immediately in the ocean of his infinite beatitude? From the

[1] *M.E.*, p. 153.
[2] *M.E.*, pp. 152–3. The concluding quotation is from St. Thomas' *De Veritate*, Q. xxiv, art. 1.

very start they would have been intrinsically impeccable, 'having in themselves a principle of stability preserving them entirely from sin'.[1]

Journet responds to this challenge as follows :

God could certainly have done this. If we look at his absolute power, or at what he could have done in conformity with his infinite wisdom and goodness, we shall say without hesitation that he *could have* created all angels and men in heavenly beatitude. And it is true that if God had chosen this course, the world would not have known sin, and would have been better.[2]

But he immediately adds a qualification which moves towards the heart of his theodicy :

At least, it would have been better from one point of view but there would not have been any room, in a world glorified from the outset, either for the forgiveness of the redemption or for the mystery of a resurrected Christ and a resurrected Church.[3]

For, as he adds on the next page,

Only *in statu viatoris*, where the ambivalence natural to every free creature is respected, so that it can either adhere or refuse to adhere to the transcendent Source of its being, can the supreme act of free option and preference, in which God is loved by the creature above all things and more than itself, be produced. This act, so dear to God, cannot be imposed, for to impose it would be to put in its place a different sort of act.[4]

God sees this free choice and love of His creatures as 'the most perfect flower of his paradise, to purchase which he is willing to run the risk of refusal by those of his creatures who wish to reject him'.[5]

We are constrained to ask at this point, in view of what has gone before, Why should, or indeed how can, the omnipotent creator 'run the risk' of a refusal by His creatures? 'Risk' implies uncertainty, as though God must wait in ignorance to see whether His creatures would cleave to Him or rebel against Him. But Journet teaches that God saw (or, as we say in our temporal mode of speech, foresaw), before He decided to put it into effect, all that would flow from the

[1] *M.E.*. p. 153, quoting *De Veritate*, Q. xxiv, art. 9.   [2] Ibid.
[3] Ibid.   [4] *M.E.*, p. 154.   [5] *M.E.*, p. 252.

particular plan of creation that He has in fact realized. 'God, who knows all things not by foresight or memory but by pure vision, only gave his plan effect once he had already made allowance from all eternity for all the free refusals of his creatures.'[1] In this case it would seem that God could instead, had He wished, have put a different plan into effect in which there would be no refusals from His creatures, because only such creatures would be included in the plan as are going always freely to worship and love their Maker.

Why, then, has God created a universe that includes beings whom He knew were going to reject Him? Journet accepts the bold and exciting suggestion of the *Exsultet*: 'O felix culpa, quae talem ac tantum meruit habere redemptorem!' (O fortunate crime which merited such and so great a redeemer.) He says, 'God permits, indeed, that evils come about so as to draw a greater good from them. . . . The world of creation in the state of innocence was good. The world of fallen and redeemed nature is good as well; following the way pointed by the *Exsultet* we have gone one step further towards affirming that it is in sum better than the original world of creation.'[2] In fact, 'the original fall was permitted only with a view to the Redemption of the world by Christ's sufferings'.[3] For God 'only allowed the irruption of the evil of sin into this first universe of creation because he foresaw the setting up of a universe of redemption which as a whole would be better'.[4] Indeed, Journet even adopts the statement of St. Francis de Sales that the redeemed state is worth a hundred times more than that of innocence.[5]

Here, then, we have a reason why God not only made fallible creatures, but fallible creatures whose fall He already foresaw 'from all eternity'. There is here a religiously and morally profound conception, whose daring matches the depth of the problem and which must surely constitute one of the few authentic flashes of light that we possess upon the mystery of evil. The value of this precious gleam of illumination is unhappily, however, largely obscured in the Catholic (as also in the Calvinist) theodicy by an eschatology that

[1] *M.E.*, pp. 231–2.   [2] *M.E.*, p. 259.   [3] *M.E.*, p. 246.
[4] *M.E.*, p. 124.   [5] *M.E.*, p. 258.

restricts to an arbitrarily selected segment of mankind this divine bringing forth of a greater good out of the evil of sin. Given the premises that 'God wishes to save all men'[1] and that He is able (although not obliged) by His irresistible grace to do so[2] one would expect the conclusion that He will in fact save all men. This logic has been followed only by a minority strand within the Christian tradition — though a minority that is today probably larger than in any previous period. According to the still dominant tradition, however, the outcome will be a double one, embodied in an eternal heaven and an eternal hell. 'God wishes to save all men: if they are saved, the glory is his, and if they are not, the fault is theirs',[3] says Journet. For God first approaches the soul in such a way that it can freely either respond or fail to respond to Him. 'At this moment, it is up to us whether we oppose his universal saving will, erect obstacles to his plans or reject his loving-kindness.'[4] If at this stage we deliberately turn away from God, and continue to do so until the moment of death, we thereby damn ourselves. For those other souls, however, who do not refuse God at the point at which He has given them their fateful freedom and responsibility, 'he can without doing them any violence or infringing on the metaphysical structure of their freedom make them, under an irresistible influence, utter that assent which raises them infinitely above themselves, draws them into his orbit and opens for them the door to his intimacy and to the beatific vision'.[5] On the one hand, then, no one can save himself, but can only permit God to save him; but on the other hand, each one can prevent his salvation, thereby bringing about his own damnation.

This represents, according to Journet, the normal or ordinary manner of God's dealings with us. It is, however, possible for Him by His divine power to approach men with 'the sovereign influence under which, admittedly, they will have to give their consent, to say yes, but no longer to deliberate or to choose between good and evil'.[6] He can do this 'without infringing on the metaphysical structure of our free

---

[1] *M.E.*, p. 155.   [2] *M.E.*, pp. 157–8.   [3] *M.E.*, p. 155.
[4] *M.E.*, p. 156.   [5] *M.E.*, p. 155–6.   [6] *M.E.*, p. 158.

will, which he alone, as its creator, can move from within'.[1] In view of this Journet later in his discussion considers the possibility that 'without destroying the metaphysical structure of their freedom [God] could from the start submit [all men] to an influence that would irresistibly draw them towards the salutary act — "though our wills resist thee press them graciously into thy service"— and so save even those who make an effort to resist him' ; and Journet replies, 'Yes, as has been said already, he both can and does do so, and quite frequently.'[2]

There now arises a question that is hypothetical but nevertheless crucial for theodicy : could God, had He wished, have exerted the irresistible influence of His grace within men's souls in such a way as to bring *all* mankind freely to salvation? Journet continues :

Could he *always* act thus? To ask this of him would be to expect from him a different world from the one he chose to make, in which the extraordinary would become the ordinary, and the exception would be changed into the rule.... But, at least, surely, the exceptions could be *more frequent*? Perhaps they could. But the real question lies elsewhere. Can we demand God to cause *even one single exception* to the ordinary rule of his subordinated power [i.e. his power within, and without altering, the structure of the universe which he had made]? Is the divine goodness, having showered its creatures with such help that, should they reject it, the fault should belong entirely to them ... bound, under pain of ceasing to be infinite, to break down the resistance of one who freely wills to rebel against it? Not this question, but another is the one we find unanswerable : why does God sometimes do what he is in no way bound to do? And why does he do it for one person rather than for another? This is where we should listen to St. Augustine : 'Do not judge, if you do not want to err'.[3]

This crucial passage is likely to leave intact, or even perhaps to intensify, the misgivings of the reader who is not committed as a matter of obedience to the Thomist solution. First, it is no adequate defence of an economy of the universe in which many creatures are damned that in order for fewer to be damned God would have to alter this economy ! For the critic's claim is that such an economy *ought* to be altered,

[1] *M.E.*, p. 158.   [2] *M.E.*, p. 168.   [3] *M.E.*, pp. 168–9.

and ought indeed never to have been instituted. Second, the proper question is not whether we have the right to *demand* of God that He should exercise His power to redeem recalcitrant sinners, or whether God is bound 'under pain of ceasing to be infinite' to save all men, but simply whether a God who does not choose to save many whom He could save can be described as perfectly good and loving.

In view of these very large difficulties it is, I think, likely that the theodicy whose outline we have now traced will seem utterly insufficient to a contemporary inquirer to whom it may be offered in defence of Christian belief. For the morally sensitive agnostic will not acknowledge as good a deity who is responsible at two levels for the sin and suffering of His creatures, in that (1) He creates personal beings who He knows will, if He proceeds to create them, fall into eternal malevolence and eternally suffer the torments of the damned, and (2) although He is able, as their creator, so to influence them from within, and without overruling their personal freedom, as to turn them from their wickedness to Himself, He does this only in some arbitrarily selected cases. Such a God is ultimately responsible for the sin and misery of the legion of lost souls, not in the sense that He is the cause of their sinning, nor in the sense that He is responsible or accountable *to* anyone for His acts, but in the sense that the existence of the whole unhappy state of affairs hangs as a necessary condition upon the thread of His omnipotent free choice. The present situation could not have come about if God had not acted in a certain way; and He acted in this way knowing what would result. In face of this theological picture it will be said by the agnostic that a God of unlimited goodness and love would either not gratuitously have created persons whose fate was to be eternal sin and misery or, alternatively, that having created them He would in His mercy have saved them, even though undeserving, and in the infinite resourcefulness of His infinite grace would have brought them at the last to be worthy and happy citizens of His Kingdom. Certainly, if we can imagine human beings possessing an analogous power of creation and prediction — perhaps in relation to persons like the test-tube babies of Aldous

Huxley's *Brave New World* — and then using this power and foreknowledge to produce people whom they know will be freely evil and deservedly miserable, we should regard such scientific gods as more devilish than divine, and should by no means be kindled to trust and love towards them. And yet they would be acting (from whatever motives) in the way in which God is said to have acted.

This is the way in which the Catholic theodicy is liable to strike an agnostic. And this is also the way in which it strikes the present writer, as a Christian who is not committed to the speculative Augustinian and Thomist doctrines. The contrast between the bad news of a God who deliberately makes creatures for whom He must also make a hell, and the good news of the God of Love heard in the parables and sayings of Jesus, is so great that I cannot regard them both as true; and every instinct of faith, hope, and charity responds to Jesus' vision of the heavenly Father rather than to that of the omnipotent Scientist who deals so cold-bloodedly with His finite creatures.

CHAPTER VI

THE PROBLEM OF EVIL IN REFORMED THOUGHT

1. AUGUSTINE AND THE REFORMERS

IT has long been recognized that the Reformation of the sixteenth century represented, theologically, a revival of Augustinianism.[1] Luther and Calvin both quoted extensively from Augustine and regarded him as presenting the best wisdom of the ancient Church, uncontaminated by the subsequent aberrations of medieval Scholasticism. But it was always the biblical and theological rather than the more speculative and philosophical sections of Augustine's writings that appealed to the Reformers.[2] They were Augustinian on the Pauline, not on the Neo-Platonic, side of his thought. This means, so far as the theodicy-problem is concerned, that the Reformers have no general theory of the nature of evil such as Augustine offered in his privative analysis, his use of

---

[1] For example, Benjamin B. Warfield said that 'the Reformation, inwardly considered, was just the ultimate triumph of Augustine's doctrine of grace over Augustine's doctrine of the Church' (*Encyclopedia of Religion and Ethics*, ed. Hastings, vol. ii. p. 224). Walther von Loewenich, in *Von Augustin zu Luther* (Wittenberg: Luther Verlag, 1959), emphasizes — especially in the essay, 'Was bedeutet uns Evangelischen Augustin?' — the continuity between Augustine and the Reformers. Calvin's numerous citations from Augustine's writings, revealing his immense respect for their author, are exhaustively studied by Luchesius Smits of Louvain in *Saint Augustin dans l'œuvre de Jean Calvin* (2 vols., Assen: Van Gorcum & Co., 1957-8). Smits concludes that 'The Reformer felt himself to be in accord with St Augustine on all the fundamental questions, and voluntarily underlined this harmony when replying to his adversaries' (i. 259).

[2] Cf. L. Smits, op. cit. i. 270. Calvin expresses his intention not to use (rather than positively to reject) Augustine's privative analysis of evil in *De Aeterna Dei Praedestinatione* (1552): 'I shall not say with Augustine, although I willingly embrace his statement as true, that in sin as in evil there is nothing positive. It is, however, an argument which does not satisfy many people.' (Calvin, *Corpus Reformatorum*, vol. vii, p. 353.) On the Reformers' view of the privative conception of evil see also Julius Müller, *The Christian Doctrine of Sin*, i. 294-5.

the principle of plenitude, and his conception of the aesthetic perfection of the universe. On the other hand, they share to the full, and even carry further, Augustine's strong doctrine of the fall of man and its paradoxical counterpoise in an equally strong doctrine of predestination.

The Reformers' lack of interest in a general philosophical theodicy is presumably due, negatively, to the absence of any contemporary heresy on the subject to be combated such as had confronted Augustine in Manichaeism; and, positively, to their passionate adherence to Scripture as the normative source of Christian truth. Thus the theology of the Reformers, faithfully built upon the *sola Scriptura* principle, reminds us by its silence that the Augustinian philosophy of evil is a work of human analysis and speculation, and that it should not be accorded the status of revealed truth. It was left to the eighteenth century to revive the more philosophical aspects of the Augustinian theodicy tradition, which had been carried over from the ancient world in Neo-Platonism and then in the Cambridge Platonism of the seventeenth century.[1]

---

[1] Ralph Cudworth, in his *True Intellectual System of the Universe* (1671), a central work of Cambridge Platonism, deals with the problem of theodicy in chap. V, sect. 5, and uses a number of the themes which later appear in Archbishop King's *Essay on the Origin of Evil* : the thought that the origin of evil lies both in the necessary inferiority of created things to their Creator and in the non-compossibility of various goods ; that God makes evils contribute to the harmony of the whole ; that we must judge the whole and not the parts ; that evil is necessary as are the dark colours in a picture, or the more plebeian roles in a drama. These latter images go back to Plotinus and before him to the Stoics. For a direct echo compare Plotinus' 'We are like people ignorant of painting who complain that the colours are not beautiful everywhere in the picture : but the Artist has laid on the appropriate tint to every spot. . . . Again, we are censuring a drama because the persons are not all heroes but include a servant and a rustic and some scurrilous clown ; yet take away the low characters and the power of the drama is gone ; these are part and parcel of it' (*En.* iii. 2, 11 quoted more fully above, p. 89), with Cudworth's 'But we are like unskilful spectators of a picture, who condemn the limner, because he hath not put bright colours every where ; whereas he has suited his colours to every part respectively, giving to each such as belongeth to it. Or else we are like those, who would blame a comedy or tragedy, because they were not all kings and heroes, that acted in it, but some servants and rustic clowns introduced also, talking after their rude fashion.' (*True Intellectual System*, ed. by Thomas Birch, Andover : Gould and Newman, 1837–8, vol. ii, pp. 339–40.) For a yet earlier use of the simile, see Marcus Aurelius' *Meditations* vi. 42.

Elements of the Neo-Platonist theodicy also appear in the writings of another Cambridge Platonist, Henry More (*Divine Dialogues*, 1688, Second Dialogue).

In eighteenth-century 'optimism' (which was a Protestant phenomenon) the Augustinian aesthetic theme, with its related principle of plenitude and its privative analysis of evil, was made central, as we shall see in the next chapter, and the more theological Augustinian doctrines were subordinated to it. This development reflects, no doubt, that change in the intellectual climate of the West that is known as the Enlightenment or the *Aufklärung*. Two centuries later, beyond the period of the Enlightenment, and indeed in reaction against it, there is, in the Neo-Reformation theology of Karl Barth an attempt, in his doctrine of *das Nichtige*, to restore one of the elements of the Augustinian tradition which the Reformers had set aside and to integrate it more closely than Augustine himself had done into the traditional theological framework. We must begin, however, with the Reformation, taking Calvin's *Institutes* as our theological source.

I. Calvin

2. Fall and Predestination in Calvin

It is not necessary to describe in detail Calvin's teaching concerning the origin of evil in the fall of free creatures, for his thought here follows closely that of Augustine as it has been described in Chapter III. It is true that Augustine gave more attention to the pre-mundane fall of the angels, whereas Calvin is for the most part content to treat the origin of evil in terms of the nearer human fall. But since Augustine does not allow the prior defection of the angels to mitigate the culpability of man's first fault, the practical difference is only that Augustine describes the fall twice, first in the heavenly places and then again on earth, whilst Calvin concentrates his attention upon the earthly fall.[1] Man, he says,

Other works of the period in which some of the same themes occur include *The Moralists* (1709) by Anthony Ashley Cooper, Earl of Shaftesbury; various versions of the *De malorum subsistentia* of the Neo-Platonist philosopher Proclus (A.D. 410–85); and (though not by a Platonist) Bishop Berkeley's *Principles of Human Knowledge* (1710), para. 153. Prior to these we find the notion of evil as non-being in Descartes' *Meditations*, iv (1641).
[1] Calvin does indeed refer briefly to the fall of the angels, and (like Augustine) carries the divine predestinating decree back beyond this first cleavage among

was created as an immortal soul in a state of original righteousness, 'so that his reason, understanding, prudence, and
judgment not only sufficed for the direction of his earthly life,
but by them men mounted up even to God and eternal
bliss. . . . Adam's choice of good and evil was free, and not
that alone, but the highest rectitude was in his mind and will,
and all the organic parts were rightly composed to obedience.'[1] However, Adam fell, as described in Genesis iii, and
'after the heavenly image was obliterated in him, he was not
the only one to suffer this punishment — that, in place of
wisdom, virtue, holiness, truth, and justice, with which
adornments he had been clad, there came forth the most
filthy plagues, blindness, impotence, impurity, vanity, and
injustice — but he also entangled and immersed his offspring in the same miseries'.[2] This is the hereditary corruption of original sin whereby 'all of us, who have descended
from impure seed, are born infected with the contagion of
sin. In fact, before we saw the light of this life we were
soiled and spotted in God's sight.'[3] And so man has lost his
original freedom and is now enslaved to sin : 'corrupted by
the Fall, [man sins] willingly, not unwillingly or by compulsion; by the most eager inclination of his heart, not by
forced compulsion from without. Yet so depraved is his
nature that he can be moved or impelled only to evil. But
if this is true, then it is clearly expressed that man is surely
subject to the necessity of sinning.[4] Further, not only has
that primeval fall deprived man of his righteousness, his
freedom, and his hope of eternal life, but it has also ruined
the world in which he lives and brought upon our race the

created beings. If the steadfastness of the one group of angels, he says, 'was
grounded in God's good pleasure, the rebellion of the others proves that the
latter were forsaken. No other cause of this fact can be adduced but reprobation, which is hidden in God's secret plan.' *Inst.* iii. xxiii. 4. Quotations from
the *Institutes* in this chapter are taken from *Calvin: Institutes of the Christian
Religion*, ed. John T. McNeill, trans. Ford Lewis Battles (London : S.C.M.
Press Ltd., and Philadelphia : The Westminster Press, 2 vols., 1961).

[1] *Inst.* i. xv. 8. Cf. Calvin's *Commentary on Genesis*, trans. John King (Edinburgh : Calvin Translation Society, 1847), on chap. 2, verse 16.

[2] *Inst.* ii. i. 5. Note, however, ii. ii. 12–16, in which Calvin explains that
whilst man's religious capacity is totally ruined, his natural capacity for science,
the arts, and human community, remains.

[3] Ibid.   [4] *Inst.* ii. iii. 5.

pains and hardships of our present existence. Adam's fall, says Calvin, 'perverted the whole order of nature in heaven and on earth', so that the creatures 'are bearing part of the punishment deserved by man, for whose use they were created'.[1]

This is the traditional Augustinian and Latin view of the fall of man as an infinitely wicked and (apart from divine grace) irreparably disastrous crime, a perverse and inexplicable fall from creaturely perfection to the most profound depth of depravity and disgrace. This cosmic calamity with all its dread consequences is wholly man's own fault, and ever since that first fatal act human beings have been born under a curse, already sentenced to eternal death. Thus far the Calvinist doctrine loads the blame for sin and suffering without remainder upon the shoulders of mankind.

But, as in Augustine's theology, there cuts across this picture a doctrine that sets the stamp of a divine predestinating decree, and therefore of ultimate divine responsibility, upon the whole chequered human situation — a situation that is so hopeful for some and (according to Calvinism) so hopeless for others. For Calvin is emphatic that divine predestination is not a matter of God foreseeing what each man will freely do, and then rewarding him accordingly, but of His so making men that they will, out of their own nature, freely follow the path for which He has predestined them, some to heaven and others to hell. 'We call predestination', Calvin explains, 'God's eternal decree, by which he determined with himself what he willed to become of each man. For all are not created in equal condition; rather, eternal life is foreordained for some, eternal damnation for others. Therefore, as any man has been created to one or the other of these ends, we speak of him as predestined to life or to death.'[2] Again, 'Since the disposition of all things is in God's hand, since the decision of salvation or of death rests in his power, he so ordains by his plan and will that among men some are born destined for certain death from the womb, who glorify his name by their own destruction.'[3] And once

---

[1] *Inst.* II. i. 5. Cf. *Commentary on Genesis*, on chap. 2, verse 8, and chap. 3, verses 17–19.   [2] *Inst.* III. xxi. 5.   [3] *Inst.* III. xxiii. 6.

more, 'Again I ask : whence does it happen that Adam's fall irremediably involved so many peoples, together with their infant offspring, in eternal death unless because it so pleased God? . . . The decree is dreadful indeed, I confess. Yet no one can deny that God foreknew what end man was to have before he created him, and consequently foreknew because he so ordained by his decree.'[1] Nor does Calvin hesitate before the implication that, if the damnation of all but 'a limited number'[2] of Adam's descendants is predestined, so also was Adam's fall, by which this was brought about : 'And it ought not to seem absurd for me to say that God not only foresaw the fall of the first man, and in him the ruin of his descendants, but also meted it out in accordance with his own decision.'[3] And combating the Thomist view that this was a matter of divine permission rather than of positive divine ordination, Calvin says, 'But why shall we say "permission" unless it is because God so wills? Still, it is not in itself likely that man brought destruction upon himself through himself, by God's mere permission and without any ordaining. As if God did not establish the condition in which he wills the chief of his creatures to be.'[4]

The fall of the first man, then, was ordained by God ; and all Adam's descendants have involuntarily inherited his guilt, acquiring with it an overwhelming predisposition to sin. And since God Himself thus set in train the fatal sequence that has led to man's present sinful state, the question arises whether men are justly punishable for their sinfulness. Calvin supposes an objector to ask, 'Why from the beginning

[1] *Inst.* III. xxiii. 7.
[2] Ibid. This is Battles' translation. The Latin is 'non multos', 'not many'. The French version of 1560 (the precise extent of Calvin's authorship of which is disputed : see John T. McNeill's Introduction to Battles' translation, I. pp. xxxviii–xxxix, n. 13) has 'quelque petite poignée de gens', 'some small handful of people'. [3] *Inst.* III. xxiii. 7.
[4] *Inst.* III. xxiii. 8. Cf. *Commentary on Genesis*, on chap. 3, verse 1, where Calvin affirms that 'Adam did not fall without the ordination and will of God'. 'It offends the ears of some,' he says, 'when it is said that God *willed* this fall ; but what else, I pray, is the *permission* of Him, who has the power of preventing, and in whose hand the whole matter is placed, but his will? . . . I hold it as a settled axiom, that nothing is more unsuitable to the character of God than for us to say that man was created by Him for the purpose of being placed in a condition of suspense and doubt ; wherefore I conclude, that, as it became the Creator, he had before determined with himself what should be man's future condition.'

did God predestine some to death who, since they did not yet exist, could not yet have deserved the judgment of death?'[1] And his answer is that, although predestined to it, men sin freely, and are therefore all personally guilty and rightly condemned. 'If all are drawn out of a corrupt mass, no wonder they are subject to condemnation! Let them not accuse God of injustice if they are destined by his eternal judgment to death, to which they feel — whether they will or not — that they are led by their own nature of itself.'[2]

Calvin is here making use of the conception of human freedom and accountability at which he had arrived in book II of the *Institutes*. We may say that, for Calvin, to have a will and to have free will are the same. Thus the sinner, whose fallen nature is such that he necessarily wills wrongly and who cannot, with his perverted nature, will rightly, remains nevertheless a free and responsible agent; for he is acting voluntarily and not from external compulsion.[3] It is to be noted that this is precisely the definition of freedom that is used by a contemporary philosopher to create difficulties for the 'free-will defence' in theodicy.[4] But whether or not this definition of human freedom ultimately helps or hinders a Christian theodicy, it does serve to render intelligible Calvin's assertion that a sinner sins necessarily, out of a fallen nature, and yet that he is morally responsible for his sins and is justly punished by his Maker.

3. PREDESTINATION VERSUS THEODICY

Calvin's is almost as extreme and uncompromising as a doctrine of predestination can be. It goes beyond Augustine's teaching in explicitly attributing reprobation as well as salvation to the positive decree of God. Augustine had taught that God's part in the damning of the reprobate is the negative act (if one may so describe it) of passing them by and leaving them to 'stew in their own juice'. God has not positively predestined them to eternal perdition, although He omits to elect them to eternal life. To Calvin's more

---

[1] *Inst.* I. xxiii. 3.   [2] Ibid.   [3] *Inst.* II. iii. 5.
[4] See pp. 303–4 below.

ruthlessly consistent mind this was an evasion of the sterner implication of the fact of divine predestination. For, 'those whom God passes over, he condemns; and this he does for no other reason than that he wills to exclude them from the inheritance which he predestines for his own children'.[1] If Calvin's position here is more repulsive than Augustine's, it is also intellectually more consistent, and more frank in its acceptance of the final conclusion of premises that Augustine and Calvin hold in common. Indeed, it is characteristic of Calvin's absolute respect for what he believes to be the message of God's Word that he makes no attempt to render predestination palatable to either the moral sense or the reason of his readers. We are, he believed, confronted here by the ultimate sovereignty and inscrutable will of our Maker. God has created us *ex nihilo* and we are His absolute property, like clay in the hands of the potter, without any rights in relation to Him; and if He chooses that some of His creatures shall enjoy eternal happiness and others endure eternal torments, we have no ground upon which to question His 'uncontrollable intent'. We can only be silent in the presence of so awful a mystery. 'The decree is dreadful [horribile] indeed, I confess. Yet no one can deny that God foreknew what end man was to have before he created him, and consequently foreknew because he so ordained by his decree.'[2]

Calvin believed that in his treatment of the tremendous theme of predestination he was eschewing all speculative reasoning, with its attendant risks of error, and was simply affirming the undoubted teaching of the Scriptures. I think it is clear, however, as we view Calvin's work form a distance, that in forming his doctrine his mind was actively at work selecting among the biblical data, drawing inferences, making speculative extensions and projections, and thus producing a theoretical construction which in its dogmatic assurance concerning the mysteries of God's righteousness and mercy goes far beyond the 'existential' standpoint of the Bible, which remains, behind its frequently mythic and parabolic use of language, ultimately reticent. Calvin was in fact

[1] *Inst.* III. xxiii. 1.   [2] *Inst.* III. xxiii. 7.

doing precisely what he criticized others for doing, namely asserting more than the Scriptures themselves reveal.

There is, however, a positive religious concern behind this doctrine, imparting to it the attraction that it has always exercised upon so many deeply Christian minds. It accentuates by contrast the wonder of the free divine grace which we have found, and which has found us, in Jesus Christ. That this positive concern was actively at work in Calvin's own use of the doctrine is evident both from his placing of his chapters concerning the double decree at the end of book III, on 'The Way in Which We Receive the Grace of Christ' (far separated, both spatially and in the theological structure of the *Institutes*, from his chapters on providence in book I and on free will in book II), and by his own description of the religious 'usefulness' and 'sweet fruit' of the doctrine. 'We shall never', he says, 'be clearly persuaded, as we ought to be, that our salvation flows from the wellspring of God's free mercy until we come to know his eternal election, which illuminates God's grace by this contrast: that he does not indiscriminately adopt all into the hope of salvation but gives to some what he denies to others.'[1] We must thus recognize a genuine religious intention behind the doctrine, even though we may believe that this intention is largely frustrated by the way in which it is carried out. Calvin so emphasizes the sovereign divine freedom, in abstraction from the total Christian conception of the divine nature, as to call God's goodness and love seriously into question. For the arbitrary saving of some and damning of others would be an act that is free not only from external constraint but also from inner moral self-direction. There would be nothing admirable, still less worthy of worship, in a free activity that consisted in creating beings whom the Creator has predetermined shall deserve and receive unending punishment.

It should be added that some scholars interpret Calvin in a way that reduces or even eliminates these objectionable features of his teaching. E. A. Dowey points out that within the structure of the *Institutes* predestination is attached to the doctrine of salvation rather than to that of creation or

[1] *Inst.* III. xxi. 1.

providence. It is not offered as a deduction from the divine omnipotence, but as a means of appreciating the wonder of God's redeeming grace; and from this point of view the idea of reprobation plays a secondary and asymmetrical role: 'the doctrine of reprobation enters into neither the relation to God nor the relation to one's fellow man, neither the worship nor the ethics, of the only one for whom it is a doctrine: the believer'.[1] We properly use the doctrine to illuminate the fact that by God's grace we know ourselves to enjoy election in distinction from reprobation; but we can make no correspondingly concrete use of the latter category.[2] For 'while it is the elect man's view of damnation, eternal reprobation is emphatically not the elect man's view of his neighbour. The "others" who are reprobate are an abstraction, not living persons with names.'[3] Thus election and reprobation do not constitute equal and balancing themes in Calvin's theology: 'good and evil, God's will and sin, election and reprobation, are never for Calvin clear parallels, because God stands in a direct and essential relation to the good and in an indirect and accidental relationship to all that is not good'.[4] In short, 'the doctrine of reprobation, when its theological locus is seen in Calvin's soteriology as a part of the knowledge of God the Redeemer, belongs to the believer's knowledge of God in faith as a limiting concept at the border of the mystery surrounding his own election'.[5]

Dowey is here emphasizing an aspect of Calvin's thought that had been insufficiently noticed by most Calvin commentators. But nevertheless it is the other aspect, in which the double decree appears as an objective metaphysical fact, that has been taken up into the theology of most of Calvin's successors and followers. Nor, surely, can Calvin himself be acquitted of a large share of responsibility for this development; for the passages in the *Institutes* that express or appear to express a metaphysical doctrine of double predestination are very striking, and bulk largest in the edition of 1559 which embodied Calvin's own final revisions. Even if Calvin would

[1] Edward A. Dowey, Jr., *The Knowledge of God in Calvin's Theology* (New York: Columbia University Press, 1952), p. 213.   [2] Cf. *Inst.* III. xxiii. 14.
[3] Dowey, op. cit., p. 214.   [4] Ibid., p. 216.   [5] Ibid., p. 215.

have regretted some of the 'Calvinism' of, for example, the Westminster Confession of 1646, nevertheless that document was fashioned within the theological movement inspired by Calvin's *Institutes*. In the Westminster Confession (which is still the official Standard of most of the English-speaking churches in the Reformed or Calvinist tradition) the chapter on God's eternal decrees has a very prominent place, prior even to that on creation, and is far separated from the chapters on salvation. 'By the decree of God,' we are told, 'for the manifestation of his glory, some men and angels are predestinated unto everlasting life, and others foreordained to everlasting death. These angels and men, thus predesti-nated and foreordained, are particularly and unchangeably designed ; and their number is so certain and definite, that it cannot be either increased or diminished.'[1]

In such a doctrine, the supreme insight and faith of New Testament monotheism, that God loves *all* His human chil dren with an infinite and irrevocable love, is lost and there is a relapse to the conception of God as the Lord of a chosen in-group whom He loves, who are surrounded by an alien out-group, whom He hates. Indeed, while the positive reli-gious concern behind the doctrine is to magnify the wonder of God's free and gracious love, its effect is to reduce that love to the proportions of our own partisan human attitudes. For we naturally think in the dualistic manner to which the polarity of heaven and hell gives ultimate scope and licence. And we equally naturally see ourselves as standing safely with the redeemed on the Godward side of the gulf. Theologians who have confidently affirmed an ultimate salvation–damna-tion dichotomy have almost invariably assumed that they themselves are among the saved. And in so far as this assump-tion is based upon a vivid sense of God's infinite and unmeri-ted grace towards themselves, it is well founded. But such is our natural human pride that, believing ourselves to be saved by God's grace, we must set up a contrast between this condition and an opposite one. We cannot be content to believe that God loves and accepts *us* unless we are assured that He hates and rejects someone else. The feeling that

[1] *Westminster Confession*, chap. 3, paras. 2 and 3.

there *must* be this contrast appears in Calvin's thinking : 'election itself', he says, 'could not stand except as set over against reprobation'.[1] But why not? Why could not God, if He wishes, elect all to salvation? There is, in Calvinist theology, no reason; for human deserts and decisions have nothing to do with God's elections, the grounds of which lie wholly within the divine will. God could, if He so desired, bring all the finite persons whom He has created into the ultimate fellowship of His kingdom. But He does not so desire. We are compelled to infer that if God had a greater love for His creatures He *would* desire the salvation of each, and therefore all, of them. For He cannot be said in any intelligible sense to love those whom He has predestined to eternal guilt and misery. Thus in its over-developed doctrine of the divine decrees Calvinism introduces a dogma which restricts God's love and thereby nullifies the attempt to present faithfully the theological structure of the Christian gospel. We see exemplified in Calvinism, no less than in Augustinianism, that failure to think of God and of His attitudes to mankind in fully personal and agapeistic terms, which is the basic defect of theodicies of the Augustinian type.

II. Karl Barth

4. BARTH'S METHOD

The theology of Karl Barth (1886–) has been characterized in a variety of ways — as dialectical, as neo-orthodox, as neo-Reformed, as the theology of the Word of God, and simply as Barthian — but none of these terms pleases Barth himself, and perhaps we should be content to say that he has written a vast and comprehensive exposition of Christian belief from a standpoint that is emphatically within the tradition of the Reformation. In the course of the more than thirty years of his work on his *Kirchliche Dogmatik* Barth has written extensively on the problem of evil. Indeed if one were to bring together the 98-page section on evil in volume III/3, paragraph 50; the 40-page exegesis of Genesis i, 1–5

[1] *Inst.* III. xxiii. 1.

in volume III/1, paragraph 41, 2 ; the 58-page discussion of the goodness of the created world in volume III/1, paragraph 42, 3 ; and the 178-page treatment of sin and the fall in IV/1, paragraph 59, one would have assembled a full-scale and even massive treatise on the subject. In the following account of Barth's contribution all these sections of his work are taken into account, although I shall leave in the background the details of his biblical exegeses and of his controversies with the various thinkers, past and contemporary, whom he takes in succession as *Gesprächspartner* for the elucidation of his own position. (In some cases, however, Barth's comments on views from which he differs are referred to elsewhere in this book in connection with the particular thinkers in question.)

At the outset of section 50, on 'Gott und das Nichtige', in volume III/3 of his great dogmatic work, Barth adopts the practical or existential as distinct from the theoretical or philosophical point of view in relation to the problem of evil. We have here, he says, 'an extraordinarily clear demonstration of the necessary brokenness of all theological thought and utterance . . . [in that] it can progress only in isolated thoughts and statements directed from different angles to the one object. It can never form a system, comprehending and as it were "seizing" the object.'[1] This is true, according to Barth, of theological language as such, but it is especially true here, because 'The existence, presence and operation of *das Nichtige*, which we are concerned to discuss, are also objectively the break in the relationship between Creator and creature.'[2] Therefore, 'Here especially theology must set an example for its procedure generally, corresponding to its object in broken thoughts and utterances.'[3] And in conformity with this conception of the appropriate method

---

[1] Karl Barth, *Church Dogmatics*, authorized English translation ed. by G. W. Bromiley and T. F. Torrance (Edinburgh : T. & T. Clark), vol. III/3, p. 293.

[2] Ibid., p. 294. The English version translates *das Nichtige* as 'nothingness'; and the editors explain in a footnote that they also considered but rejected 'the nihil', 'the null', 'the negative', and the 'non-existent'. Geraint Vaughan Jones translates *das Nichtige* as 'the inimical principle of Negation' ('God and Negation', *Scottish Journal of Theology*, vol. vii, Sept. 1954). However, rather than adopt any of these terms I have retained Barth's own word in all quotations.

[3] *C.D.* III/3, p. 294.

Barth does not hesitate to utter paradoxes: *das Nichtige* is to be taken with the utmost seriousness as a mortal threat, and yet it has been completely overcome and abolished by Christ on the cross; it is the work of God 'on His left hand', and has no power which He has not given it, and yet it is utterly opposed and inimical to Him.

What, then, does Barth mean by *das Nichtige*? 'There is', he says, 'opposition and resistance to God's world-dominion. . . . This opposition and resistance, this stubborn and alien factor, may be provisionally defined as *das Nichtige*.'[1] But can it be said that such a thing 'is'? Not in the sense in which God and His creatures are. For *das Nichtige* has nothing in common with God and His creatures. Nevertheless, it is not nothing. ('Das Nichtige *ist* nicht das Nichts.'[2]) It 'is' as something of which God takes account; for He is known as 'the God who is confronted by *das Nichtige*, for whom it constitutes a problem, who takes it seriously, who does not deal with it incidentally but in the fulness of the glory of His deity, who is not engaged indirectly or mediately but with His whole being, involving Himself to the utmost'.[3]

5. The 'Shadowside' of Creation

Barth is anxious to distinguish *das Nichtige*, by which he means evil in the strongest possible sense, namely that which is utterly and essentially inimical to God and His creation, from the shadowside (*Schattenseite*) or negative aspect of the universe, which has often been mistakenly identified with evil in the strong sense, but which really has a quite different status as a necessary antithesis and contrast within that total creaturely world which the Creator saw to be 'very good'.

Barth is not as precise as one could wish in indicating the borders, and thus showing the content, of this shadowside. In his fullest account of it he says that,

in creation there is not only a Yes but also a No; not only a height but also an abyss; not only clarity but also obscurity; not

[1] *C.D.* iii/3, p. 289.
[2] *Kirchliche Dogmatik*, iii/3, p. 403 (*C.D.*, p. 349).
[3] *C.D.* iii/3, p. 349.

only progress and continuation but also impediment and limitation; not only growth but also decay; not only opulence but also indigence; not only beauty but also ashes; not only beginning but also end; not only value but also worthlessness. It is true that in creaturely existence, and especially in the existence of man, there are hours, days and years both bright and dark, success and failure, laughter and tears, youth and age, gain and loss, birth and sooner or later its inevitable corollary, death. It is true that individual creatures and men experience these things in most unequal measure, their lots being assigned by a justice which is curious or very much concealed. Yet it is irrefutable that creation and creature are good even in the fact that all that is exists in this contrast and antithesis.[1]

And at another point he derives this shadowside from the creation's essential finitude and variety.

The diversities and frontiers of the creaturely world contain many 'nots'. No single creature is all-inclusive. None is or resembles another. To each belongs its own place and time, and in these its own manner, nature and existence. What we have called the 'shadow side' of creation is constituted by the 'not' which in this twofold respect, as its distinction from God and its individual distinctiveness, pertains to creaturely nature.[2]

Thus Barth's shadowside corresponds to what has traditionally been called metaphysical evil, namely finitude, imperfection, impermanence, and the fact of having been created *ex nihilo* and being thus ever on the verge of collapsing back into non-existence; but Barth is in agreement with the traditional rejection by Catholic writers of the view that this necessary limitation and imperfection are to be accounted *evil*. 'It belongs to the essence of creaturely nature, and is indeed a mark of its perfection, that it has in fact this negative side, that it inclines not only to the right hand but also to the left, that it is thus simultaneously worthy of its Creator and yet dependent on Him, that it is not "nothing" [*Nichts*] but "something," yet something on the very frontier of nothingness [*Nichts*], secure, and yet in jeopardy.'[3] Accordingly this shadowside of the world does not stand in opposition or resistance to God's creative will. 'On the contrary, this

---

[1] Ibid., pp. 296–7. On the *Schattenseite*, as well as *C.D.* III/3, para. 50, 2, see *C.D.* III/1, para. 42, 3.   [2] *C.D.* III/3, pp. 349–50.   [3] Ibid., p. 296.

will is fulfilled and confirmed in it.'[1] And — as Barth heard in the musical microcosm of Mozart's harmonies — even in its negative aspects, and hence as a totality, creation praises its Maker, and is perfect.[2] We know this because God in Christ has made His home in the creaturely world, participating in both its light and its shadow, 'and has again and expressly claimed the whole of creation as His work, adopting and as it were taking it to heart in both its positive and negative aspects'.[3]

It is essential, according to Barth, not to confuse *das Nichtige* with the shadowside of creation, because we might then be deluded into thinking that the former is a special case of the latter, and that it is thus covertly good and contributes to the ultimate perfection of the universe. When this happens, its 'nature and existence are attributed to God, to His will and responsibility, and the menacing and corruption of creation by *das Nichtige* are understood as His intention and act and therefore as a necessary and tolerable part of creaturely existence'.[4] But this would be a fatal delusion, under cover of which *das Nichtige* would be able further to establish its malignant power over us.

6. 'DAS NICHTIGE'

Whereas the negative side of creation is to be faced with only a relative seriousness, true irreconcilable evil, *das Nichtige*, can only be the object of unqualified fear and loathing. The full horror and peril of *das Nichtige* are disclosed to us when we see it in its relation to Jesus Christ. Through him there is revealed 'the true *Nichtige* which is utterly distinct from both Creator and creation, the adversary with whom no compromise is possible, the negative which is more than the mere complement of an antithetical positive, the left which is not counterpoised by any right, the antithesis which is not merely within creation and therefore dialectical but which is primarily and supremely to God Himself and therefore to the totality of the created world'.[5] But when we see the true

[1] *C.D.* III/3, p. 296.   [2] Ibid., p. 299.   [3] Ibid., p. 301.
[4] Ibid.   [5] Ibid., p. 302.

menace and horror of *das Nichtige* as it is locked in mortal
combat with God's Christ, we also see it overcome and
abolished. As Christians we can look back to its defeat in
Christ. But nevertheless we still live in the interim between
his resurrection and coming again, in which *das Nichtige*
seems as powerful as ever even though we know by faith
that it has already been decisively defeated. Hence the
paradoxical status of *das Nichtige* : it is utterly inimical
to God and yet God 'comprehends, envisages and controls
it'.[1]

Indeed, *das Nichtige* is primarily the enemy of God Himself
and only secondarily of God's creature. 'The controversy
with *das Nichtige*, its conquest, removal and abolition, are
primarily and properly God's own affair.'[2] *Das Nichtige* is
'the "reality" on whose account (i.e. against which) God
Himself willed to become a creature in the creaturely world,
yielding and subjecting Himself to it in Jesus Christ in order
to overcome it. . . . The true *Nichtige* is that which brought
Jesus Christ to the cross, and that which He defeated
there.'[3]

Relating *das Nichtige* to other and more familiar concepts
Barth says that the most important of all its forms is human
sin. 'When seen in the light of Jesus Christ, the concrete
form in which *das Nichtige* is active is the sin of man as his
personal act and guilt. . . . In the light of Jesus Christ, it is
impossible to escape the truth that we ourselves as sinners
have become the victims and servants of *das Nichtige*, sharing
its nature and producing and extending it.'[4]

But although sin is the concrete form of *das Nichtige*,

Yet *das Nichtige* is not exhausted in sin. It is also something
under which we suffer in a connection with sin which is sometimes palpable but sometimes we can only sense and sometimes is
closely hidden. . . . Contrary to his will and expectation, the sin of
man is not beneficial to him but detrimental. . . . Sin as such is not
only an offence to God ; it also disturbs, injures and destroys the
creature and its nature [for] it is attended and followed by suffering, i.e. the suffering of evil [Übel] and death. It is not merely

---

[1] Ibid.   [2] Ibid., p. 354.
[3] Ibid., p. 305.   [4] Ibid., pp. 305–6.

attended and followed by the ills which are inseparably bound up with creaturely existence in virtue of the negative aspect of creation, but by the suffering of evil [*Übel*] as something wholly anomalous which threatens and imperils his existence and is no less inconsistent with it than sin itself. . . .[1]

As a result of the triumph of *das Nichtige* in our own sinful rebellion against God, death — which in itself belongs only to the innocent shadowside of existence — becomes 'the intolerable, life-destroying thing to which all suffering hastens as its goal, as the ultimate irruption and triumph of that alien power which annihilates creaturely existence and thus discredits and disclaims the Creator'.[2] There are thus 'real evil [*Übel*] and real death', caused by *das Nichtige* infecting the shadowside of the world; and there are likewise 'a real devil with his legions, and a real hell'.[3] ('Real', says Barth, means here: 'in opposition to the totality of God's creation'.[4]) Thus *das Nichtige* is not simply sin but something more comprehensive — the enemy of God, which takes the forms of sin and pain, suffering and death; 'in the physical evil concealed behind the shadowy side of the created cosmos we have a form of the enemy and no less an offence against God than that which reveals man to be a sinner'.[5]

Accordingly if we were to ask Barth whether disease and bodily pain fall within the shadowside of creaturely existence, or represent hostile acts of *das Nichtige*, he would presumably reply that whilst ideally they are merely part of the negative, but innocently negative, aspect of creaturely life, yet concretely, as they occur within our sinful human experience, they have become manifestations of *das Nichtige*. Accordingly it was as the power of God's Kingdom overturning the work of the devil that Jesus combated disease in His healing miracles.

The Gospel records of the miracles and acts of Jesus are not just formal proofs of his Messiahship, of His divine mission, authority and power, but as such they are objective manifestations of His character as the Conqueror not only of sin but also of evil [*Übel*]

---

[1] *C.D.* III/3, p. 310.   [2] Ibid.   [3] Ibid.   [4] Ibid.
[5] Ibid., p. 315. In this passage Barth uses *Böse* (*physich Bösen*) where we might expect *Übel*, to emphasize that behind *Übel* there lurks *Böse*, i.e. *das Nichtige*.

and death, as the Destroyer of the destroyer, as the Saviour in the most inclusive sense. He not only forgives the sins of men; He also removes the source of their suffering. He resists the whole assault. To its power He opposes His own power, the transcendent power of God. He shows Himself to be the total Victor.[1]

7. The Origin of 'das Nichtige'

What is the ultimate origin of *das Nichtige*? Barth's answer is as follows:

> Grounded always in election, the activity of God is invariably one of jealousy, wrath and judgment. God is also holy, and this means that His being and activity take place in a definite opposition, in a real negation, both defensive and aggressive. *Das Nichtige* is that from which God separates Himself and in face of which He asserts Himself and exerts His positive will. . . . God elects, and therefore rejects what He does not elect. God wills, and therefore opposes what He does not will. He says Yes, and therefore says No to that to which He has not said Yes. He works according to His purpose, and in so doing rejects and dismisses all that gainsays it. Both of these activities, grounded in His election and decision, are necessary elements in His sovereign action. He is Lord both on the right hand and on the left. It is only on this basis that *das Nichtige* 'is', but on that basis it really 'is'.[2]

And then he proceeds to the following statements:

> As God is Lord on the left hand as well, He is the basis [Grund] and Lord of *das Nichtige* too. Consequently it is not adventitious. It is not a second God, nor self-created. It has no power save that which it is allowed by God. It, too, belongs to God [ist von Gott]. It 'is' problematically because it is only on the left hand of God, under His No, the object of His jealousy, wrath and judgment. It 'is', not as God and His creation are, but only in its own improper way, as inherent contradiction, as impossible possibility.[3]

It is thus Barth's contention that *das Nichtige* is brought into 'existence' by God's decision to create a good universe. By willing a good creation, faithful to Himself, God has unwilled, or willed not to will, its opposite; and this divine act of negation and rejection has as its object *das Nichtige* as

[1] Ibid., p. 311.   [2] Ibid., p. 351.   [3] Ibid.

a malignant opposition to God and to His creatures. 'That which God renounces and abandons in virtue of His decision is not merely nothing [nichts]. It is *das Nichtige*, and has as such its own being, albeit malignant and perverse.'¹ Again, referring to the primitive chaos that was 'without form and void', of Genesis i, 2, Barth says, 'This negation of [God's] grace is chaos, the world which He did not choose or will, which He could not and did not create, but which, as He created the actual world, He passes over and set aside, marking and excluding it as the eternal past, the eternal yesterday.'²

Finally, as the object of God's *opus alienum*, His work on the left hand, which is His jealousy, wrath, and judgement, *das Nichtige* has no perpetuity but is doomed to final extinction. It exists only in tension with God's *opus proprium*, the work of His right hand, which is the work of grace. But, 'As God fulfils His true and positive work, His negative work becomes pointless and redundant and can be terminated and ended'.³ For 'The purpose of His *opus proprium* is the termination of His *opus alienum* and therefore the elimination of its object',⁴ and this has been accomplished in principle and in prolepsis by the death and exaltation of Jesus Christ. Accordingly, 'In the light of Jesus Christ there is no sense in

---

¹ *C.D.* iii/3, p. 352.
² Ibid., p. 353. Barth's exegesis of Genesis i, 2–3 is a classic example of theological *eis*egesis, or reading into the text; for he equates the *tohu wa-bohu* ('without form and void') of verse 2 with *das Nichtige*, and thus discovers his own theory of evil in the biblical account of creation. He argues that verse 2 must be regarded as 'a portrait, deliberately taken from myth, of the world which according to His revelation was negated, rejected, ignored and left behind in His actual creation, i.e. in the utterance of His Word'. (*C.D.* iii/1, p. 108.) Again, he interprets the 'darkness' of verse 3 as referring to the primitive chaos which God put behind Him and which now exists only as the rejected, hostile *Nichtige*, in contrast to God's good creation. Accordingly, 'a true and strict analogy to the relationship between light and darkness is to be found only in the relationship between the divine election and rejection, in the eternal Yes and No spoken by God Himself . . .' (Ibid., pp. 123–4.) It should be added that Barth's discussion of the Genesis text is characteristic of much of his writing in that a reader to whom Barth's explicit thesis is quite unacceptable is nevertheless likely to find his pages illuminating and suggestive in the highest degree. (On the relation between the Hebrew *tohu* and the Greek τὸ μὴ ὄν (non-being), see Thorlief Boman, *Hebrew Thought Compared with Greek*, 1954, trans. by Jules Moreau, London: S.C.M. Press, 1960, pp. 56–58.)
³ *C.D.* iii/3, p. 361.   ⁴ Ibid., p. 362.

which it can be affirmed that *das Nichtige* has any objective existence, that it continues except for our still blinded eyes, that it is still to be feared, that it still counts as a cogent factor, that it still has a future, that it still implies a threat and possesses destructive power'.[1] The fact that *das Nichtige* has been finally done away with will, however, only be universally revealed in the return of Jesus Christ in glory. In the meantime God permits it to retain a semblance of power in order that He may use it to serve His own ends. 'In this already innocuous form, as this echo and shadow, it is an instrument of His will and action.'[2] Thus in the end even *das Nichtige* is included among the 'all things' of which it is said that they 'work together for good to those who love God'.[3]

8. CRITICISM: (a) THE ORIGIN OF 'DAS NICHTIGE'

Barth's treatment of the problem of evil is open to criticism at at least three main points.

One of these, namely Barth's decision to call evil *das Nichtige* (instead of *Böse*) and so to involve himself in the ancient but questionable meontic tradition of philosophy, will be discussed in connection with that tradition as a whole in Chapter VIII.[4]

A second criticism centres upon Barth's conception of the origin or ground of *das Nichtige* in the act of divine rejection that was involved in His creation of a good universe. In willing and affirming a good creation God has unwilled, or willed against, a contrary possibility, and has thereby given that which was rejected and excluded a negative but nevertheless virulent power over against the real creaturely world.[5] This view may be criticized both from within Barth's own thought world, as an infringement of his ban upon speculative theorizing, and from outside that thought world, as a naïvely mythological construction, which cannot withstand rational criticism.

---

[1] Ibid., p. 363.   [2] Ibid., p. 367.
[3] Ibid., p. 368. Cf. Romans viii. 28.   [4] See below, pp. 185 f.
[5] See above, pp. 139–40 f.

Barth has himself insisted upon the necessarily 'broken' and unsystematic character of valid Christian discourse concerning evil.[1] All that we can properly do, he suggests, is to affirm the apparently contradictory truths that we learn from Scripture — the utterly malignant character of evil as God's enemy, and yet its subjection to His control — without professing to weave these together into an intelligible unity. But in his account of the coming to 'be' of *das Nichtige* he offers a boldly speculative theory which would enable us to understand how and why *das Nichtige* 'exists'. He is thus doing what he has criticized others for doing, namely going beyond the data of faith and becoming entangled in the dangers of philosophical construction.[2]

Further — and this could be taken as either mitigating or as adding to the offence — the particular speculation in which Barth indulges is mythological rather than rational. As such it may well serve, like the biblical creation myth, to feed the religious imagination; and if Barth had offered it as a myth (in the tradition of Plato, who consciously used myths concerning matters transcending human knowledge) it might have been able to stand as such without incurring philosophic criticism. But if Barth's suggestion is intended as a contribution to the science of theology, its anthropomorphic character becomes a liability. When a human being makes a choice it is presupposed that the alternatives between which he chooses exist as facts or as possibilities independently of himself. Thus, in choosing good we must reject evil. But it is far from clear that this necessity applies to God, creating *ex nihilo* and in absolute freedom. Might we not more properly think of God as able, if He wished, to create a good universe that is not accompanied by the threatening shadow of rejected evil? At any rate this is a question to be explicitly asked and answered rather than to be settled by primitive picture thinking. When it is explicitly asked a further important question arises for Barth, which will be pursued in the next section.

[1] See above, p. 133.
[2] Cf. G. C. Berkouwer, *The Triumph of Grace in the Theology of Karl Barth*, trans. Harry R. Boer (Grand Rapids, Mich. : 1956), pp. 221-3.

9. CRITICISM : (*b*) THE STATUS OF 'DAS NICHTIGE'

This further question stands on the boundary of Barth's own consideration of the subject; for in conformity with his declared method he did not feel called upon to raise and resolve it. But the reader whose mind is focused upon the theodicy-problem and who, in studying its history, has noted the main alternatives and traced the entailment routes linking its various aspects, will not be content to leave unexplored the further implications of Barth's position.

In formulating the question that arises at this point it is understood (1) that by evil we do not here merely mean creaturely finitude and limitation and whatever else comprises that which Barth describes as the shadowside of the created world, but the real opposition to God in sin and evil, which he calls *das Nichtige*; (2) that evil in this sense does not exist independently and in its own right, but only as the object of God's rejection, denial, and condemnation, only as that to which God has irrevocably said no ; and (3) that God has defeated this enemy in Jesus Christ and is able to control and use it until it ceases at his return in glory to exist in any sense whatever. When therefore one speaks of the 'existence' of evil, one is referring to the fact that evil 'is' in the peculiar negative but virulent fashion described by Barth. All this being granted, we must now ask whether, in Barth's system, God *wanted*, in creating a good universe, to bring evil into 'existence' also, or whether the creation of good *entailed* the bringing of evil into 'existence' by some necessity not subject to the divine will? Did God want (for some reason of His own) to bring evil into 'existence' in the course of creating a good world, or could He, if He was going to create a good world, not avoid at the same time giving rise, in this negative fashion, to evil? In other words, could God, if He had wished, have created a good universe without thereby also causing evil to 'be' as a real though negative threat to that creation? Or is there, on the contrary, a necessary connection between good and evil such that good inevitably drew evil in its wake, whether God wished this or not?

Suppose, first, we say that God's creation of a good universe gave this negative but dangerous mode of 'existence' to its obverse, evil, not because God so desired but because of some ontological necessity outside His control. Barth's language seems to suggest some such necessity when he says that '*Das Nichtige* is that from which God *separates Himself* and in face of which He asserts Himself and exerts His positive will',[1] or again that the Chaos of Genesis i, 2, which he identifies with *das Nichtige*, 'is the *unwilled and uncreated reality* which constitutes as it were the periphery of His creation and creature'.[2] The suggestion of such passages seems to be that prior to God's creative activity there was a double potentiality, both for good and for evil, so joined together that in realizing the potentiality for good God could not avoid activating the contrary potentiality of evil. He had specifically to will against it, and could not help thereby giving it a peculiar negative but active status as the object of His aversion, rejection, and denial.

Now such a view would be open to an objection similar to one levelled against Leibniz's position:[3] namely, that it denies God's unlimited power or (in Barth's terminology) freedom. According to Leibniz, God does not want there to be evil in His universe, but because of certain eternal and irrefragable links of compossibility, which are independent of His will, He is unable to create a universe without evil. According to Barth (in this first interpretation of his teaching) God does not want *das Nichtige* to 'be'. But He cannot create the universe which He *does* want without at the same time conjuring up *das Nichtige* as its attendant obverse; for the link between them holds independently of, and indeed contrary to, His will. In the case both of Leibniz and of Barth, the objection, from the standpoint of theodicy, is that when faced with the demand to reconcile the absolute goodness and power of God with the fact of evil, they jettison the latter element in the Christian understanding of God. God is limited by something that holds independently of Him

[1] *C.D.* III/3, p. 351. My italics.
[2] Ibid., p. 352. My italics.
[3] See below, pp. 170 f.

144

and of which He is obliged to take account. In Leibniz's system this 'something' is an eternal realm of logical possibilities; in Barth's it is an ontological possibility or potentiality, which is not indeed eternal but which is of unchangeably malignant character. God can defeat it — at a great cost — but He cannot avoid having to defeat it. On either view God is ultimately defined as a limited God.

This conclusion, however, would presumably be wholly unacceptable to Barth, one of the cardinal points of whose theology is the absolute divine freedom. It would further involve the peculiarity that although God cannot — if He is going to create a world — prevent the presence and operation of evil within it, yet once evil arises He can and does combat and exterminate it. He does not have the power to prevent it from arising, but he does have the power, when it has arisen, to slay it! This paradox naturally prompts us to consider the other answer that Barth might give to our question, namely that God has deliberately created the conditions under which evil has arisen and permits its presence in His world for a good purpose.

On such a view the good purpose for which evil (still in the radical sense of *das Nichtige*) 'exists' is to make possible the supreme good of redemption. Starting from the Christological centre of theology, and above all from the death and resurrection of Christ, which has 'opened the Kingdom of Heaven to all believers', one might ask, What conditions are presupposed by this divine act of salvation? If the whole creation centres upon this great event, is it not implied that man's need for salvation was envisaged in God's creative plan, the presence of evil being a necessary precondition of redemption, and the fall accordingly serving ultimately the high purpose of setting God as Saviour at the centre of His creation? Thus instead of 'O felix culpa . . . we might sing 'O felix Nihil, quae talem ac tantum meruit habere redemptorem!'

There are passages in which Barth seems to be taking such a position, as for example when he says:

In this intention of the Creator and therefore this final goal of the creature as manifested in the divine revelation, there is implied from the very outset, as far as the nature and essence of the creature are concerned, a twofold determination : on the one hand an exaltation and dignity of the creature in the sight of God (for otherwise how could it be His partner, or be accepted by Him?) ; and on the other hand the equally clear need and peril of the creature before Him (for otherwise how could it be so exclusively referred to His lordship and help in the covenant, and to reconciliation with God in the person of His Son?). God created man to lift him in His own Son into fellowship with Himself. This is the positive meaning of human existence and all existence. But this elevation presupposes a wretchedness of human and all existence which His own Son will share and bear. This is the negative meaning of creation. Since everything is created for Jesus Christ and His death and resurrection, from the very outset everything must stand under this twofold and contradictory determination.[1]

It is not entirely clear to what extent Barth is referring here to *das Nichtige* and to what extent only to the shadowside of creation. The section as a whole from which this quotation is taken seems to deal primarily with the latter. But on the other hand it is, according to Barth, *das Nichtige* that has been overcome by Christ on the cross, and it would therefore seem to be *das Nichtige* to which he is referring here. However, it is not easy, amidst these conflicting indications, to be sure how Barth wishes to be understood.

We cannot, then, assert that Barth has given either of the two alternative answers to the question that we have posed — the question, namely, whether God could have created a good world without at the same time giving a negative but powerful 'existence' to evil. Barth seems instead deliberately to refrain from choosing either possibility. And if he were called upon to defend his silence at this point, he would no doubt appeal to the method which he announced at the threshold of his main discussion of evil. He said there, it will be recalled, that because evil represents a fracture in created existence it cannot be brought under any general principle, not can we hope to attain a coherent and systematic doctrine concerning it. One can only affirm the different truths that Scripture affirms or implies concerning it, without insisting that there must be a humanly intelligible relation between

---

[1] *C.D.* iii/1, pp. 375-6.

them. In his substantive discussion Barth carries out this programme.

Indeed, I think we must go further than this and say that not only does Barth not teach that evil 'exists' either ultimately by necessity or ultimately by divine ordinance, but he positively *rules out* each of these two possible ways of relating his account of evil to the doctrine of God

On the one hand, Barth's repeated emphasis upon the absolute Lordship and freedom of God excludes the thought that there is a necessity independent of the divine will, such that in electing good God cannot but give rise to evil as a real alternative which He rejects and combats. This is a pervasive theme of the entire *Dogmatik*; and within section 50 itself Barth says that God is the *Grund* and *Herr* of *das Nichtige*, which is not there by accident (*nicht von ungefähr*) but is from God (*von Gott*).[1] If God is the ground and the lord of evil, evil can hardly 'exist' by some necessity independent of Him.

On the other hand the contrary thought that the strange negative but threatening factor of *das Nichtige* is ultimately God's own work, and contributes in its own strange but pre-ordained way to the fulfilment of the divine purpose, is again and again repudiated by Barth as a disastrous failure to take the reality of evil seriously. 'It is', he says, 'obvious that [*das Nichtige*] neither can nor may be understood as something which [God] Himself has posited or decreed, and that it cannot be subsumed under any synthesis.'[2] And in another place he says of sin (which he has characterized as the most important form of *das Nichtige*)[3] that God confronts it in Jesus Christ with His unconditional 'No'. 'It is a No to which there is no hidden Yes, no secret approval, no original or ultimate agreement.'[4] And concerning the actual occurrence of sin: 'We say too much even if we say that this event may take place according to the divine will and appointment.'[5] Further, that sin and evil arise ultimately

[1] *Kirchliche Dogmatik*, III/3, p. 405 (*C.D.* p. 351).
[2] *C.D.* III/3, pp. 304–5.
[3] Ibid.   [4] *C.D.* IV/1, p. 409.
[5] Ibid., p. 410. Kurt Luthi, in his comparison between Schelling and Barth on the problem of evil, points out that in this aspect of his thought Barth is

by divine ordination is a feature of Schleiermacher's theodicy which Barth emphatically rejects.[1] Can we perhaps find a middle way for Barth between these two alternatives by saying that evil does stand in relation to God's will, rather than to some independent cosmic necessity, but only to His *negative* will : it is a work of God, but only a work on His left hand? In this connection it may be significant that instead of saying without qualification that evil has no basis in God and no part in His will and work Barth says at one point that sin (as a form of *das Nichtige*) 'has no *positive* basis in God . . . no *positive* part in His life, and therefore no *positive* part in His will and work.'[2] And so it may be that Barth would wish to affirm a basis for *das Nichtige* in God's will, and would reconcile this with the divine enmity against evil by relating it, not to God's positive will, but to His negative and contrary will, His willing *against* it. Many of Barth's statements would be compatible with such a position. However, one hesitates to attribute it to him, because it would so manifestly offer a merely verbal and apparent solution. For the original alternatives — divine will or independent necessity — present themselves again concerning that which God does 'on His left hand', by willing against it. We have to ask whether this peculiar negative occurrence 'on God's left hand' is something that He intends and plans, or something that takes place by a necessity which is independent of His intention? When it is said that evil comes to 'be' as the object of God's wrath, rejection, and denial, we have to ask whether its so coming to 'be' is desired and planned by God, or something occurring by a necessity that is contrary to His wish? Thus the question is

highly selective in his use of the biblical evidence. For in the Old Testament there is (as Schelling had noted) a strong strand of *Indienststellung* in respect of evil, a view of evil as God's servant and instrument. 'Indeed the witness of the Old Testament, which is emphatically a witness to *one* God, does not shun the consequence that certain influences and activities, which otherwise are ascribed to the demonic realm, must in order to avoid any trace of dualism be ascribed to this one God ; but Barth does not draw this conclusion, probably because his starting-point and his aim forbid him to feel this question as a question.' *Gott und das Böse: Eine biblischtheologische und systematische These zur Lehre vom Bösen, entworfen in Auseinandersetzung mit Schelling und Karl Barth* (Zürich: Zwingli Verlag, 1961), p. 261.   [1] See below, pp. 237 f.
[2] *C.D.* iv/i, p. 409.   My italics.

not resolved, but only postponed, by invoking the image of the left hand and the related notion of willing-against.

We are left in a strange position indeed! Evil (in the sense of *das Nichtige*) came to 'be' as that which God repudiated when He created His good universe. But Barth does not feel obliged to suppose that it thus came to 'be' either by a necessity independent of the divine will, or by the divine will itself. He not only refuses to choose between these possibilities, but by implication he repudiates both! For both, he would no doubt claim, are ruled out by God's own revelation of Himself. We know from the Scriptures that God is absolute and not subject to any necessities which He has not Himself established. We also know from the Scriptures that God is the implacable enemy of *das Nichtige*, which therefore cannot in any sense be His own work. We have, then, as theologians, to leave the problem hanging in the air, without presuming to settle it. Having formulated it, we can only turn aside from it as something that is beyond our ken and not our proper business.

But can we be content to leave the matter at this point? We should not be expected to adopt Barth's own distinctive account of the nature and origin of evil as a revealed truth which, although incomprehensible, must be accepted in faith. For it does not represent revealed truth at all. It is a product of Barth's own fertile and fascinating mind. The notion that evil has been brought into 'existence' as that which God rejected when he elected His good creation, is not among the data of Christian faith. It is not a part of, nor is it an unproblematic deduction from, the biblical revelation. Barth's claim to find it in Genesis i is a quite patent imposing of his own speculations upon the text. And if such a theory were to have a protected status, absolved from responsibility for what it logically entails, theology would cease to be an intellectually respectable or interesting subject and would degenerate into the mere proclaiming of arbitrarily adopted positions.

The consequence of Barth's teaching is to conceal the final alternatives facing any theodicy. For every position that maintains the perfect goodness of God is bound either to let go the absolute divine power and freedom, or else to hold that evil exists ultimately within God's good purpose. Theodicies of the Augustinian type tend, in various ways, to adopt the first alternative, whilst the Irenaean tradition of theodicy tends to adopt the second. Barth, whose doctrine of evil belongs in general to the Augustinian family, declines this ultimate choice — but not, unfortunately, on grounds that can either excuse others from facing it or give them any help when they do face it.

But although Barth's treatment of the problem of evil only accentuates that problem, he does shed valuable light in the course of his discussion. Perhaps the chief constructive value of Barth's thought on this subject lies in his distinction between the shadowside of creaturely existence, and evil in the much stronger sense of enmity against God of which the primary expression is sin. This is a valuable distinction, and while it corresponds roughly to the traditional division between metaphysical evil on the one hand and sin and suffering on the other, Barth with his dramatic nomenclature has drawn it with a new vividness. There is likewise great value in his suggestion that the 'evils' (as they generally have been called) of the shadowside, although not in themselves deadly and inimical to God's purposes, have become so in man in his state of alienation from God.

CHAPTER VII

EIGHTEENTH-CENTURY 'OPTIMISM'

1. A Product of the Augustinian Tradition

THE eighteenth century was the golden age of theodicies, when the problem of evil was often at the centre of discussion and when comprehensive solutions to it were being confidently offered. In this century of the Enlightenment the conception of evil as serving a larger good — a conception derived, as we have seen, from Plotinus and Christianized by Augustine — was developed to the ultimate conclusion that despite all that is bad within it our universe is nevertheless the very best that is possible. This was the thesis of a work that was widely read in Britain during the eighteenth century, the *Essay on the Origin of Evil* by the Anglican divine, William King (1650–1729), who became Archbishop of Dublin. King's book was first published in Latin in 1702 and appeared in 1731 in an English translation by Edmund Law, fortified both by Law's own notes and by extracts from King's posthumous papers in defence of his doctrines. This version went through five editions during the eighteenth century and was widely read.[1] It was, for example, probably King's formulation of the so-called optimistic thesis, in the original Latin edition, that provided the philosophical content of Pope's *Essay on Man*.[2] In prose this type of theodicy was expounded in England by John Clarke, in his *Enquiry into the Cause and Origin of Evil*

---

[1] Written in 1697, *De Origine Mali* was first published in Latin in London and Dublin in 1702, with a reprint of the London edition at Bremen in 1704. Edmund Law's translation was published in five editions; London, 1731; London, 1732; Cambridge, 1739; Cambridge, 1758; London, 1781. (On the origin of the book see King's 'Quaedam Vitae Meae Insigniora', reprinted in Sir Charles Simeon King, *A Great Archbishop of Dublin* (London: Longmans, Green & Co., 1906).)

[2] See Arthur Lovejoy, *The Great Chain of Being*, lectr. vii, n. 9, and F. Billicsich, *Das Problem des Übels in der Philosophie des Abendlandes*, vol. ii, p. 179, n. 14.

(1720)[1] and by Soame Jenyns in his *Free Inquiry into the Nature and Origin of Evil* (1756).[2] But it was, of course, on the continent that the greatest expression of eighteenth-century 'optimism' appeared in Leibniz's *Theodicy* (1710),[3] which has been described as 'without doubt the greatest effort that philosophy has made to solve the problem of evil'.[4] Pope's *Essay on Man* (1732), as an optimistic theodicy in versified form, had its German counterparts during the eighteenth century in the poetry of Barthold Heinrich Brockes, Albrecht von Haller (who wrote a poem called 'Ueber den Ursprung des Uebels'), Friedrich Hagedorn, Johann Peter Uz, and C. M. Wieland. However, this way of thinking never lacked its critics. Dr. Samuel Johnson, for example, wrote a crushing review of Soames Jenyn's *Free Enquiry*;[5] and Johnson's own philosophical novel *Rasselas* (1759) is itself a sustained rejection of the optimistic thesis. But the most famous criticisms of philosophical optimism are to be found, in literature, in Pierre Bayle's *Dictionnaire historique et critique* (1697), Voltaire's novel, *Candide ou l'optimisme* (1759), and David Hume's *Dialogues Concerning Natural Religion* (1779) and more decisively, in the events of the world, in the great Lisbon earthquake of 1755 — a 'great' upheaval of nature not so much in its physical proportions as in its tremendous psychological impact upon mid-eighteenth-century Europe.[6]

Eighteenth-century 'optimism' (I set the word between quotation marks because there is an important sense in which the traditional designation is entirely inappropriate) is a variation of the Augustinian type of theodicy, and belongs firmly within the tradition which we are studying in Part II. This has not always been recognized. Roman-Catholic

---

[1] John Clarke, D.D., was Chaplain in Ordinary to King George I, and his book contained the Robert Boyle Lectures for 1719.
[2] Soame Jenyns was for some years Member of Parliament for Cambridge.
[3] G. W. Leibniz, *Theodicy* (*Essais de théodicée, sur la bonté de Dieu, la liberté de l'homme, et l'origine du mal*), trans. by E. M. Huggard (London: Routledge & Kegan Paul Ltd., 1952).
[4] Charles Werner, *Le Problème du Mal dans la Pensée Humaine*, op. cit., p. 23.
[5] Samuel Johnson, *Works*, ed. Arthur Murphy (1792), vol. 8.
[6] For a study of the contemporary reactions to the Lisbon earthquake, see T. D. Kendrick, *The Lisbon Earthquake* (London: Methuen & Co., Ltd., 1956).

writers sometimes treat the work of Leibniz (and by implication eighteenth-century optimism in general) as though it were opposed to the traditional Catholic solution to the problem of evil. But in fact Leibniz departed from the Augustinian–Thomist theodicy at only one significant point : Aquinas had believed that although this world is a product of infinite goodness and power, its Maker could, if He had wished, have created yet better worlds; whereas Leibniz believed that an omnipotent and infinitely good Being, in creating a world, could make only the best that is possible. This disagreement turns out on analysis not to represent as direct a clash as appears at first sight, since each party is interpreting the notion of a 'best possible world' in a different way.[1] There does, however, lie behind this difference of interpretation the deeper difference that — as will be shown below — whereas Thomism in effect compromises the divine love rather than the divine power, eighteenth-century optimism in effect sacrifices God's power rather than His love, to achieve their respective theodicies. Yet even this important difference occurs within a very wide context of agreement. In his privative account of evil; in his use of the principle of plenitude to account for natural imperfection; in his reliance upon the aesthetic conception of evil as part of a larger good; and in his retention of the heaven and hell dichotomy in the final fulfilment of the divine purpose, Leibniz is in direct continuity with the Augustinian tradition. Basically, as a learned Catholic historian of theodicies has said, Leibniz's *Theodicy* 're-edited the Augustinian and Thomist defence, traditional since the time of these two geniuses, of the divine goodness and wisdom manifested in the general order of the universe even if not always in each phenomenon considered in isolation'.[2] The same is true to an equal or perhaps to a greater extent of the work of William King; and indeed before turning to Leibniz's great classic it will be instructive to look at King's *Origin of Evil*. King was not an original thinker; but precisely for this reason his clear and uncritical exposition presents the

[1] See pp. 166 f. below.
[2] A. D. Sertillanges, *Le Problème du Mal*, vol. i, pp. 230–1.

Augustinian themes as they had become detatched from the
more creative minds that first produced them and refined
through a long tradition into the form in which they were
the common currency of educated discussion.[1]

2. KING'S 'ORIGIN OF EVIL'

Archbishop King uses the three-fold division of the phenomena of evil as imperfection ('the absence of those perfections or advantages which exist elsewhere, or in other
beings') ; natural evil ('pains and uneasinesses, inconveniences and disappointments of appetites') ; and moral evil
('vicious elections').[2]

Concerning the evil of defect or imperfection — corresponding to Leibniz's metaphysical evil — King's position
represents the Augustinian–Thomist theodicy in its fully
developed form. Created beings must, in the nature of the
case, be less perfect than their Creator ; they must, then, in
varying degrees be imperfect ; hence, as King's editor,
Edmund Law, says, 'Granting therefore this one principle,
which cannot be denied, (viz. that an effect must be inferior
to its cause) it will appear that the evil of imperfection,
supposing a creation, is necessary and unavoidable ; and
consequently, all other evils which necessarily arise from
that, are unavoidable also'.[3] Given this inevitable finitude
and imperfection, God has chosen in the present system of
the world the very best that could be : 'It might have been
better perhaps in some particulars, but not without some new,
and probably greater inconveniences, which must have
spoiled the beauty either of the whole, or of some chief part.'[4]
Thus King arrives at the distinctive notion of eighteenth-
century optimism, that of the best possible world ; and Law

[1] On the historical antecedents of eighteenth-century optimism, see p. 122
n. 1 above.
[2] William King, *An Essay on the Origin of Evil*, trans. and ed. by Edmund
Law (4th ed. 1758), p. 92. (This edition was sometimes bound in 2 vols., the
second volume beginning with p. 311, but having a new title-page.) King
even sees the very imperfection of created things as evidence of the infinite
goodness of their Maker : 'Had not God been infinitely good, perhaps he might
not have permitted imperfect beings ; but have been content in himself, and
created nothing at all'. (*Essay*, p. 107.)
[3] *Essay*, p. 103, n. 18.   [4] *Essay*, pp. 108–9.

explains that by 'the very best system' is meant 'one that is fitted for, and productive of the greatest absolute general good'.[1]

Following Augustine and the long tradition which stems from him, King links the finitude and imperfection of created beings with the 'non-being' out of which they have come. This idea is very clearly formulated in another work of that period, Scott's *Christian Life*, which Law quotes:

God is the cause of perfection only, but not of defect, which so far forth as it is natural to created beings hath no cause at all, but is merely a negation or non-entity. For every created thing was a negation or non-entity before ever it had a positive being, and it had only so much of its primitive negation taken away from it, as it had positive being conferred upon it; and therefore, so far forth as it is, its being is to be attributed to the Sovereign Cause that produced it; but so far forth as it is not, its not being is to be attributed to the original non-entity out of which it was produced. For that which was once nothing, would still have been nothing, had it not been for the cause that gave being to it, and therefore that it is so far nothing still, i.e. limited and defective, is only to be attributed to its own primitive nothingness.[2]

Thus, as King says,

Whatever arises from *Nothing* is necessarily imperfect; and the less it is removed from nothing . . . the more imperfect it is. There is no occasion therefore for an evil principle to introduce the evil of defect, or an inequality of perfections in the works of God: for the very nature of created beings necessarily requires it, and we may conceive the place of this malicious principle to be abundantly supplied from hence, that they derive their origin from *Nothing*.[3]

In response to the question, Why has God created beings below the highest species of dependent creature? King appeals to the principle of plenitude, which Augustine had long ago baptized into the service of Christian theology.[4] It is in accordance with this principle that King says,

[1] *Essay*, p. 108, n. 19.
[2] Quoted in King, *Essay*, p. 116, n. 21.   [3] *Essay*, pp. 115–16.
[4] Law, in his edition of King's *Essay*, quotes Joseph Addison's perfect formulation of this principle: 'Infinite goodness is of so communicative a nature, that it seems to delight in the conferring of existence upon every degree of perceptive being' (*Spectator*, no. 519).

If the production of a less perfect being were any hindrance to a more perfect one, it would appear contrary to divine goodness to have omitted the more perfect and created the less; but since they are no manner of hindrance to each other, the more the better.[1]

The basic criticism that has been suggested here of other theodicies of the Augustinian family — that they conceive of God's relation to His creation, including mankind, in predominantly impersonal terms — applies emphatically to King's work. An ingenious physical model, which he uses in connection with the principle of plenitude, well illustrates the impersonal tendency of his thinking:

If any one had a mind to fill a certain vessel with globes of various magnitudes, and had distinguished them into their several degrees, so that those of the second degree might have place in the interstices left by those of the first; and those of the third order in the interstices of the second, and so on. 'Tis evident that when as many of the first magnitude were put in as the vessel could contain yet there would be room for those of the second. . . . [Thus] when as many creatures were made of the superior order as the system of the world was able to contain . . . nothing hindered but that there might be room for others of a lower degree: as when as many globes of greater magnitude were put into the vessel as it could hold, yet there was still a space for others of a less dimension; and so on *in infinitum*.[2]

In this discussion of the evil of imperfection and defect King lays the foundation for his treatment of the other kinds of evil. Turning to the natural evil of pain, he sees this as arising inevitably from the existence of matter in motion. For matter would be useless to compose animal bodies and their common environment if it lacked the capacity of motion: 'And if this be once admitted in matter, there necessarily follows a division and disparity of parts, clashing and opposition, comminution, concretion and repulsion, and all those evils which we behold in generation and corruption'.[3] It is true that we cannot prove in detail that the laws according to which nature moves, and the various shocks and

[1] *Essay*, p. 117.   [2] *Essay*, pp. 119–20, n. F.
[3] *Essay*, pp. 133–4.

clashes which occur in a world of matter regulated by these laws, constitute the best possible system :

> You'll say that some particular things might have been better. But, since you do not thoroughly understand the whole, you have no right to affirm thus much. We have much greater reason to presume that no one part of it could be changed for the better, without greater detriment to the rest, which it would either be inconsistent with, or disfigure by its disproportion.[1]

Further, given that souls are united with bodies, which are in turn organic to their material environment, the sensation of pain operates as a warning to the soul to react to a danger that threatens the body : 'nor ought it to cease urging, till what was hurtful be removed : without this importunity perhaps the strongest animal would not last even a day. The sense then of pain or uneasiness produced in the soul upon the mutilation or dissolution of the body, is necessary for the preservation of life in the present state of things.'[2]

In summarizing his conclusions concerning natural evil King gives repeated hints of the ultimately dualist implication of eighteenth-century optimism. For in affirming that the world is as good as it possibly could be, in that God (given His initial free decision to create)[3] was pulled by certain necessities and restricted by certain incompatibilities to form precisely the universe that now exists, even though it contains much that displeases Him, the optimistic theodicy in effect saves God's goodness by denying His omnipotence. So King affirms that,

> if the mundane system be taken together, if all the parts and periods of it be compared with one another, we must believe that it could not possibly be better ; if any part could be changed for the better, another would be worse ; if one abounded with greater co...eniences, another would be exposed to greater evils ; and that necessarily from the imperfection of all creatures. . . . Nor is it any argument against the divine omnipotence, that it could not free a creature in its own nature necessarily imperfect, from that native imperfection, and the evils consequent upon it. He might, as we have often said, have not created mortal inhabitants, and

---

[1] *Essay*, p. 137.   [2] *Essay*, pp. 154-5.
[3] Cf. *Essay*, chap. v, sect. i, 4.

such as were liable to fears and griefs . . . but with regard to the system of the whole 'twas necessary that he should create these or none at all. . . . If therefore we could compare the good things with the evil; if we could view the whole workmanship of God; if we could thoroughly understand the connection, subordinations, and mutual relation of things, the mutual assistance which they afford each other; and lastly, the whole series and order of them; it would appear that the world is as well as it could possibly be; and that no evil in it could be avoided, which would not occasion a greater by its absence.[1]

Here — as at many other points throughout his book — King depicts God as doing the best that He can in face of unalterable circumstances. He is not free to do what He would ideally like to do; but within the limits of the eternal laws of compatibility and incompatibility among finite things, which are brought into play by the pressure towards a maximization of being, He has made the best universe that could be made. The most fundamental of the necessities thus postulated by King is undoubtedly that created by the operation of the principle of plenitude. If God is going to create a universe at all, He is apparently obliged to create one that contains every possible kind of dependent creature. This axiom recurs in the pages of King's *Essay*; for example, when speaking of the creation of animals, which habitually sustain themselves by the gross and unspiritual method of eating, King says, 'The infinite power of God was able to produce animals of such capacities; and since the creation of them was no inconvenience to other beings who might exercise themselves in a more noble manner, may not the infinite goodness of God be conceived to have almost compelled him not to refuse or envy those the benefit of life?'[2] It is as though the infinite possibilities of being challenge God to create as many of them as He can, even if the resulting world necessarily contains all manner of evils.

When he turns, as his last topic, to the problem of moral evil, King finds his solution (though with minimal emphasis upon the traditional fall doctrine) in what has come to be called the free-will defence: morally responsible creatures must be free to sin, and in allowing them to sin God is but

---

[1] *Essay*, pp. 195–7. [2] *Essay*, pp. 172–4.

respecting their status as persons. The question is, then, why God should have wanted to create creatures endowed with this dangerous faculty of free will. Here we find two answers in King, one stemming from the old principle of plenitude, and the other prophetic of a quite different type of theodicy. On the one hand, King says that God need not have made persons : 'But such a monstrous defect and hiatus would have been left in nature by this means, *viz.* by taking away all free agents, as would have put the world into a worse condition than that which it is in at present. . . .'[1] But a little later he adds another reason, which had within it — unobserved by King — the capacity to subvert and supersede the principle of plenitude :

any one may understand how much a work which moves itself, pleases itself, and is capable of receiving and returning a favour, is preferable to one that does nothing, feels nothing, makes no return, unless by the force of some external impulse : any person, I say, may apprehend this, who remembers what a difference there is between a child caressing his father, and a machine turned about by the hand of the artificer. There is a kind of commerce between God, and such of his works as are endowed with freedom ; there's room for covenant and mutual love.[2]

But, given that there are to be personal creatures, why does not God intervene to prevent them from so disastrously misusing their freedom? Because 'more and greater evils would befall the universe from such an interposition, than from the abuse of free-will'.[3] God would have to do violence both to the natural order and to the finite wills that He has created, and would thereby reduce personal beings to the level of the animals. But 'You may urge, that you had rather want this pleasure [of the consciousness of freely acting rightly] than undergo the danger ; that is, you had rather be a brute than a man'.[4] In response to this suggestion King sinks back to the impersonal level of the principle of plenitude : 'But supposing it were convenient to you to be a brute, yet it would not be convenient for all nature : the system of the universe required free agents : without these

[1] *Essay*, p. 340. [2] *Essay*, p. 342.
[3] *Essay*, p. 356. [4] *Essay*, pp. 367–8.

the works of God would be lame and imperfect; his goodness chose the benefit of the universe rather than that of yourself. . . .'[1] Further, in abolishing human freedom God would be depriving Himself of that aspect of His creation in relation to which He exercises His attribute of wisdom; for 'the divine wisdom seems to have set apart the government of free agents as its peculiar province', and 'God made the world in order to have something wherein to exercise his attributes externally'.[2]

We may leave King on this characteristic note, which reflects the pervasively impersonal standpoint of the tradition of theodicy deriving from Augustine.

3. LEIBNIZ'S 'THEODICY'

As in the case of Archbishop King's *Essay*, one feels in reading Leibniz's *Theodicy* that the problem of evil was for its author an intellectual puzzle rather than a terrifying threat to all the meaning that he had found in life. Karl Barth has said of Leibniz that 'at bottom he hardly had any serious interest (and from the practical standpoint none at all) in the problem of evil';[3] and I for one do not feel inclined to dispute this dictum.[4] Having shown to his own satisfaction that we are living in the best possible world, Leibniz was content to enjoy his own comparatively comfortable lot, leaving it to those who were less fortunate to make the best they could of it.

The sense in which this is for Leibniz the best possible world (or universe, for he uses 'world' in a comprehensive sense) reveals the predominantly aesthetic, as distinct from personal, motif which presides over his theodicy. The best possible world is that which permits a maximization of being. This maximum is defined in terms not only of quantity but also of variety. As Leibniz says in the *Theodicy*, the actual world consists of those possibles 'which, being united, produce most reality, most perfection, most significance';[5]

---

[1] *Essay*, p. 369.   [2] Ibid.   [3] Karl Barth, *C.D.* III/1, p. 392.
[4] For a contrary view, however, see Friedrich Billicsich, *Das Problem des Übels in der Philosophie des Abendlandes*, ii. 112. Billicsich points out that Leibniz had been concerned with the theodicy problem from his youth.
[5] Leibniz, *Theodicy*, para. 201.

and as he says in a letter to Louis Bourget, 'the actual universe is the collection of those [existents] which form the richest composition'.[1]   A merely quantitative maximum would, in comparison, be dull and uninteresting : 'To multiply one and the same thing only would be superfluity, and poverty too.   To have a thousand well-bound Vergils in one's library, always to sing the airs from the opera of Cadmus and Hermione, to break all the china in order to have cups of gold, to have only diamond buttons, to eat nothing but partridges, to drink only Hungarian or Shiraz wine — would one call that reason?'[2]

Given, then, that the divine purpose in creating, which is to manifest God's goodness beyond the borders of His own being, is better served by the production of a richly varied realm rather than of only a single type of existence; and granting further that not all of the things that are individually possible are mutually compossible, the Creator had to choose one particular coherent set of possibilities upon which to bestow existence.   His choice was made from an infinity of different universes which were present in idea to the divine mind.   Each constituted a complete possible history from creation onwards, and each formed a systematic whole such that to alter the least feature of it would be to change it into a different universe.   It was these comprehensive possibilities that God surveyed, and from among which He summoned one into existence by His creative power :

One may say that as soon as God has decreed to create something there is a struggle between all the possibles, all of them laying claim to existence, and that those which, being united, produce most reality, most perfection, most significance carry the day.   It is true that all this struggle can only be ideal, that is to say, it can only be a conflict of reasons in the most perfect understanding, which cannot fail to act in the most perfect way, and consequently to choose the best.[3]

Leibniz thinks of the possibilities between which God chooses as objective to the divine Mind in the way in which mathematical truths are objective to our own minds.   The

[1] *Philosophische Schriften*, ed. Gerhardt, iii. 573.
[2] *Theodicy*, para. 124.   [3] Ibid., para. 201.

eternal verities, 'which are altogether necessary, so that the opposite implies contradiction',[1] and which determine the cosmic possibilities, are 'in the understanding of God, independently of his will'.[2] There are here both a continuity with a long tradition and at the same time a modification of it. The Scholastics had located the realm of Platonic Ideas within the mind of God, and had assumed that these are the eternal archetypes upon which created things are modelled.[3] Hence one may say that, for the Scholastics, God's creative activity was conditioned by the existence within the divine Mind of a fixed inventory of kinds of creatable things. But they did not think of this ideal realm as involving conflict and contradiction, and accordingly they did not make use of the notion of compossibility. But it is, in Leibniz's system, the fact of compossibility, or rather the negative fact of noncompossibility, that compels God to choose as the best possible world one which nevertheless contains a great deal of evil. For His choice is limited by inherent compatibilities and incompatibilities. And so Leibniz offers a kind of cybernetic myth in which the divine intellect, like an infinite calculating machine, sets up every possible combination of existents, surveys them exhaustively, and selects the best:

The wisdom of God, not content with embracing all the possibles, penetrates them, compares them, weighs them one against the other, to estimate their degrees of perfection or imperfection, the strong and the weak, the good and the evil. It goes even beyond the finite combinations, it makes of them an infinity of infinites, that is to say, an infinity of possible sequences of the universe, each of which contains an infinity of creatures. By this means the divine Wisdom distributes all the possibles it had already contemplated separately, into so many universal systems which it further compares the one with the other. The result of all these comparisons and deliberations is the choice of the best from among all these possible systems, which wisdom makes in order to satisfy goodness completely; and such is precisely the plan of the universe as it is. Moreover, all these operations of the divine understanding, although they have among them an order and a priority of nature, always take place together, no priority of time existing among them.[4]

[1] *Theodicy*, para. 2.   [2] Ibid., para. 20. Cf. 380.
[3] See above, p. 85   [4] *Theodicy*, para. 225.

This means in the case of our own world that,

> God, before decreeing anything, considered among other pos-
> sible sequences of things that one which he afterwards approved.
> In the idea of this is represented how the first parents sin and
> corrupt their posterity; how Jesus Christ redeems the human
> race; how some, aided by such and such graces, attain to final
> faith and to salvation; and how others, with or without such or
> other graces, do not attain thereto, continue in sin, and are
> damned. God grants his sanction to this sequence only after hav-
> ing entered into all its detail, and thus pronounces nothing final
> as to those who shall be saved or damned without having pondered
> upon everything and compared it with other possible sequences.
> Thus God's pronouncement concerns the whole sequence at the
> same time; he simply decrees its existence.[1]

This last point provides the basis for Leibniz's reconciliation of human freedom with the prior divine choice of a
complete world history. For it has often been pointed out
that the Leibnizian universe, despite its inventor's advertise-
ments to the contrary, is as rigidly determined as Spinoza's.
In selecting the best possible universe for creation God has
decreed a complete sequence of events, filling the whole of
time; and accordingly every phase of the world throughout
its entire temporal extent will fulfil the complex specification
selected by God in His original choice. But then, as Charles
Werner says, 'If it is true that our universe was already
there, fully formed in the divine understanding, and that
God's creative action simply consisted in making it pass from
the realm of possibility into real existence, without any
change ever intervening, does not this world, thus frozen as
though in ice, present to us a perfect and devastating image
of necessity?'[2]
Leibniz's reply is that the necessity constituted by the
completeness of the universe in the divine mind does not
exclude genuine human freedom and responsibility. For
when our world first came as a total complex possibility
under the Creator's scrutiny, it included all the free actions
of free beings throughout its history:

[1] Ibid., para. 84.
[2] Charles Werner, *Le Problème du Mal dans la Pensée Humaine*, p. 33. Cf. Arthur
Lovejoy, *The Great Chain of Being*, pp. 166 f.

Since . . . God's decree consists solely in the resolution he forms, after having compared all possible worlds, to choose that one which is the best, and bring it into existence together with all that this world contains, by means of the all-powerful word *Fiat*, it is plain to see that this decree changes nothing in the constitution of things : God leaves them just as they were in the state of mere possibility, that is, changing nothing either in their essence or nature, or even their accidents, which are represented perfectly already in the idea of this possible world. Thus that which is contingent and free remains no less so under the decrees of God than under his prevision.[1]

In this reconciliation of human freedom with divine sovereignty Leibniz stands within the broad Augustinian tradition. His argument parallels the reply made by Augustine to the charge that divine predestination takes away man's liberty : namely, that God's foreseeing of our free actions does not render them any less free.[2]

Returning to the main line of Leibniz's argument, it is clear that he is so far from denying the existence of evil that he asserts its indispensability ! He acknowledges its reality — though always as a privation of goodness and being[3]— under the three heads of metaphysical evil, which consists in finitude and imperfection ; physical evil, which consists in pain and suffering ; and moral evil, which is sin.[4] For Leibniz, as for the Augustinian tradition as a whole, metaphysical evil is fundamental, and the other modes of evil flow from it. And the principle which accommodates these evils into the best possible world is the aesthetic principle that a good whole may contain parts that would in isolation be bad. 'Not only does [God] derive from [evils] greater goods, but he finds them connected with the greatest goods of all those that are possible : so that it would be a fault not to permit them.'[5] Leibniz offers in support various examples of the way in which evils may contribute to a more comprehensive good. For instance, in directing a battle a general may make a mistake that nevertheless turns out for the best ; and 'a little acid, sharpness or bitterness is often more pleasing than sugar ; shadows enhance colours ; even a dissonance

[1] *Theodicy*, para. 52. [2] See above, pp. 74–75 [3] *Theodicy*, paras. 30–33.
[4] Ibid., para. 21. [5] Ibid., para. 127.

in the right place gives relief to harmony';[1] and there are the words which they sing 'on the eve of Easter, in the churches of the Roman rite:

O certe necessarium Adae peccatum, quod Christi morte deletum est!
O felix culpa, quae talem ac tantum meruit habere Redemptorem!'[2]

In short, 'all the evils of the world contribute, in ways which generally we cannot now trace, to the character of the whole as the best of all possible universes'; so that 'if the smallest evil that comes to pass in the world were missing in it, it would no longer be this world; which, with nothing omitted and all allowance made, was found the best by the Creator who chose it'.[3]

Leibniz does not attempt to demonstrate from the appearances of nature that this is indeed the best possible world. 'I cannot show you this in detail', he says, 'For can I know and can I present infinities to you and compare them together? But you must judge with me *ab effectu*, since God has chosen this world as it is.'[4] Leibniz's conclusion is thus an article of faith derived from the prior faith that this world has been created by an all-powerful and perfectly good deity. Since God is good it would be inconsistent with His own nature to have chosen any other world than the best: for 'supreme wisdom, united to a goodness that is no less infinite, cannot but have chosen the best'.[5] God could not create any other kind of universe. He is not, however, moved or urged in this by anything external to Himself. His original decision to create a dependent universe was a decision of the absolute divine freedom. His further decision to create the best out of all the possible worlds arose from an inner necessity of His own nature. Such an inner necessity is not however to be confused with constraint, and we must not 'imagine that since God cannot help acting for the best he is thus deprived of freedom'.[6] For it is not a defect or an unfreedom in a saint or an angel to have a nature that tends

[1] Ibid., para. 12.   [2] Ibid., para. 10.   [3] Ibid., para. 9.
[4] Ibid., para. 10.   [5] Ibid., para. 8. Cf. paras. 117 and 145.
[6] Ibid., para. 168. Cf. paras. 45, 175, and 191.

to good; and likewise it is not a defect or unfreedom in God that, being perfectly good, He always and unerringly acts for the best.

4. THE 'BEST POSSIBLE WORLD'

The proposition that the perfect and omnipotent God will, if He elects to create a dependent universe, create the best possible such universe, is Leibniz's central thesis. It is at this point that he departs from the Thomist and Catholic development of the Augustinian theodicy. Abbé Charles Journet attacks Leibniz's axiom, finding in it a double incoherence: that of the Absolute morally bound to create and that of the best of all possible worlds'.[1]

Under the first heading, Journet assumes that Leibniz 'presupposes in God an inclination which necessarily carries him to create',[2] and says, 'We impugn the transcendence of God and ignore his absolute indifference in regard to all created things, when we assume in him a moral necessity to create'.[3] But does Leibniz in fact teach this? I cannot find that he does. The passages that Journet cites (paras. 233, 230, 234, and 175) refer to God's choice of the best possible world, given that He has already decided to create; and they specifically say of that prior decision that 'The decree to create is free.'[4] It is true that there is something in God's nature which in fact leads Him to create; otherwise He would not have done so. But this 'something' is the same for Leibniz as for Aquinas, namely God's overflowing goodness expressing itself in the free production of a dependent realm. Thus Leibniz says that 'it is goodness which prompts God to create with the purpose of communicating himself',[5] and Aquinas says that 'if natural things, in so far as they are perfect, communicate their good to others, much more does it appertain to the Divine Will to communicate by likeness its own good to others, as much as is possible'.[6] There is

---

[1] Charles Journet, *The Meaning of Evil*, p. 118.   [2] Ibid., p. 120.
[3] Ibid., n. 149.
[4] *Theodicy*, para. 230.
[5] Ibid., para. 228.
[6] *S.T.*, pt. 1, Q. xix, art. 2. Cf. also the following passage from the same place: 'Hence, although God wills things apart from Himself only for the sake

here no real difference between Aquinas' and Leibniz's
accounts of the divine 'motive' in creating; according to
each of them, creation flows from the outgoing goodness of
God.

Journet develops his second charge as follows:

> The notion of the best of all possible worlds is by definition un-
> realizable — like that of the fastest possible speed — for 'what-
> ever things he has made, God could make a better one', and so on
> indefinitely. To demand that God, to be above reproach, must
> make the best of all possible worlds is to demand him to make
> what is not feasible, and to give existence to something absurd.[1]

The debate is at this point, I believe, based upon a confusion, which resolves itself when we make explicit the differ-
ing definitions of 'best possible world' that are being used.
Aquinas' position, which Journet is supporting, is valid in
relation to one definition, and Leibniz's in relation to
another; and when these two definitions are made explicit
there is no direct clash between the two views. The Thomist
doctrine assumes a scale of possible universes rising from
non-being to God, who is treated for this purpose as the
terminal member of the series. This scale contains an infinity
of steps (as does the series of fractions between 0 and 1), so
that between any two universes it is always theoretically
possible for there to be another. Thus there is always a
possible universe which is superior to any given actual
universe, but less than God. In other words, the scale of
universes approaches asymptotically to the divine perfection
without ever reaching it, and consequently there is always
room for yet better ones. Within this framework of thought
it is correct to reject as meaningless the notion of the best
possible world. Leibniz, on the other hand, was assuming
an unlimited range of possible universes, among which there
is one that contains more reality than any other. This one
will not necessarily be, and in fact is not, particularly close
to the level of the divine perfection; but nevertheless there

of the end, which is His own goodness, it does not follow that anything else
moves His will, except His goodness. So, as He understands things apart from
Himself by understanding His own Essence, so He wills things apart from
Himself by willing His own goodness.'   [1] *M.E.*, pp. 117–18.

is no other compossible system that is as a totality superior to it. The infinite mind of the Creator has surveyed the infinite realm of world-possibilities and has selected the best. Defined in this way the notion of the best possible world is coherent and feasible.

Thus, on the one hand, Thomism is right in holding that there could not be a best possible world, understood as the closest possible approximation to perfection within an infinite series of such approximations. But, on the other hand, Leibniz was right in holding that there could be a best possible world in the sense of that member of an infinite class of possible worlds that best satisfies a stated criterion of excellence. Accordingly, the best possible world that the Thomist tradition denies is not that which Leibniz affirms. So far as this difference between Thomism and Leibnizian optimism is concerned, both might be right. Each position, however, generates its own grave difficulties, which can best be brought to light by further reflection upon the notion of a best possible world.

The difficulty entailed by the Thomist position is as follows. It is affirmed that God could, had He wished, have made a better world. As this contention is developed by Journet, it is clear that it does not amount simply to the abstract mathematical truth that within an infinite but bounded series there can always be a yet closer approximation to the limit. For this could entail a merely infinitesimal lack in our present world; the world might for all practical purposes be perfect, although it would still be true that a yet closer approximation to perfection remains conceivable. But Journet holds that God could have made a better world in the much more substantial sense that He could have made a world free from many of the gross evils of the present order. Some of these evils, to be sure, are logically inseparable from some of the world's goods, as Journet, following Aquinas, points out. But he does not profess that in view of these irrefragable linkages between good and evil the present arrangement is (as Leibniz claims) the best possible. Journet allows quite explicitly that God could, had He wished, have decreed a different arrangement in which there would have been proportionately more good and less evil. 'God could', says Journet, 'according to his absolute power, have made better worlds than ours.'[1] And in a passage about the sufferings of animals, which I have quoted above, he says, '. . . granted the existence of our animal world, and hence of suffering, could not the sum of suffering be less? The only answer, of course, is that it could be less. And also it could be greater . . .'[2] But, Journet adds (again representing the traditional Thomist teaching) whatever the character of the world it would still be infinitely distant from the glory and goodness of God :

An unbridgeable gap separates the infinite, uncreated good from the whole universe of created and creatable things. . . . Even for God it will never be possible to pour out the fullness of uncreated being into the vessels of created beings, or to enclose the infinite within the finite. Whatever world he decides to make, what will be manifested of his infinite fullness will never be equivalent to what remains to be manifested. There will always be an infinite margin in which other worlds could occur.[3]

This argument moves on a highly abstract plane and should be tested by being brought into relation with concrete reality. Suppose that the world was as it is at present, with the single exception that it did not include cholera germs. The plain untheological man in the street will say that it would to that extent have been a better world. Journet will agree, but will insist that this better world would still be infinitely remote from the splendour of the divine perfection, and that, whatever improved world we may imagine by thinking away this evil or that, it will still be true that there could have been yet better ones. The suggestion is thus that when regarded from the divine vantage point one world is virtually as good as another, since all are, alike, infinitely remote from God's illimitable perfection.

There is clearly evident in these reasonings the tendency to think of God and of His relation to the world in nonpersonal terms. Conceived as eternal Being overflowing in the creation of more and more being, God can doubtless not

[1] *M.E.*, p. 113.   [2] Ibid., p. 139. See pp. 108-9 above.
[3] Ibid., pp. 110-11.

only tolerate but find satisfaction in the existence of the cholera germ. For a fully abundant range of creatures cannot exclude any particular form of life; and even if, for the sake of man, this germ were omitted, the world would still be infinitely removed from the glory of its Maker. The world is very good; and whilst it must always be true that it could conceivably be still better, there was no reason why God should have made the world other than He did. But if, on the other hand, we think of God's goodness and love in more personal terms, we cannot so easily reconcile the existence of cholera, or more generally what might be called the anti-human aspects of nature, with infinite love and power in the Creator. For from man's point of view the presence or absence of cholera germs makes a very poignant difference. And as the loving heavenly Father of mankind, He must surely see disease as an enemy to His purpose and as incompatible with His rule. Accordingly to say that God could have made the world free from cholera, but did not, is to impugn His love for His human creatures. This is the horn of the dilemma upon which the Thomist reaction to the concept of the best possible world impales itself, led thereto by its tendency to think of God's goodness and love in impersonal and emanationist terms.

The Leibnizian position, on the other hand, in effect denies the infinite power of God, and thus arrives upon the other but equally uncomfortable horn of the dilemma. For in Leibniz's *Theodicy* the place that had been accorded to matter as the source of evil in Neo-Platonic dualism is occupied by fixed possibilities and compossibilities having an eternal existence within God's intellect but not subject to His will. Rejecting the Manichaean form of dualism, and at the same time setting up his own form of it, Leibniz says, 'But as matter is itself of God's creation, it . . . cannot be the very source of evil and of imperfection. I have already shown that this source lies in the forms or ideas of the possibles, for it must be eternal and matter is not. Now since God made all positive reality that is not eternal, he would have made the source of evil, if that did not rather lie in the possibility of things or forms, that which alone God did not make, since

he is not the author of his own understanding.'[1]   Hence,
'the source of evil lies in the possible forms, anterior to the
acts of God's will'.[2]   Thus the two ultimate and co-ordinate
realities that stand over against each other within the Godhead are, on the one hand, the creative divine will, and, on
the other hand, the eternal compossibilities, in conformity
with which the divine will must operate.   Given the limitation imposed by these compossibilities, God has made the
best world that He could — indeed, within the given restrictions, the best possible world.   But, as Schleiermacher pointed
out, 'the whole productive activity of God is assumed to be
selective and therefore secondary'.[3]   For, as John Stuart Mill
rightly perceived, 'In every page of the work [Leibniz]
tacitly assumes an abstract possibility and impossibility,
independent of the divine power :   and though his pious
feelings make him continue to designate that power by the
word Omnipotence, he so explains that term as to make it
mean, power extending to all that is within the limits of that
abstract possibility'.[4]   Mill's description of the dualism that
he finds in Leibniz is entirely just :

> [Such writers] have always saved [God's] goodness at the ex-
> pense of his power.   They have believed, perhaps, that he could,
> if he willed, remove all the thorns from their individual path, but
> not without causing greater harm to some one else, or frustrating
> some purpose of greater importance to the general well-being.
> They have believed that he could do any one thing, but not any
> combination of things :   that his government, like human govern-
> ment, was a system of adjustments and compromises ;   that the
> world is inevitably imperfect, contrary to his intention.[5]

Much can be said for such a dualism as we find it in Mill
and in other writers.   But one can hardly dispute that it
represents a radical departure from the conception of God
that has operated in all the main branches of Christendom ;
and from this point of view we must say that Leibniz's doctrine of a realm of eternal possibilities, which does not permit

---

[1] *Theodicy*, para. 380.   [2] Ibid., para. 381.
[3] *The Christian Faith*, p. 241.
[4] *Three Essays on Religion* (London :   Longmans, Green, Reader, & Dyer,
4th ed., 1875), p. 40 n.
[5] Ibid., p. 40.

there to be a world free from evil, is sub-Christian in the restriction that it sets upon the divine sovereignty.[1] Arthur Lovejoy, who remarks that eighteenth-century optimism had much in common with Manichaean dualism, makes the excellent comment:

The very ills which Bayle had argued must be attributed to the interference of a species of extraneous Anti-God, for whose existence and hostility to the good no rational explanation could be given, were by the optimist attributed to a necessity inhering in the nature of things; and it is questionable whether this was not the less cheerful view of the two. For it was possible to hope that in the fullness of time the Devil might be put under foot, and believers in revealed religion were assured that he would be; but logical necessities are eternal, and the evils which arise from them must therefore be perpetual.[2]

It is indeed true that a metaphysical system in which evil is attributed to an eternal necessity, and is thereby accorded an ultimate status, constitutes a highly ambivalent form of optimism. It could almost equally appropriately be regarded as an extreme form of pessimism. For as Voltaire's Candide asks in dismay, 'Si c'est ici le meilleur des mondes possible, que sont donc les autres?'[3] If this is the best *possible* world, there can be no hope either for its improvement or for an eventual translation to a better, and despair or resignation could well be the appropriate reaction.[4]

Thus with the development of the thesis that this is the best of all possible worlds the Augustinian tradition of theodicy faces a painful dilemma. If it insists (with Leibniz) that this is indeed the best possible world, it thereby implies that God was powerless to make a better one, and so denies His omnipotence; but if on the other hand it asserts (with Thomism) that this is not the best possible world, it calls in question God's goodness and love, which were not sufficient to induce Him to will a better one.

---

[1] Cf. A. D. Sertillanges, *Le Problème du Mal*, i. 233–4.
[2] *The Great Chain of Being*, p. 209.
[3] *Candide*, chap. 6.
[4] Cf. Basil Willey, *The Eighteenth Century Background* (London: Chatto & Windus, 1940), p. 55. Willey's chap. 3, on 'Cosmic Toryism', is a study of eighteenth-century 'optimism'.

5. 'BEST POSSIBLE'— FOR WHAT PURPOSE?

Thus Thomism and eighteenth-century optimism answer in opposite but equally unacceptable ways the question whether this is the best possible world. Perhaps their common failure results from certain mistaken presuppositions that they hold in common. Both Aquinas and Leibniz think in terms of the great chain of being consisting of all the modes of existence from the highest to the lowest, with man about mid-way on the scale. And when they speak of the goodness of the universe they are referring to the adequacy of this complex whole considered as an expression of the overflowing creativity of God. The goodness of the universe is not conceived by them in instrumental terms, as its suitability to fulfil some specific divine purpose, but as an intrinsic goodness which is ultimately equivalent simply to existence itself. Being as such is good; and hence the more there is of it, in diversity as well as in quantity, the better. Thinking of the universe in this way, Thomism holds that in being less than God it is bound to be less than perfect, and that the precise extent to which it is imperfect is a matter for divine decision. God's power is unlimited, but He did not choose to make a better world than this; and even if He had, it would still have been infinitely below the ultimate perfection which exists in Himself alone. On the other hand, Leibniz says that God's power is in effect limited by the eternal possibilities, and that the world which He has made to reflect His goodness is the best practicable one. Thus either God's goodness or His power is impugned. But as an alternative to this way of conceiving of the world we may take up the thread which has been represented so far in these pages by Hugh of St. Victor's teaching that the world has been created for the sake of mankind. The question whether this is the best possible world will then depend upon a prior question concerning God's purpose in creating man and setting him within the kind of world in which he finds himself. The best possible world will be that which best serves the purpose that God is seeking to fulfil by means of it. But if the divine intention for this world relates to man, two very different conceptions of that intention are still possible. One has generally been presupposed by agnostics and the other by Christians — with the result that they have often been at cross-purposes. Hume, Mill, and Russell, for example, assume that the purpose that an infinitely good and powerful deity must have in creating an environment for finite persons is to produce for them a maximum of pleasure and a minimum of pain. The alternative view, which is found in many Christian works, particularly outside the Augustinian tradition, is that the world is a 'vale of soul-making',[1] designed as an environment in which finite persons may develop the more valuable qualities of moral personality. Thus Leibniz's conception of the best possible world leads us back to the prior question concerning the function of the world within God's purpose. This question will be taken up again in Chapter XIII.

[1] See below, pp. 289 f.

CHAPTER VIII

DIVIDING THE LIGHT FROM THE DARKNESS

1. THE MAIN FEATURES OF THE AUGUSTINIAN TYPE OF THEODICY

THE purpose of this final chapter of Part II is to summarize the main features of the Augustinian type of theodicy, and then to ask concerning each of these features whether it should be carried over into our attempt in Part IV to confront the problem of evil directly, independently of the claims of tradition.

Although they are closely intertwined in the thought of Augustine himself, we can nevertheless distinguish between the more theological and the more philosophical strands in his theodicy. The theological themes are : the goodness of the created world, pain and suffering as consequences of the fall, the 'O felix culpa', and the final dichotomy of heaven and hell. The philosophical themes are : evil as non-being, metaphysical evil as fundamental, the principle of plenitude, and the aesthetic conception of the perfection of the universe.

These two sets of ideas continued together — with variations in their relative prominence — through the medieval period. At the Reformation, however, they fell apart, for the Reformers accepted the theological but not the more philosophical aspects of the Augustinian theodicy. Two centuries later, in the glow of the Enlightenment, the eighteenth-century optimists partly reversed this discrimination by adopting the philosophical strands while de-emphasizing the theological tenets of Augustinianism. In our own day, however, in the thought of Karl Barth, we have a continued Reformed use of the theological themes of the Augustinian tradition, accompanied by a revised version of its philosophical themes.

In criticizing the various aspects of this complex tradition we shall accordingly sometimes be criticizing primarily the Catholic theodicy, which stems from St. Augustine and St. Thomas, sometimes the more theological aspects of this as they are held in common with Reformed theologians, and sometimes the more philosophical aspects of it as they are held in common with the eighteenth-century optimists. And generally the criticism will take the form of a proposed distinction between a valid and an invalid use of the idea in question.

I. *The Theological Themes*

2. THE GOODNESS OF THE CREATED UNIVERSE

The Augustinian theodicy is built upon the doctrines of God and creation. God is good; or in the Greek terms that Catholic thought has always freely interspersed with the language of the Bible, God is the Good. And all that He has created is likewise, in its own derivative way, good. Everything, in so far as it represents the divine creativity, is by nature good. For its existence is willed by God and expresses His approval and positive valuation. This fundamental position can be regarded only as a direct and valid inference from the biblical affirmation that God viewed the world which He had made and 'behold, it was very good'.[1]

But at this point a discrimination must be made. Augustine and Aquinas were not content to affirm this biblical teaching. They put alongside it, and mingled and confused with it, the Neo-Platonic equation of being with goodness. They taught that to exist is *ipso facto* to be good — understanding the doctrine as entailing that the more existence there is the more goodness there is, and that even a thoroughly evil being is still, *qua* existent, good.

It is perhaps not easy to maintain a clear dividing line between, on the one hand, the biblical-theological affirmation that the universe is good in that God wills that it should exist, and, on the other hand, the philosophical doctrine that

---

[1] Genesis i. 31.

being and goodness are identical. Indeed, the distinction could be minimized to the point at which the latter notion is construed as simply a translation of the former. However, as it occurs in the Augustinian-Thomist tradition the identity of being and goodness has its own separate philosophical root in Neo-Platonism and must be regarded as an independent philosophical position.

Augustine's own defence of it amounts to the claim that certain characteristics, which are necessarily present in different degrees in every existent thing — principally 'measure, form, and order', — are intrinsically good.[1] To possess these characteristics is to be a part of the continuum of entities constituting the created universe, so that to exist is, as such, to be good. This is Augustine's contention. But on what grounds is it proposed? No philosophical arguments accompany Augustine's use of the idea, which stands simply as an inheritance from the Neo-Platonic vision of reality, in which the degrees of being and goodness together correspond to descending emanations of the ultimate divine One. There appears to be no basis within Christian theology for affirming the intrinsic goodness of existence in any other than the biblical sense that God wills and values the world that He has created. The philosophical identification of being with goodness at worst suggests an emanationist view of the status of the created world, and is at best an indirect and opaque way of saying that the universe is a creation of God, willed and affirmed by its good Creator.

Aquinas proceeds on different and more Aristotelian grounds to his own equation of being with goodness. He says even more explicitly than Augustine that 'goodness and being are the same as regards things ('Bonum et ens idem sunt secundum rem'), and differ only logically'.[2] As we might say in another terminology, 'being' and 'goodness' differ in connotation but share the same denotation. Aquinas provides an argument for this position based upon Aristotle's definition that 'good is what all desire'.[3] Everything desires its own continued existence and the fulfilment or perfection

---

[1] See pp. 55 f. above. [2] *S.T.*, pt. I, Q. v, art. I.
[3] *Nicomachean Ethics*, i. 1.

of its own nature. Since each existing thing thus clings to existence and resists dissolution, existence is always desired and is accordingly always good; for 'since every nature desires its own being and its own perfection, it must be said also that the being and the perfection of any nature is good'.[1] Now this argument establishes, on the basis of the initial Aristotelian definition, that each creature's existence is a good to that creature. But it does not establish that each creature's existence is intrinsically good, or that it is good in the sight of God, or indeed in the sight of any but the individual creature itself. So far as this argument is concerned, the universe might consist of a multitude of beings who are each evil and abhorrent to God, and such that it would be better if they did not exist, each nevertheless wholeheartedly desiring its own continued existence. Thus Aquinas' reasoning does not warrant a general identification of the existent with the good.

The conclusion of the matter would seem to be that from the standpoint of Christian theism only God can be said in an absolute and unqualified sense to be good (for 'No one is good but God alone'[2]), and everything else that is good is so only in a secondary sense, as an object of His approval or love. When we affirm that the creation is good we are claiming that it is willed and valued by God. It exists because God has willed that it should exist. This is the value pertaining to it objectively, beyond the circumstance that this or that aspect of it is desired or enjoyed by this or that finite creature. But this does not entail any metaphysical doctrine of the identity of being and goodness; nor does there appear to be any adequate reason to adopt such a doctrine.

3. HUMAN SUFFERING AS A PUNISHMENT FOR SIN

The theodicy-tradition, which has descended from Augustine through Aquinas to the more tradition-governed Catholic theologians of today, and equally as we find it in the Re-

---

[1] *S.T.*, pt. 1, Q .xlviii, art. 1.
[2] Mark x. 18.

formers and in Protestant orthodoxy, teaches that all the evil that indwells or afflicts mankind is, in Augustine's phrase, 'either sin or punishment for sin'.[1]

That an individual's sufferings are a divine punishment for his sins is a very ancient and a very natural theory, which is liable to appeal both to the conscience of the sufferer and to the judgement of the onlooker. Indeed, the punitive theory is built into our language, for 'pain' is derived from the Latin *poena*, which bears the meaning of penalty or punishment. We still use 'pain' as a synonym for 'penalty' in such phrases as 'No smoking in the engine room on pain of dismissal'. The theory is also strongly affirmed in Ecclesiasticus xxxix. 29 : 'Fire and hail, famine and pestilence, all these have been created for vengeance'. It is, however, repudiated in the Book of Job, and in the New Testament by our Lord himself.[2] There does, of course, lie behind the punitive view of suffering the fact that every sufferer (beyond infancy) is also a sinner; but so far as observation can reveal the truth to us it is not in general the case that a man's sufferings are proportioned to the degree of his sinfulness. On the contrary, the relatively innocent sometimes suffer grievously whilst the wicked sometimes seem to remain comparatively untroubled.

However, the disproportion between guilt and pain does not undermine the Augustinian theodicy, for this speaks not only of the 'actual' sins of individuals but also of the 'original' sin of the race as a whole, of which the evils afflicting mankind generally, and having to be borne unequally by different individuals, are a divinely appointed consequence. Thus the sufferings of the apparently innocent are met by the doctrine that, properly speaking, no one is innocent—for no one, not even a new-born child, is innocent in respect of Adam's crime.

There is, however, a fundamental objection to this aspect of the Augustinian tradition, considered from the point of view of the theodicy-problem : namely, that no alleviation of the dark mystery of evil is obtained by pointing back to

[1] *De Genesi Ad Litteram*, Imperfectus liber, chap. i, para. 3.
[2] Luke xiii. 1–5 ; John ix. 1–5.

its origin in a fall of man, or even beyond this in a fall of Satan and his angels. Whether or not the traditional doctrine of an historical fall (i.e. the fall of man as an event that took place on this earth some definite, though not necessarily ascertainable, number of years ago) is true, considered as a contribution to the solution of the problem of evil it only explains *obscurum per obscurius*. In place of the honest mystery of man as by nature frail, fallible, vulnerable, and sinning, it presents the wanton paradox of man (or the angels) being placed as finitely perfect creatures in a finitely perfect environment and then becoming the locus of the self-creation of evil *ex nihilo*. To say that an unqualifiedly good (though finite) being gratuitously sins is to say that he was not unqualifiedly good in the first place; and to infer that he was created as a morally imperfect being is to suggest that God, who gave him this imperfect nature, should not blame him too severely when further evidences of imperfection flow from it. But this line of thought points towards an alternative type of theodicy which, instead of seeking a solution by looking to the past and finding its clue to the meaning of evil in a heinous original crime, seeks for light by looking in faith to the future, to an eventual triumphant bringing of good out of evil.

Further, the doctrine that, in the Abbé Journet's words, 'All the trials in our human lives are due to sin, though not all in the same way: some are attributable to original sin, others to our personal sins',[1] treats the ancient myth of the fall of man — which expresses a fitting awareness of our immense distance from the end set for us in God's purpose — as though it were an historical account of an actual state of perfection followed by a first moral offence and man's precipitation into his present existence amid hardship and suffering. According to the older conception of the fall of man, based upon a literal interpretation of the Genesis narrative, our mortality, our sad vulnerability to disease and natural disaster, as well as all the forms and consequences of 'man's inhumanity to man', are continuing products of a disastrous primeval apostasy, apart from which human life

[1] *M.E.*, p. 218.

would still be free from such penalties. But the time has long been with us when Christians can not only see, but must frankly say, that the Genesis story is not history but myth. For the past century evidence has been available concerning the earlier states of mankind, before the brief span of recorded history, and none of this evidence lends any support to the theory that the human race is descended from a single original pair, or that mortality and liability to disease and disaster are other than natural to the human animal in his place within the larger system of nature. On the contrary, the hypothesis of a pre-fallen paradisal stage of history in which man enjoyed a serene consciousness of God and in which, but for some first sin, he would have remained in perfect peace both with his own kind and with the other species, conflicts with so much of our modern scientific understanding of the structure and unity of nature that it is no longer possible to combine biblical literalism with a responsible attitude to scientifically acquired knowledge. Only a drastic compartmentalization of the mind could enable one to believe today in a literal historical fall of man from paradisal perfection taking place in the year x B.C. It is not necessary to go back over the biological, anthropological, geological, and paleontological work of the past century and a half and to relive the battles between advancing science and retreating theological dogmatism. Let us instead simply say without equivocation that the fall is a mythic conception which does not describe an actual event in man's history or prehistory. It is not a happening in the chronological past, whether six thousand, or sixty thousand, or six hundred thousand, or any other number of years ago. Its relation to human history is that of mythic presupposition rather than constituent event. The story of the fall does not describe genetically how our human situation came to be as it is, but analyses that situation as it has always been. Man has never lived in a pre- or un-fallen state, in however remote an epoch. He has never existed in an ideal relationship with God, and he did not begin his career in paradisal blessedness and then fall out of it into sin and guilt. From his first emergence from the lower forms of life he has been in other than perfect fellowship with his Maker. We cannot speak of a radically better state that *was*; we must speak instead in hope of a radically better state which *will be*. The implications of this for theodicy will be developed in Chapter XIV.

4. 'O Felix Culpa . . .' versus Eternal Torment

Augustine's affirmation that 'God judged it better to bring good out of evil than to suffer no evil to exist'[1] is as far-reaching in its implications for theology as it is unavoidable as an *ex post facto* recognition of the cosmic situation. Since evil exists, and since God is sovereign, it must be that God has permitted it for some good purpose. And since the basic avoidable and therefore permitted evil is sin, which is an occurrence within free rational creatures, the good for the sake of which sin is permitted also pertains to free rational creatures and consists in their redemption. The special good of the salvation of sinners is such as to outweigh both the evil of their sin and the cost involved in their redemption. As Journet says, God 'only allowed the irruption of the evil of sin into this first universe of creation because he foresaw the setting up of a universe of redemption which as a whole would be better'.[2]

Speculative and daring as this statement may seem, it represents a position that is almost inevitably reached when one contemplates the sovereignty of God in its relation to the fact of evil. It would be an intolerable thought that God had permitted the fearful evil of sin without having already intended to bring out of it an even greater good than would have been possible if evil had never existed. It must, then, be the case that sin plus redemption is of more value in the sight of God than an innocence that permits neither sin nor redemption, otherwise God would have created innocents, and would have so protected and guided and inwardly graced them that they would never fall. This is the clear presupposition of the 'O felix culpa'[3] — an insight which must, I believe, be one of the cornerstones of Christian

---

[1] *Ench.* viii. 27.   [2] *M.E.*, p. 124.   [3] See below, p. 280, n. 1.

theodicy. An attempt will be made to use it as such in the constructive sections of this book.

But within the Augustinian tradition this insight is accompanied, and annulled, by the doctrine that in the case of many souls (generally assumed to be a majority) God will not in fact bring the good of redemption out of the evil of sin, but that on the contrary sin will continue without end, accompanied by unending punishment. Whatever there may be to be said for and against the doctrine of hell from the point of view of biblical exegesis and doctrinal theology (and these matters will be discussed in Chapter XVII), the point that has to be made here is that this doctrine works directly against the theodicy suggested by the 'felix culpa' idea. If God has allowed sin only in order to bring out of it the greater good of redemption, then to the extent that redemption fails to occur the divine purpose has been frustrated. And yet God's freedom and sovereignty have been so defined that the divine purpose cannot be frustrated; for God Himself determined in the first place, by what the Thomists call His absolute, in distinction from His subordinated, power, all the factors and conditions in relation to which He has a purpose. It is He who has chosen, out of an infinite range of possibilities, to create a universe in which He foresees (or timelessly knows) that some free beings will fall. And yet His intention in relation to this situation is — according to the Augustinian tradition — to redeem by His free grace some, but not others; and those others will then have the stamp of eternity placed upon the fallen and miserable state which, within the divine foreknowledge, they have freely chosen. Now this doctrine of God's creation of beings whom He knows will fall and whom, having fallen, He does not intend to raise up again, is directly at variance with the thought that He permits sin only in order to bring out of it the greater and more glorious good of redemption. We can only claim that the second-order good of a redeemed humanity justifies the first-order evil of sin to the extent that sinners are in fact finally redeemed. So far as the rest of fallen humanity are concerned, no justification has been offered, but on the contrary the problem has been exacerbated by the appalling doctrine that God creates persons who He knows will merit damnation and who He is content should be damned.

Catholic teaching on this point has been defended as follows by Dr. Richard Downey, writing in the composite work, *The Teaching of the Catholic Church* :

> the further objection is raised that if God foresaw, even though he did not positively forewill, the damnation of the wicked, nevertheless, as an infinitely good God, he ought to have abstained from creating such souls. A little reflection, however, makes it clear that if God could be influenced in that way by a condition outside himself, the condition would be greater than he ; he would be limited, constrained from without, and therefore not infinite, and not God. It would be as though the damned soul, whilst still only a mere possibility, could defy the Omnipotent to create it.[1]

This argument rests upon the curious premise that God is only infinite and free if He does something the result of which He knows will be displeasing to Him ! But why should it be considered a limitation upon God to refrain from self-frustrating activity? Surely it is a power rather than a weakness to be able to avoid defeating one's own ends. Indeed, presumably God has, in choosing to create the present universe, already chosen not to create other possible universes which would have been unacceptable to Him. But thus to choose the better rather than the worse does not indicate a restriction upon God's power. On the contrary, there would be limitation in being obliged to create a universe that the Creator did not want. And likewise it would be a limitation for God to be obliged to produce, simply because He *can* produce, free beings who He knows will, if created, fall away from Him and constitute a permanent stain upon His universe.

A more promising response to the difficulty might be the following : It does not make sense to speak of God foreseeing the free actions of beings whom He has not yet created, and then on the basis of such foreseeing deciding whether or

---

[1] *The Teaching of the Catholic Church*, ed. by George D. Smith (London : Burns Oates & Washbourne, 1948), vol. i, chap. vii, 'Divine Providence' by Richard Downey, p. 245. Cf. Dom Bruno Webb, 'God and the Mystery of Evil', *The Downside Review*, vol. 75, no. 242 (Autumn 1957), p. 354.

not to create them.[1]   For only *existing* beings can be said to
have a future for God to foresee.   It is true that in the case
of unfree creatures, whose actions were divinely predetermined, God could consider, without creating them, what
they would do if He created them.   However, the actions of
free creatures can be known to God 'in advance' only by
His seeing their contingent future actions in His own eternal
present.   But if He has not created them, or even decided
whether to create them, there are no free beings to have any
future actions.   Freedom exists only in its actual exercise,
and a being who may never be created cannot exercise freedom and cannot have a divinely foreseeable future.   What
we must therefore say — according to this argument — is
that God creates free beings, and having done so He can at
once foresee their future obedience or disobedience.   However, having once created them God does not go back upon
His own creative act; He respects His creatures' freedom
and responsibility even when He now sees that they are
going to misuse it.   This might meet the particular difficulty
under consideration.   It would, however, only open the door
to another difficulty.   For we must ask whether it makes
sense to speak of God creating free beings as such, without
any determinate nature which would lead them to act either
rightly or wrongly.   This is a problem that is to be taken up
more fully in Chapter XIV, when we consider some contemporary philosophical contributions to our problem.

II. *The Philosophical Themes*

5. Evil as Non-being

Turning now to the more philosophical aspects of the
Augustinian theodicy, and to the idea of evil as non-being,
we have, I believe, to distinguish between a valid theological insight arising out of the Christian revelation itself,
and a questionable philosophical conceptuality by means of
which this insight has often been presented.   The distinction

[1] Cf. Charles Werner, *Le Problème du Mal*, p. 32.

is between, on the one hand, the understanding of evil as the going wrong of something good, which follows from the Christian understanding of God and creation, and, on the other hand, the various philosophical theories of evil as nothingness or non-being, whether in the negative sense of a *privatio boni* or in the more positive and dynamic sense of *das Nichtige*. I shall argue that these categories are not required by the theological insight to which they have attached themselves and that they are in addition open to serious philosophical objections.

Let us note first the theological insight in question, which has already been discussed in connection with St. Augustine. Against the background of their belief in the goodness of the created world, Augustine and Aquinas taught that the fact that there is evil in the universe does not entail that there are entities or 'substances' whose essential nature is evil. On the contrary, every existing thing is a good creation of a good God, and evil accordingly consists in the partial loss by anything of the particular mode of goodness that constitutes its own special form of being. Evil is thus loss and lack, a deprivation of good, and instead of having any positive goal or function of its own it tends, by its inherently negative character, towards nullity and non-existence.

As a characterization of evil, within the framework of Christian theology, this privative definition must be accepted as wholly sound. It represents the only possible account of the ontological status of evil in a universe that is the creation of an omnipotent and good God. From this standpoint evil cannot be an ultimate constituent of reality, for the sole ultimate reality is the infinitely good Creator. Evil can only consist in a malfunctioning or disorder that has somehow come about within an essentially good creation.

The privative view of the status of evil thus follows inevitably from various prior positions of Christian faith and is valid within this context.[1] It is an inference from the Christian doctrines of God and creation. Apart from this theological framework, however, an affirmation of the privative character of evil would be as arbitrary as a contrary

[1] See pp. 59 f. above.

affirmation of the privative character of good. Either would represent an optional way of thinking about the relation between good and evil, one seeing good as primary and evil as its shadow, and the other seeing evil as positive and good as filling only the interstices of an evil universe. Neither view can claim to be read off unambiguously from the facts of human experience. As experienced, good and evil are equally real, equally positive, equally insistent upon recognition as forces to be reckoned with. Only a wider metaphysical framework of belief impels us to believe that the true status of evil is that of negation and lack within a universe whose positive nature is good.

We are therefore not authorized to draw any empirical conclusions from the doctrine of the negative character of evil, taken by itself. It does not entail that evil is other than a real fact and a grievously oppressive problem; and its major proponents have never supposed that it did. It is therefore not permissible to dismiss the privative analysis of evil as 'a philosophical theory which explains that experience away by denying that evil is real. [According to this theory] Evil exists only in semblance; it is the negation or privation of good';[1] or to say that 'Some theists seek a solution by denying the reality of evil or by describing it as a "privation" or absence of good. They hope thereby to explain it away as not needing a solution.'[2] The privative doctrine is not offered by Augustine, Aquinas, Leibniz, or others in the same tradition, as a solution to the problem of evil.[3] All that it does is to rule out a dualist solution and thereby to advance the definition of the problem a stage by posing the question, How does privation of good come about in a universe that is created and ruled by a good God?

[1] J. S. Whale, *The Christian Answer to the Problem of Evil* (London: S.C.M. Press Ltd., 4th ed., 1957), p. 58. Leonard Hodgson similarly incorrectly describes the traditional Thomist analysis of evil as privation of being and goodness as 'the explanation away of evil as unreal'. *For Faith and Freedom* (Oxford: Basil Blackwell, 1956–7), ii. 49–50. Cf. i. 199–201.
[2] H. J. McCloskey, 'God and Evil', *The Philosophical Quarterly* (April, 1960), p. 100. Reprinted in Nelson Pike ed., *God and Evil*.
[3] However, it must be admitted that sometimes writers expounding sympathetically the privative analysis of evil have ill-advisedly applied the term 'solution' to it: e.g. François Petit, O.P., *The Problem of Evil*, trans. Christopher Williams (New York: Hawthorne Books, 1959), p. 68.

Rightly understood, then, this element in the Augustinian response to our problem is to be accepted. It is an affirmation, in the presence of the mystery of evil, of the sole ultimacy and perfect goodness of God, and of His creation of the world *ex nihilo* — an affirmation which, however, so far from resolving the problem, only intensifies it by setting a 'No Thoroughfare' sign over the first and simplest of the solutions that suggest themselves, namely dualism.

However, this legitimate theological insight became entangled in the minds of Augustine and his successors with a philosophical conception of non-being. This conception has a long and eventful history, stretching from the first bold speculative movements of pre-Socratic Greek thought, through various exotic realms of medieval mysticism and theology, to the critique of the idea offered by twentieth-century logic and the warm hospitality extended to it in contemporary existentialism. Non-being has been postulated as a necessity of thought, and rejected as a verbal deceit; it has been hailed as the divine depth and abyss, and reviled as the deadly enemy and source of all evil. The term itself is thus something of an umbrella, covering both a fact of experience and a logical or philosophical concept by means of which this has been expressed. This conjunction of a questionable philosophical category with an 'existential experience' has been a prolific source of confusion.

In terms of the Greek distinction, which has been made familiar again in our own time by Nicholas Berdyaev[1] and Paul Tillich,[2] we are concerned with 'meontic' non-being (τὸ μὴ ὄν), that is to say non-being conceived as some kind of negatively positive or positively negative reality or influence, in distinction from mere blank nothing (τὸ οὐκ ὄν). In the literature in which it appears meontic non-being is presented either as a necessity of thought, or as something given in distinctively human experience, or indeed as both. It is important to distinguish these two claims, even though

---

[1] See, for example, *The Destiny of Man*, trans. Natalie Duddington (London : Geoffrey Bles, 1937), pp. 33 f.

[2] See, for example, *Systematic Theology*, vol. i (London : Nisbet & Co. Ltd., 1953), p. 209, and vol. ii (London : Nisbet & Co. Ltd., 1957), pp. 22 f.

they were often both made by the same writer; for it is arguable — as I shall be arguing here — that non-being as an existential experience is real enough, but that it has become wrongly identified in terms of an alleged necessity of thought which is in fact merely verbal.

We may begin with the fact of experience. It is the existentialist writers who, under such names as *le néant* (Jean Paul Sartre),[1] *das Nichts* (Martin Heidegger),[2] Non-being (Paul Tillich),[3] *das Nichtige* (Karl Barth),[4] and Nothingness,[5] have faced the threatening, oppressive, or disturbing aspects of our human finitude. Because we are finite, mortal, and insecure we not only from time to time fear this and that specific danger, but we also (according to existentialists from Søren Kierkegaard[6] onwards) suffer chronically from *Angst*, a generalized anxiety or dread, which cannot be appeased by guarding against particular concrete perils. This pervasive *Angst* arises inevitably from man's precarious and estranged situation. He is 'thrown into the world'; he finds himself in exile from birth onwards; living as though under the incomprehensible rule of an unapproachable despot; inhabiting a Wasteland. The universe is to him *unheimlich*, stony, and alien. Behind this sense of hollowness, which existentialists believe to be the typical mode of experiencing life in our twentieth century, we encounter nothingness as a terrifying gulf of meaninglessness.

For the atheistic existentialists the non-being that thus engenders *Angst* within us constitutes our ultimate, inescapable situation, and we can only try to learn to live in it with courage. For Christian existentialists, on the other hand, the non-being that pervades our lives represents our

---

[1] See *L'Être et le néant* (1943), trans. Hazel E. Barnes (London: Methuen & Co. Ltd., 1957), pt. i, chap. 1.
[2] See *Was ist Metaphysik?* (1929), trans. R. F. C. Hull and Alan Crick in *Existence and Being* (London: Vision Press Ltd., 1949).
[3] See *Systematic Theology I* (Chicago University Press, 1951), pt. ii, chap. 8, and vol. ii (1957), pt. iii, chap. 12.
[4] See *Church Dogmatics*, vol. iii/3, para. 50.
[5] See Helmut Kuhn, *Encounter with Nothingness* (London: Methuen & Co. Ltd., 1951).
[6] See especially *The Concept of Dread*, trans. Walter Lowrie (Princeton University Press, 1944).

estrangement from God as the ground of our being, and points to our need for a salvation from beyond ourselves.

Existentialist literature 'speaks to the condition' of some more than of others; and it can hardly be accidental that twentieth-century Existentialism has come out of situations of intense social strain and catastrophe — Germany after the First World War, and France during and after the Second World War. It has been born out of agony, both individual and collective, and it apparently makes its appeal within the highly industrialized sections of humanity, whose life it depicts as the spiritual nightmare that it can be for minds acutely sensitive to the decay of tradition, the collapse of established cultural forms, and the threat of a nuclear holocaust. It expresses the neuroses of an age that finds itself being carried into the unknown on the wheel of immense and bewildering changes.

As an expression of the strains and contradictions of contemporary megalopolitan life much of the existentialist literature must be held to be both valid and highly valuable. But with our special problem in mind we must ask, What has led the existentialists to use the notion of non-being to express these particular aspects of our human experience? Partly, perhaps, the fact that the term was available to them from a long mystical tradition (where it had however a very different meaning[1]), and partly a certain poetic appropriateness in non-being as a characterization of the negative realities of death and meaninglessness. Death is non-life, and meaninglessness is non-meaning; and so generically they can be subsumed under 'non-being'. This is an apt figure of speech, evoking imaginative reverberations. But it must not be taken as other than poetic diction. Death, in the sense of the state of having died, can be thought of as nothingness; but the fear of death, which is the existential reality, is something positive, namely the fear of death! And meaninglessness is likewise perhaps in a sense a form of nothingness; but the sense of meaninglessness, the feeling that one's life lacks any

---

[1] On the differences between the Nothingness of existentialist thought, and the positive 'rich nought' of the divine in Christian mysticism, see Helmut Kuhn, *Encounter with Nothingness*, pp. 90 f.

point or significance, which is once again the existential
reality, is as real and positive a state of mind as any other.
Thus nothingness, or non-being, used to express the awareness of death or the sense of futility and spiritual emptiness,
is an imaginative image, like 'height', 'depth', and 'light'
used as religious terms. Such language is valid as metaphor,
but becomes deprived of its symbolic power and reduced to
a mere factual mis-statement if it is understood literally.

This evocative metaphor has been identified by some
writers with a supposed necessity of thought. For it has
seemed to a number of thinkers that contingent being,
created *ex nihilo*, must be a precarious mixture of being and
non-being. Thus Tillich says that 'The root meaning of "to
exist", in Latin, *existere*, is to "stand out". Immediately one
asks: "To stand out of what?" . . . The general answer to
the question of what we stand out of is that we stand out of
non-being. . . . If we say that everything that exists stands
out of absolute [*oukontic*] non-being, we say that it is both
being and non-being . . . it is a finite, a mixture of being
and non-being.'[1] The element of non-being in the creature
produces an instability and defectibility that accounts for the
malfunctioning that is the essential nature of evil. We have
met this idea in Augustine himself, in Aquinas and contemporary Thomism, and in eighteenth-century optimism, from
which the following very matter-of-fact account of it is
drawn:

every created thing was a negation or non-entity before ever it
had a positive being, and it had only so much of its primitive negation taken away from it, as it had positive being conferred upon
it. . . . For that which was once nothing, would still have been
nothing, had it not been for the cause that gave being to it, and
therefore that it is so far nothing still, i.e. limited and defective, is
only to be attributed to its own primitive nothingness.[2]

It has seemed self-evident or logically necessary to writers
in this tradition that there is such a thing as non-being. Just
as there cannot be an inside without an outside, so there

[1] *Systematic Theology*, vol. ii, p. 22. Cf. vol. i, p. 209.
[2] King, *Essay*, vol. i, p. 116. The longer passage from which these two
sentences are taken appears above, on p. 155.

cannot be a finite being without the infinite non-being out of which it stands or has been taken or from which it is divided. And this non-being, which surrounds being, threatening to overwhelm it as darkness threatens to black out light, is the unavoidable source of evil, playing a role similar to that of ἀνάγκη (necessity) in much Greek thought, or of matter in Gnosticism and Neo-Platonism.[1]

Karl Barth's notion of *das Nichtige* as a categorization of evil is also sometimes apparently presented as a necessity of thought, arising out of the concept of divine choice or election. For God to choose to create a good universe is for Him to pass by and reject evil and, by the very act of rejecting it, to give it the status of an enemy that is hostile to the good creation and must be combated and eventually destroyed. The basic logical principle involved is that choice presupposes a duality, so that the divine choice of a good creation presupposes the 'existence' (the peculiarly negative but nevertheless dangerous 'existence') of something else which He does not choose, but on the contrary rejects, namely evil.

What are we to say of these alleged necessities? We must, I think, say that they are either (in the case of Augustine) untenable in logic, or (in the case of Barth) of very dangerous tendency in their theological presuppositions and implications.

From the point of view of twentieth-century logic, the notion of meontic non-being is an example of the inveterate tendency of the human mind to hypostatize or reify language. The term 'being' generates the cognate term 'non-being'; but it does not follow that there in any sense is or exists anything of which this is the name. In modern logical theory the negative existential proposition has been very fully analysed. Russell's theory of descriptions[2] shows that to put 'exist' after a general name (e.g. 'Cows exist') is to assert

[1] Julius Müller (*The Christian Doctrine of Sin*, i. 290 n.) reminds us that this thought is often to be met with in the Theosophists and theosophic sects.

[2] Bertrand Russell, 'The Philosophy of Logical Atomism', sect. v, in *Logic and Language*, ed. by R. C. March (London: George Allen & Unwin, 1956); *Introduction to Mathematical Philosophy* (London: George Allen & Unwin, 1919), chap. 16; *History of Western Philosophy* (London: George Allen & Unwin, 1946), pp. 859–60.

that the description for which the name stands has a referent
or referents ; and conversely that to say of some kind of thing
that it does not exist (e.g. 'Unicorns do not exist') is not to
locate it in any metaphysical realm of non-being, but is
simply to deny that some particular description has a referent.
From this point of view there is no basis for the hypostatiza-
tion of non-being. The situation is simply that we have the
generally useful habit of presuming an entity of some kind
corresponding to a noun ; but sometimes the language generates words that have no denotation — and non-being is a
case in point.

Barth's argument — that in creating, God chooses good
and rejects evil, which henceforth has the character of being
denied and opposed by God — would be in order if this were
a human choice. But when applied to the Godhead the
argument becomes highly questionable. It requires the
premise that God, in creating, must choose between realities
which already stand in some way before Him (or within
Him), seeking His election. But such a premise ignores, and
by implication denies, the distinctively Christian doctrine of
creation *ex nihilo*. By postulating a previously existing situation within which God acts, and of whose character He must
in acting take account, Barth is half-way towards a Manichaean dualism. It is from one point of view the saving
grace of his thought on this subject, and from another point
of view its final breakdown, that he does not draw any of the
conclusions to which his arguments lead.[1]

The upshot of the matter is thus that non-being can be
useful as a piece of poetic diction, suggesting certain aspects
of our human experience of evil, but that, as an ontological
or metaphysical concept it represents a mistaken hypostati-
zation of language, and can be of no positive help in relation
to the theodicy problem.

6. METAPHYSICAL EVIL AS FUNDAMENTAL

In discussing this aspect of the Augustinian theodicy-
tradition we need not be primarily concerned about the

---

[1] See above, pp. 143 f.

propriety or otherwise of the phrase 'metaphysical evil'. It may be (as e.g. Père Sertillanges argues[1]) that the creaturely finitude, limitation, and consequent imperfection which the term denotes should not be described as 'evil'. For our present purpose, however, it does not greatly matter whether the basic structural characteristic of creaturely existence to which evils are traced is or is not regarded as being itself an evil. The important point is that, according to the Augustinian theodicy, certain imperfections are inevitable in a created and dependent universe, and these inevitable imperfections are the source of many or all the (other) evils that occur in it.

The argument is as follows. In creating, God must either produce another reality on the same level of perfection as Himself — in fact another God — or else a creaturely realm which is inferior to its Maker. The first alternative is ruled out as absurd; whence it follows that the creation must be less perfect than its Creator. That is to say, it must be imperfect.

The argument can now take any number of different forms — or it may combine several or all of them.

1. The argument can at this point become extremely vague and tenuous, moving without precise steps of reasoning straight from the necessary fact of imperfection in general to the presumed necessity for all the specific imperfections and evils that we find in the world.[2] But this — by itself at any rate — is a very unsatisfactory procedure, since the general necessity that a creaturely realm should be less than God is by no means equivalent to a positive necessity that it should contain, for example, hatred, cruelty, and insanity, or cholera, earthquakes, and starvation.

2. An attempt may be made to deduce the specific evils of the natural order from the existence and properties of matter. Given parcels of matter interacting in space, there must needs be clashes, injuries, decomposition, and destruction.[3] But this also is an argument of only limited value,

[1] *Le Problème du Mal*, vol. ii, pp. 7–8.
[2] E.g. Edmund Law in King's *Essay*, p. 103, n. 18.
[3] E.g. King's *Essay*, pp. 133–4.

since it does not show why there need be matter at all, or why matter (taking this now as a name for the basic 'stuff' of existence) must have the properties that we find it to have.

3. The oldest strand of the Augustinian theodicy-tradition invoked the principle of plenitude, which explains why there must be not only imperfections in general but all the specific forms of imperfection that currently obtain. The suggestion is that in making a dependent world God has given being to a maximum range of forms of existence, constituting a hierarchy that descends from the highest archangel to the lowest amoeba or virus. Such a world must include all those forms of life and organizations of matter whose interactions cause the mingled good and evil which we observe and experience. In discussing this theme in connection with St. Augustine we have already noted (*a*) that historically it arose from the Neo-Platonist picture of reality, and has no roots in specifically Christian ground, and (*b*) that the principle is not in fact exemplified in the actual universe (which by no means contains all possible species), and that we cannot explain this incompleteness without either abandoning the principle of plenitude, or supposing a limit to God's power.[1]

It should be noted that from the point of view of theodicy the principle of plenitude is closely related to the aesthetic theme of the perfection of the universe when seen as a totality. For on the basis of the principle of plenitude it is claimed that whilst some of the lower forms of existence, considered in isolation, appear to be evil, they are nevertheless necessary links in the great chain of being, which is good as a totality and which expresses as adequately as possible the creative fecundity of God. Thus the principle of plenitude, as an aid to theodicy, presupposes that evils can indeed be justified in this kind of holistic way; and the discussion of this point below is thus relevant to the present topic.[2] In the meantime we must conclude that in the absence of any positive reason for accepting the principle of plenitude the counter-considerations are sufficient to warrant us in leaving it aside.

[1] See above, pp. 85 f.   [2] See below, pp. 197 f.

4. The Augustinian-Thomist theodicy likewise traces the fall of angels and men to a basic and inevitable metaphysical weakness of their nature. They have been made out of nothing and are for that reason inherently mutable, lacking the eternal poise of the uncreated self-existent Being. There is thus an ineradicable weak point in every created thing, and it is virtually certain that this weakness will sooner or later manifest itself in some form of evil. As Père Sertillanges puts it,

> The glass is fragile : it is indeed possible that in a whole city, on a certain day, no glass breaks ; but that none ever breaks, however long the city lasts and however often they are used, is not possible. Considering the degree of fragility, it will most certainly break. It is the same with fragile substances, fragile animals, fragile humans, fragile consciences. *There* lies evil.[1]

Thus even moral evil arises with virtual inevitability from the inherent imperfection of the universe as a created and dependent order. For a diversified universe, containing as wide a variety of beings as possible, must include creatures that can fail ; and Aquinas taught that 'what can fail sometimes does'.[2]

Defending this Thomist principle that failure is inevitable among contingent beings, P. M. Farrell, O.P., says :

> It is of the definition of contingent being that it should be such that it can not-be — i.e. it must be capable of defection from being. If it cannot not-be, it is not contingent but necessary. But to defect from being is to involve a privative absence of good which is the 'definition' of evil. Thus evil is involved in the very concept and definition of contingent being. Evil, i.e. is a necessary consequence of contingency.[3]

There are, however, two reasons for being dissatisfied with this argument. First (as Marvin Zimmermann has pointed out in a note on Farrell's article[4]), Farrell slides from contingency as entailing the possibility of ceasing to exist, to contingency as entailing the possibility of (existing but)

[1] *Le Problème du Mal*, vol. ii, p. 36.
[2] *S.T.*, pt. 1, Q. xlviii, art. 2.
[3] P. M. Farrell, 'Evil and Omnipotence', *Mind*, lxvii (July 1958), p. 401.
[4] Marvin Zimmermann, 'A Note on the Problem of Evil', *Mind*, lxx (April 1961), p. 253.

ceasing to be good. He establishes a point concerning contingency in the first sense, and then assumes, illegitimately, that he has established it concerning contingency in the second sense. And second, Farrell slides from the necessary *possibility* of defection from good to the necessary *fact* of defection from good, and concludes that evil (and not only the possibility of evil) is a necessary consequence of contingency. So he is able to proceed a few paragraphs later, 'Since contingent being necessarily involves defection from being and therefore evil . . .'[1]

As against Farrell, then, and likewise against the Thomist tradition which stands behind him, we must insist that the fact that created beings exist contingently does not entail that they *must* fail. It would be possible for God to create contingent beings which, though in principle capable of failing, are in fact so constituted and sustained that they never fail. In other words, we cannot as monotheists blame the existence of evil upon a metaphysical necessity. On the contrary, to bring sin (or any other form of evil) under the two principles that a created being must in the nature of the case be fallible and that what can fail sometimes does, is in effect — but contrary to the intention of the Augustinian-Thomist tradition — to lay upon the Creator the ultimate responsibility for the existence of evil. For if He chose to make creatures who are bound sooner or later to fail (even though they do so without external compulsion), He cannot reasonably complain when they do fail. He must have foreseen that they would fail if He made them, and He must nevertheless have decided to make them. This consideration points to a fatal contradiction within the Augustinian-Thomist theodicy.

7. The Aesthetic Perfection of the Universe

There is one further important element that has always been present, with greater or less prominence, in theodicies of the Augustinian type. This is the aesthetic theme, which was absorbed into Christian thought from Neo-Platonism, with

[1] Farrell, 'Evil and Omnipotence', op. cit., p. 401.

Augustine as the chief point of contact between the two worlds of thought.

The traditional analogy was based upon the visual arts (excluding of course the cinematograph), in which the picture or the piece of sculpture stands complete before us, so that change and development are not constitutive of it and the time dimension remains external. In a painting, for example, contrasts arising from the presence of dark as well as light colours, and even of elements that are in themselves ugly and repellent, may contribute to the beauty of the whole; and this fact may suggest by analogy a positive function for the dark patches of evil within a universe which in its totality is morally harmonious, well-ordered, and beautiful.

The danger, however, of this aesthetic analogy is that it is heavily weighted towards monism. Evil tends to disappear, its terrible reality concealed within the larger pattern. For example, as Sertillanges expatiates on the organic and aesthetic perfection of the creation, evil seems to diminish in importance and we are invited to see the universe from a standpoint from which it is lost to view in the sublimity of the whole.

[The Creator] intends that everything harmonises, and if there is evil, as is inevitable in so large a plan, he sees it in its relation to the whole, just as does the artist, who is also a friend of sacrifice — of forms or of value — and in order that the totality shine forth. The more man raises himself to an intuition of all and its infinite secondary complexity, the more the notion of evil disappears.[1]

This aesthetic theme is, however, capable of being reconstructed in terms of a more complex analogy based upon art forms in which the element of duration and change plays an essential part, as it does in music. Here we have evil in the forms of dissonance and conflict and noises that are in themselves grating and disagreeable, contributing to a beauty that resides in the auditory sequence as a whole. Some music includes unfulfilled beginnings and even an element of clash and disharmony at one stage of the musical development in order to make possible a later triumphant resolution in which the dissonant notes are worked into a complex

---

[1] Sertillanges, *Le Problème du Mal*, vol. ii, p. 50.

harmony that would not be possible without them. On this analogy the aesthetic perfection of the universe is no longer that of its state at any one moment, but — adopting the Augustinian picture — that of the entire panorama of existence from the initial overflowing of the divine goodness in the creation of a secondary and dependent realm with its inherent limitations and imperfections, through the drama of the virtually inevitable failure of free creatures and their later triumphant redemption, and completing itself in the gathering up of the blessed to enjoy for ever the Beatific Vision, whilst the damned remain in rigid self-exclusion from God's presence and by their plight preserve eternally the dreadful distance between good and evil.

However, even this improved version of the aesthetic analogy, making use of the added dimension of time, and pointing to an eschatological resolution of the interplay between good and evil in the final fulfilment of God's purposes, is still open to a fundamental objection which, since it operates against other aspects of the Augustinian tradition of theodicy as well, will be presented in a separate section.

8. A Basic Criticism

Perhaps the most fundamental criticism to be made of the Augustinian type of theodicy concerns a pervasive presupposition within it. This is the impersonal or subpersonal way in which God's relationship to His creation is prevailingly conceived. Such a way of thinking shows itself at a number of points :

1. God's goodness and love are understood — as they are typically understood throughout medieval theology — primarily as His creative fecundity, His bestowing of the boon of existence as widely as possible.[1]

2. Accordingly the 'motive' for creation is the self-diffusing nature of the ultimate divine Being. As Père Sertillanges puts it in a contemporary Catholic restatement, 'The divine Good seeks to overflow [cherche à se répandre]

---

[1] Cf. pp. 83–84 above.

as if God were too full of God',[1] and consequently 'the world has been made in order to express and communicate the sovereign good outside itself'.[2] The Neo-Platonist picture of the emanating divine One has clearly inspired these first two aspects of the Augustinian tradition.

3. In accordance with the principle of plenitude and the related idea of the universal scale of nature, man has been created because without him there would be a gap in the great chain of being:

> Where all must full or not coherent be,
> And all that rises, rise in due degree;
> Then, in the scale of reas'ning life, 'tis plain,
> There must be, somewhere, such a rank as Man.[3]

Or, as Aquinas said more succinctly, 'God wishes man's existence for the sake of the perfection of the universe.'[4] The emphasis is not that man is valued and loved for his own sake as finite personal life capable of personal relationship with the infinite divine Person, but upon the thought that man is created to complete the range of a dependent realm which exists to give external expression to God's glory.

4. Again, the tracing of sin back to the metaphysical circumstances of man's contingency, on the Thomist principle that 'what can fail sometimes does',[5] invites the kind of impersonal analogy that was offered by Sertillanges. Glass, being by nature fragile, breaks sooner or later: 'it is the same with fragile substances, fragile animals, fragile humans, fragile consciences. *Le mal est là.*'[6] From such a standpoint human personality is readily thought of as a kind of metaphysical substance, and sin as a personal act is lost sight of.

5. The notion of evil as non-being is essentially an impersonal conception. The Protestant theologian Jean Cadier, in an important article on 'Calvin et saint Augustin', has said that,

By this definition of evil as non-being St. Augustine threw into the process of theological reflection a principle which was to lead to a particular conception of grace, salvation, the Christian life,

---

[1] *Le Problème du Mal*, vol. ii, p. 36.   [2] Ibid., p. 40.
[3] Alexander Pope, *Essay on Man*, epistle i.   [4] *S.c.G.* i. 86.
[5] Cf. pp. 102–3 and 196–7 above.   [6] *Le Problème du Mal*, vol. ii, p. 38.

and the Church. In effect, if sin is a privation, the sinner is *un déficient*. Consequently the grace which saves him will fill up this deficiency, and will be an irresistible grace [un don de force]. The instrument of this infusion of supernatural life will be the sacrament. The Church will have the treasury of these sanctifying graces at its disposal and will distribute it by means of its priests. Medieval theology would develop all the consequences of this principle . . .[1]

It may be that Cadier's thesis is overstated; but nevertheless there is a significant coherence between the conception of evil as *privatio* and the Augustinian conception of grace as a metaphysical force that repairs the defect in being, rather than as (in John Oman's famous phrase[2]) a gracious personal relationship.

6. The whole aesthetic or quasi-aesthetic understanding of the perfection of the universe is sub-personal in character. The universe, including the finite personal life within it, is seen as a complex picture or symphony or organism whose value resides in its totality, and whose perfection is compatible with much suffering and sin in some of the constituent units.

7. And finally, the principle of moral balance, which was invoked no less by Calvin and Protestant orthodoxy than by Augustine and the Catholic tradition, is impersonal at the point at which an impersonal principle is most out of place. The assumption that a universe in which there are sin and appropriate punishment is as acceptable in God's sight, because it is as morally well audited, as one in which there is neither sin nor punishment, is juridical precisely where an ethical and personal principle is called for. Sin is regarded here as a quantity rather than as a breach of personal relationship.

At all these points the central stream of the Augustinian theodicy-tradition operates within a framework that is seriously inadequate for the consideration of God's relationship to His creation as this has been revealed to us in the

[1] *Augustinus Magister*, Communications du Congrès International Augustinien (Paris, 1954), ii, p. 1055.
[2] See John Oman, *Grace and Personality* (Cambridge: The University Press, 4th ed., 1931).

person and work of Jesus Christ. Twentieth-century theologians of all communions are more explicitly conscious than were their predecessors in most other periods that our positive knowledge of God and of His manner of dealing with His creation is derived from the Incarnation. And the category that inevitably dominates a theology based upon God's self-disclosure in Christ is the category of the personal. Because God has revealed Himself to us in and through a human life we must think of Him and of His attitude to ourselves in personal terms. Further, if God is personal we must see man as standing in a quite different relationship to Him from that in which the material universe stands to its Creator. Instead of thinking of God's goodness and love as creative power inexhaustibly producing new modes of being, we must think of them as personal qualities which have their most intense focus in God's relationship to the finite persons whom He has created. Instead of seeing human life as a link in the great chain of being, whose value in God's sight consists in its completeness as comprehending every grade of existence, we must see finite personality as made in God's image and as of unique significance to Him. Instead of seeing the creation of man as determined by the exigencies of the universal chain of nature, we must see it as determined by God's free desire to create beings for fellowship with Himself. Instead of construing evil as metaphysical non-being, we must see it primarily as a failure in personal relationship. Instead of upholding the perfection of the universe as an aesthetic whole, we must think of it as perfect in the rather different sense that it is suited to the fulfilment of God's purpose for it. And instead of invoking the juridical principle of moral balance, we must see God's free forgiveness at work, ultimately transforming sinners into citizens of His Kingdom.

For if God is the Personal Infinite, man alone among God's creatures is, so far as we know, capable of personal relationship with Him. This fact singles man out and gives him a special place in the created order. In view of this it seems more probable that God initially willed to create beings who could live in relationship with Himself and created a material universe as the appropriate environment for them, than that

He initially willed to create a material world as such, crammed with the utmost variety of forms of life, and that this world happens to include personal beings within it. Instead, then, of thinking of the origin and fate of human personality as a function of an aesthetically valued whole, we should see the great frame of nature, with all its sources of evil, as the deliberately mysterious environment of finite personal life. This approach will be developed further in Part IV.

Two qualifications must, however, immediately be added. First, in speaking of mankind as having been made as personal beings in the image of God and as occupying a unique place in creation, we do not exclude the possibility, or even the probability, that there are other forms of personal life than man — higher spiritual beings in the heavenly places, and intelligent creatures on other planets of our solar system or on planets of other stars. These possibilities remain fully open. Nevertheless we cannot make any positive use of them in our theology. For if — as seems from the standpoint of Christian faith to be altogether likely — there are higher beings such as the Bible calls angels, we do not know enough about them to draw them within the scope of rational discussion; and if there is intelligent life on other planets we do not yet have any positive evidence concerning it. But neither the existence nor the non-existence of such beings would affect what is proposed here concerning the special place of finite personal life in the divine economy. The status of personal life can be considered independently of any definitive knowledge as to what this class in fact contains. And as a second qualification, it may well be that the material universe and the ranges of sub-human life within it have very important further significances to their Maker than simply as an environment for personal life. We can set no limit whatever to the possibly multi-dimensional complexity of the divine purpose. It may be that God fulfils many different intentions at once and that innumerable strands of the divine will intersect in the universe in which we live.[1] But while we cannot, and have no wish to, exclude such

[1] See further, pp. 295 f. and 351 f. below.

possibilities, neither can we properly build any specific theological conclusions upon them.

Whatever realms of life and dimensions of meaning there may be beyond our present awareness and concern, our positive knowledge of God's nature and purpose still derives from His incarnation in Jesus Christ. Here we see the divine love for persons expressed in activities of healing, teaching, challenging, forgiving — activities that are wholly personal in character. To say that 'God was in Christ reconciling the world to Himself'[1] is to say that Jesus' attitudes to the people with whom He had to do in first-century Palestine were God's attitudes to those same individuals, expressed in and through the attitudes of a particular historical personality. The actions constituting Jesus' impact upon the world were the actions of an agape which was continuous with, and directly revelatory of, the eternal agape of God. And a theology based upon the Incarnation will point above all to the great redemptive reality that is known in Christian experience and to which the New Testament documents are the primary witness — the active agape of God at work in human life.

In the light of the Incarnation, then, any justification of evil must (I suggest) be a justification of it as playing a part in bringing about the high good of man's fellowship with God, rather than as necessary to the aesthetic perfection of a universe which, in virtue of its completeness, includes personal life. A Christian theodicy must be centred upon moral personality rather than upon nature as a whole, and its governing principle must be ethical rather than aesthetic.

[1] II Corinthians v. 19.

CHAPTER IX

SIN AND THE FALL ACCORDING TO THE HELLENISTIC FATHERS

1. THE BIBLICAL BASIS OF THE FALL DOCTRINE

WITHIN the Augustinian tradition, which has dominated the thought of Western Christendom since
the fifth century, the doctrine of a fearful and calamitous fall of man long ago in the 'dark backward and abysm
of time', and of a subsequent participation by all men in the
deadly entail of sin, is, as we have seen, deeply entrenched.
According to this conception in its developed form, man was
created finitely perfect, but in his freedom he rebelled against
God and has existed ever since under the righteous wrath and
just condemnation of his Maker. For the descendants of
Adam and Eve stand in a corporate unity and continuity of
life with the primal pair and have inherited both their guilt
and a corrupted and sin-prone nature. We are accordingly
born as sinners, and endowed with a nature that is bound to
lead us daily into further sin; and it is only by God's free,
and to us incomprehensible, grace that some (but not all)
are eventually to be saved.

It is helpful to distinguish two separable elements within
this tradition : namely, the assertion of an inherited *sinfulness*
or tendency to sin, and the assertion of a universal human
guilt in respect of Adam's crime, falling upon us on account
of a physical or mystical presence of the whole race in its
first forefather. As we shall see, the former idea is common
to all Christian traditions — whether in the form of a
physiologically or of a socially transmitted moral distortion
— whilst the latter idea is peculiar to Augustinian and
Calvinist theology.

The Augustinian picture is so familiar that it is commonly

thought of as *the* Christian view of man and his sinful plight. Nevertheless it is only *a* Christian view. As F. R. Tennant pointed out at the outset of his valuable study of *The Sources of the Doctrines of the Fall and Original Sin*,[1] 'S. Paul's teaching as to the connexion of human sin and death with Adam's transgression is but one of the various possible interpretations of this narrative [Genesis iii], slowly and tentatively reached after some centuries of Jewish exegesis and reflection. S. Augustine's fuller and more definite doctrine is but a developed form of one of the possible interpretations of the statements of S. Paul, arrived at after the preparation of further centuries of Jewish speculation.'[2] That this is so has been established not only in Tennant's book but in a number of others, including Julius Müller's classic work, *The Christian Doctrine of Sin*,[3] and N. P. Williams's *The Ideas of the Fall and of Original Sin*.[4] The historical picture which has come to be widely accepted can be summarized as follows :

1. Probably the first of the ancient Hebraic stories to be used to account for the sinful state of the world was that of the angel marriages in Genesis vi. 1–8. This is used by the Jahwist redactor of Genesis to explain why God found it necessary to destroy mankind by means of a great Flood. The story is that the sons of God mated (by implication, unlawfully) with the daughters of earth, thus mixing the divine and human essences ; the children of these marriages (again, by implication) being the giants (*nephilim*), whom the writer also confusingly identifies with the 'mighty men of old'. Surveying this situation, God 'saw that the wickedness of man was great in the earth, and that every imagination of the thoughts of his heart was only evil continually. And the Lord was sorry that he had made man on the earth. . . .'[5] During the last two centuries before Christ, when these older legends came into their own and were put to theological use, this brief passage was elaborated into

---

[1] Cambridge University Press, 1903.
[2] Tennant, op. cit., p. 1.
[3] Op. cit., bk. iv, chap. iii, sect. 5.
[4] London : Longmans, Green & Co. Ltd., 1927.
[5] Genesis vi. 5–6.

the story of the Watchers in the Book of Enoch[1] ('I Enoch'),
a composite work whose writing probably covered much of
the last two centuries B.C.

2. However, towards the end of the pre-Christian era the
Watcher legend, based on Genesis vi, lost ground and was
gradually replaced by the story of the fall of Adam, based
on Genesis iii. In Jubilees (written somewhere between 135
and 105 B.C.) the Adamic story alone reveals the origin of
sin; and in Slavonic Enoch ('II Enoch', written around
the time of the birth of Jesus) the paradise-fall story plays
the more prominent role, with only an echo of the Watchers
theme remaining in Satan's seduction of Eve. In II Esdras
(*c.* A.D. 100) the Watcher legend has disappeared altogether,
leaving the Adamic myth in sole possession of the field. This
was the status of the story within Judaism at the time when
St. Paul was writing his epistles.

3. But the Adamic myth itself, as we find it in its primitive
simplicity in the early pages of the Book of Genesis, must be
carefully distinguished from the later interpretation of it
adopted by St. Paul, and the further development of this by
St. Augustine, and of course from its literary presentation in
the greatest epic poem in our language, telling

> Of man's first disobedience, and the fruit
> Of that forbidden Tree, whose mortal taste
> Brought Death into the world, and all our woe . . .

In the Augustinian elaboration of the Genesis story Adam's
pre-fallen state is an exalted condition of 'original righteousness', the snake is Satan in disguise, and the fall results in
our mortality and the inheritance by the whole subsequent
human species of both an imputed guilt for the first crime
and an inherited moral taint or disease. But none of this is
to be found in the text of Genesis iii. There man's first
condition is one of primitive simplicity; he is not set in a
heavenly or paradisal state but in an earthly garden which
he must tend; the snake is a snake and not a fallen angel;

---

[1] Chaps. 6 f. This work, as well as the other apocryphal and pseudepi-
graphical writings referred to in this section, appear in Eng. trans. with critical
introductions in *The Apocrypha and Pseudepigrapha of the Old Testament*, ed. R. H.
Charles, 2 vols. (Oxford: Clarendon Press, 1913).

and there is no suggestion either of an inherited guilt or of a congenital tendency to sin.

4. These dramatic fall stories — Genesis iii and Genesis vi — lived most vividly within the popular and apocalyptic religion of the more ordinary Jewish people (the 'people of the land') outside the priestly and rabbinical circles of Jerusalem during the two closing centuries of the pre-Christian era and the first century or so of the Christian era, the Genesis iii story largely displacing that in Genesis vi, as has already been mentioned, by about the turn of the eras. But in official rabbinical theology yet another view of the nature and origin of evil was current. This is the doctrine of the 'evil imagination' or the evil inclination or impulse (*yecer ha-ra*) referred to in Genesis vi. 5: 'God saw that the wickedness of man was great in the earth, and that every imagination (*yecer*) of the thoughts of his heart was only evil continually.' What the rabbinical schools derived from this passage was not perhaps so much a doctrine, in the sense of a systematic body of teaching, as a key idea which was applied in a variety of ways. The *yecer ha-ra* is sometimes thought of as ruling a man and determining his actions (as apparently in Genesis vi. 5 itself), but more usually as one of two mutually contrary influences, the 'evil imagination' and the Torah, between which man stands as a free being exercising responsible choice. Sometimes the origin of the evil imagination is located in man himself, as though it represented a culpably acquired habit of the will. More usually it is taught that God Himself has implanted the *yecer ha-ra* in each human heart — yet not so as to take away the individual's personal responsibility in face of its promptings; for although the *yecer ha-ra* is an *evil* imagination, its energy can always be sublimated to good ends in the life of piety. (N. P. Williams sees this aspect of the doctrine as akin to the notion of the morally neutral *libido* in Jungian psychology.[1])

By the time of Christ, however, the 'evil imagination' doctrine was coalescing with the Adamic myth, and it had become established rabbinical teaching (which St. Paul took with him from Judaism to Christianity) that Adam's trans-

[1] Williams, op. cit., pp. 67–69.

gression has affected his descendants as well as himself by generating an evil imagination within them.[1] It was also held that Adam would not have been mortal but for his sin, which has thus brought death into the world.[2]

5. Our Lord himself, whilst he clearly implied the universality of sin among men (e.g. in the words of the Lord's Prayer, 'Forgive us our trespasses . . .'[3] and in such sayings as, 'If ye then, being evil, know how to give good gifts to your children . . .'[4]), gave no endorsement in his recorded teachings either to the Adamic or the Watcher legend or to the rabbinical conception of the evil imagination. Neither do any of the New Testament writers, other than St. Paul, enter into the question of the origin of sin.[5]

If, then, we distinguish between, on the one hand, God's self-revelation in and through the *Heilsgeschichte* recorded in the Bible, centring upon the person and life of Jesus the Christ, and, on the other hand, the successive attempts to interpret the significance of these events in the theology of the Church, we must say that the Christian revelation itself emphatically points to the reality of sin as a universal human condition, but that it does not offer any specific theory as to its origin — not even that sponsored by the Church's first theologian, St. Paul.

2. FROM PAUL TO AUGUSTINE

When we seek the New Testament basis of the doctrine of the fall of mankind in Adam and Eve it is thus to Paul that we turn, and especially to Romans v. 12–21 and I Corinthians xv. 21–22. N. P. Williams suggests, as a probable reconstruction of the transmission of the fall theme from Judaism to Christianity, that the Adamic story was current in the

---

[1] This is reflected in II Esdras (*c.* A.D. 100) iii. 21–22 : 'For the first Adam, burdened with an evil heart, transgressed and was overcome, as were also all who were descended from him. Thus the disease became permanent. . . .'
[2] E.g. II Esdras (*c.* A.D. 100) iii. 7 : 'And thou didst lay upon [Adam] one commandment of thine ; but he transgressed it, and immediately thou didst appoint death for him and for his descendants'.
[3] Matthew vi. 12.   [4] Matthew vii. 11.
[5] There are, however, lingering traces of the Watchers legend in Jude vi–vii and II Peter i. 4.

Galilean religious culture, which was nourished by the inter-testamental apocalyptic literature, and that Paul, meeting this idea among the Galilean followers of Jesus and being struck by the Adam–Christ parallel, proceeded to employ it with such force as to impress it (after the formation of the New Testament canon) upon the mind of the developing Church.[1]

Paul's teaching is notoriously difficult to interpret; but nevertheless the following relatively general points represent a wide measure of agreement among Pauline commentators:

1. Death (which is perhaps to be understood in a spiritual as well as a physical sense) is a consequence of sin (Romans v. 12; I Corinthians xv. 21).

2. Mankind forms a corporate whole in relation to God, and mortality came upon this racial unity as a result of the sin of our first ancestor, Adam (Romans v. 18–19).

3. This conjunction of sin and death has descended from Adam to all his children through a tendency to sin which is part of our inherited psycho-physical make-up (Romans v. 18–19).

4. The inherited sinful tendency then produces actual sins, which are branded as such by the prohibitions of the law (Romans v. 20 and vii. 7–12).

5. As sin and death came thus through one man, Adam, so they will be abolished through one man, Christ (I Corinthians xv. 22; Romans v. 15 and 17).

6. As well as evil men there are evil spirits (Ephesians vi. 12); these exercise great power in the world — massively in the realms of pagan life outside Judaism and Christianity (I Corinthians ii. 8 and x. 20; II Corinthians iv. 4), but also occasionally even within the Church itself (II Corinthians ii. 11 and xii. 7; I Corinthians vii. 5; I Thessalonians ii. 18).

There is thus clearly present in St. Paul the root idea of what was later to be called 'original sin',[2] namely the idea of a sinful bias or tendency which operates in all human

---

[1] Williams, op. cit., pp. 118–22.
[2] The term 'original sin' (*originale peccatum*) was apparently first used by Augustine in his early work, *De diversis quaestionibus ad Simplicianum*. See N. P. Williams, op. cit., p. 327.

beings, but which is nevertheless not created *de novo* by each individual for himself. Paul probably assumed that the mode of transmission of this taint was by physiological inheritance. This likewise represented the predominant view in the early Church, although some writers seem to have thought more in terms of what we might call a social heredity operating by parental example and general environmental influence.[1] Certainly by the time of Augustine the notion of the physiological inheritance of sinfulness was predominant ; and in the thought of Augustine himself it is the sinful, because concupiscent or lustful, nature of the sexual intercourse by which a child is begotten that causes the child's character to be wicked from the moment of birth.

Another important development during this same period in the writings of the Latin Fathers was a growing sense of the heinousness of the fall. There is throughout the fourth century a heightening of the mythological perfection of Adam in the Garden of Eden, and a corresponding lengthening of the extent and deepening of the tragedy of his fall. The fall ceases to be thought of — as generally by the Greek-speaking theologians — as a *deprivatio*, the loss of something good, and is seen instead as a *depravatio*, a wicked corruption.

An element that is absent in St. Paul but had become widely accepted in Latin theology by the time of Augustine, and was then securely built by him into the continuing tradition of the West, is the idea that Adam's descendants somehow share the guilt of Adam's sin. This is made intelligible either by the thought of the mystical unity of the human race or by the doctrine of the seminal presence of Adam's descendants in his loins. On either basis the sin of the first man is the sin of all men, and all men are accordingly guilty and deserving of eternal death. To quote Augustine :

For we are all in that one man [Adam], since we all were that one man who fell into sin by the woman who was made from him before the sin. For not yet was the particular form created and distributed to us, in which we as individuals were to live, but already the seminal nature was there from which we were to be propagated and this being vitiated by sin, and bound by the

---

[1] E.g. Clement of Alexandria. See N. P. Williams, op. cit., p. 206.

chain of death, and justly condemned, man could not be born of man in any other state. And thus, from the bad use of free will [by Adam], there originated the whole train of evil, which, with its concatenation of miseries, convoys the human race from its depraved origin, as from a corrupt root, on to the destruction of the second death, which has no end, those only being excepted who are freed by the Grace of God.[1]

With this quotation we are back in the midst of the Augustinian theodicy. But the reason for this compressed sketch of the development from St. Paul, through the Latin or Western Fathers, to Augustine has been to provide a background and contrast for the presentation of another strand of thought which co-existed with it. This is the characteristically different point of view of the Hellenistic or Eastern Fathers, in tracing which we must now return to the earliest period of Christian thought and to a situation before the Augustinian theology had become established.

3. THE BEGINNINGS OF THE HELLENISTIC POINT OF VIEW

So far as Christian teaching concerning sin and the fall is concerned, the 'sub-apostolic age' (the period between the work of the biblical writers at about the end of the first century and that of Irenaeus towards the end of the second century) was a time of great fluidity and of the absence of established dogmas. This fact is significant for us today as we seek to discern the shape of a viable Christian theodicy, for it shows what wide freedom of interpretation and speculation existed among those Christian thinkers who were closest in time to the life and teaching of our Lord. As we have already noted, the Christian revelation itself, as it occurred in the saving events centring upon the life of Christ, presupposes the dread reality of sin and suffering but does not provide us with any divinely authorized theory as to their origin. The first few generations of Christians were very largely 'starting from scratch' when they speculated on this subject, and we are therefore free, if there seems good reason to do so, to return behind the later dogmas to the freedom

[1] *C.G.*, bk. xiii, chap. 14.

of that earlier period — learning something, however, we may hope, from the difficulties in which the later theologians sometimes became entangled.

In the surviving Christian literature of the hundred years or so after the apostles there is, despite Paul's use of the figure of Adam, no common way of thinking about sin and evil. On the whole, wrongdoing is straightforwardly blamed upon the individual sinner (as it is in the teaching of Jesus in the synoptic gospels), and man's God-given freedom and responsibility are strongly emphasized.[1] There is also, however, a constant reference to the evil spirits who invisibly infest this world and who are the source of temptations, heresies, diseases, and the oppressive cruelties of the pagan world.[2] Indeed, in so far as there is any widely agreed way of tracing sin to a source beyond the free will of the individual sinner it is referred to the swarming hordes of demons under the leadership of Satan, who are man's enemies and whom Christ somehow defeated on His cross.[3] Today we are more inclined, without necessarily discounting the possibility of evil spiritual beings, to reinterpret in terms of psychological forces most of the events that former ages attributed to the work of devils. The demons that tempt us or confuse us or lure us into evil are complexes and psychoses and libidinous pressures and the like in our own minds. They are just as real, and just as much in need of being cleansed by the redemptive power of Christ, as the demons of former popular imagination ; but it is clear that the problem of evil is not solved, but only restated, by referring to them. For the demonic realities, whether they be conceived as external beings or as elements within our own unconscious nature, are as troublesome to theodicy as any other instance of evil.

In addition to this sense of the pervasive activity of demons, and the no less emphatic sense of human freedom and responsibility, there are in the sub-apostolic literature a few

[1] E.g. Justin Martyr, *First Apology*, chap. 43 ; *Dialogue with Trypho*, chap. 141.
[2] E.g. Justin Martyr, *First Apology*, chaps. 14 and 56–58.
[3] Gustav Aulén's *Christus Victor* (E.T. 1931) presents this 'classic' view of atonement.

215

scattered references to the Adamic story,[1] but no more exclusive use is made of it than of the rival Watchers story or of the doctrine of the 'evil imagination'.[2] Paul's strong advocacy of the imagery derived from Genesis iii would seem to have had little effect in the Church as a whole for about a hundred years after his writing of I Corinthians and the Letter to the Romans. But in the mid-second century, in the course of the Church's mortal struggle against Gnosticism, the New Testament canon was formed, with Paul's letters as part of it. From about this time onwards his teaching about the fall was regarded as authoritative, and virtually all subsequent Christian thought concerning sin and evil has been set within the framework of the Adamic myth.

But within this framework two significantly different developments have taken place, the one (sketched in the previous section) going through Augustine and the Western Church, the other going through Irenaeus and the Eastern Church. The former — thanks to the massive theological effort of Augustine himself — contained a developed theodicy; the latter tradition did not, but has nevertheless provided the foundation for a radical Christian alternative to the Augustinian theodicy. Within this early period, however, whose literature we are surveying there are only the first tentative hints of such a non-Augustinian theodicy. Thus Tatian, in his *Oration against the Greeks* (c. 175), says that God did not make men already good, but made them free to become good by obedience to Himself.[3] And Theophilus, in his book *To Autolycus* (c. 175), regards Adam and Eve as children, placed in the Garden of Eden in order that they should mature and become perfect by obedience, and so

[1] E.g., *Epistle of Barnabas* (c. 130), chap. 12; *Epistle to Diognetus*, 12; Justin Martyr, *Dialogue with Trypho*, chaps. 79 and 88.
[2] During this period there are references to the Watchers story in II Peter ii. 4 (c. 150), and Athenagoras' *Plea for the Christians* (c. 180), chap. 24.
[3] Chap. 7. 'And each of these two orders of creatures [angels and men] was made free to act as it pleased, not having the nature of good, which again is with God alone, but is brought to perfection in men through their freedom of choice, in order that the bad man may be justly punished, having become depraved through his own fault, but the just man be deservedly praised for his virtuous deeds, since in the exercise of his free choice he refrained from transgressing the will of God. . . .' (Trans. by Marcus Dods in the Ante-Nicene Library.)

Sin and the Fall

come to share God's immortality.[1] However, man has disobeyed God and now has to come to his perfection through the long detour of a life outside the Garden.[2]

Here are the first tentative beginnings of themes that were to be developed by Irenaeus, and some of which were to emerge again in the modern world in Schleiermacher and then again in more recent theologians, especially in Britain.

4. IRENAEUS

It was in Irenaeus (c. 130–c. 202), Bishop of Lyons and author, in response to Gnosticism, of the Church's first systematic theology, that there comes clearly to light the point of view that was to characterize the Greek as distinct from the Latin Fathers. Irenaeus distinguishes between the image (εἰκών) of God and the likeness (ὁμοίωσις) of God in man. The 'imago', which resides in man's bodily form, apparently represents his nature as an intelligent creature capable of fellowship with his Maker, whilst the 'likeness' represents man's final perfecting by the Holy Spirit. For 'the man is rendered spiritual and perfect because of the outpouring of the Spirit, and this is he who was made in the image and likeness of God. But if the Spirit be wanting to the soul, he who is such is indeed of an animal nature, and being left carnal, shall be an imperfect being, possessing indeed the image [of God] in his formation, but not receiving the likeness through the Spirit.'[3] This distinction can, I think, not unfairly be presented in more contemporary terms by saying that man's basic nature, in distinction from the other animals, is that of a personal being endowed with moral freedom and responsibility. This is the divine εἰκών in him; he is made as person in the image of God. But man, the finite personal

[1] Bk. ii, chaps. 24–25.   [2] Bk. ii, chap. 26.
[3] *Against Heresies*, v. vi. 1. Translation in the Ante-Nicene Library. The Greek Orthodox scholar P. Bratsiotis, describing this distinction in the early Eastern Fathers as a whole, says, 'The εἰκών [imago] is related, according to these Church Fathers, to man's spiritual nature as a rational and free being. But the ὁμοίωσις [similitudo] means, according to the same Church Fathers, man's longing and positive striving toward God, and at the same time man's destiny, which is to come into the likeness of God.' 'Das Menschenverständnis in der griechisch-orthodoxen Kirche', *Theologische Zeitschrift* (1950), p. 378.

217

creature capable of personal relationship with his Maker, is as yet only potentially the perfected being whom God is seeking to produce. He is only at the beginning of a process of growth and development in God's continuing providence, which is to culminate in the finite 'likeness' of God. Thus whilst the image of God is man's nature as personal, the divine likeness will be a quality of personal existence which reflects finitely the life of the Creator Himself.

Irenaeus accordingly thinks of man as originally an immature being upon whom God could not yet profitably bestow His highest gifts : 'God had power at the beginning to grant perfection to man; but as the latter was only recently created, he could not possibly have received it, or even if he had received it could he have contained it, or containing it, could he have retained it.'[1] Again :

If, however, any one say, 'What then? could not God have exhibited man as perfect from the beginning?' let him know that inasmuch as God is indeed always the same and unbegotten as respects Himself, all things are possible to Him. But created things must be inferior to Him who created them, from the very fact of their later origin ; for it was not possible for things recently created to have been uncreated. But inasmuch as they are not uncreated, for this very reason do they come short of the perfect. Because, as these things are of later date, so are they infantile ; so are they unaccustomed to, and unexercised in, perfect discipline. For as it certainly is in the power of a mother to give strong food to her infant, [but she does not do so], as the child is not yet able to receive more substantial nourishment ; so also it was possible for God Himself to have made man perfect from the first, but man could not receive this [perfection], being as yet an infant.[2]

In his *Proof of the Apostolic Preaching* Irenaeus pictures Adam and Eve in the Garden of Eden as children ;[3] and their sin is accordingly not presented as a damnable revolt, but rather as calling forth God's compassion on account of their weakness and vulnerability.[4] There is even a hint of the 'O felix culpa' theme in such passages as this :

---

[1] *A.H.* IV. xxxviii. 2.　　　[2] *A.H.* IV. xxxix. 1.　　　[3] Chap. 16.
[4] *A.H.* III. xx. 1.　Cf. *Proof of the Apostolic Preaching*, chap. 12, 'But the man was a little one, and his discretion still undeveloped, wherefore also he was easily misled by the deceiver.' (Trans. by Joseph P. Smith, S.J., *Ancient Christian Writers*, vol. xvi, London : Longmans, Green & Co., 1952.)

This, therefore, was the [object of the] long-suffering of God, that man, passing through all things, and acquiring the knowledge of moral discipline, then attaining to the resurrection from the dead, and learning by experience what is the source of his deliverance, may always live in a state of gratitude to the Lord, having obtained from him the gift of incorruptibility, that he might love Him the more; for 'he to whom more is forgiven, loveth more' (Luke vii. 43).[1]

There is also to be found in Irenaeus the thought, correlative to that of man's finitude and weakness, of his cognitive freedom in relation to God, which is safeguarded by the ambiguities of God's self-revealing activity in history and by the corresponding need for an uncompelled response of faith on man's own part. 'Not merely in works', Irenaeus says, 'but also in faith, has God preserved the will of man free and under his own control.'[2] And in the next chapter: 'It was for this reason that the Son of God, although He was perfect, passed through the state of infancy in common with the rest of mankind, partaking of it thus not for His own benefit, but for that of the infantile stage of man's existence, in order that man might be able to receive Him.'[3]

Our present life is accordingly pictured as a scene of gradual spiritual growth:

By this arrangement, therefore, and these harmonies, and a sequence of this nature, man, a created and organized being, is rendered after the image and likeness of the uncreated God — the Father planning everything well and giving His commands, the Son carrying these into execution and performing the work of creating, and the Spirit nourishing and increasing [what is made], but man making progress day by day, and ascending towards the perfect, that is, approximating to the uncreated One. . . . Now it was necessary that man should in the first instance be created; and having been created, should receive growth; and having received growth, should be strengthened; and having been strengthened, should abound; and having abounded, should recover [from the disease of sin]; and having recovered, should be glorified; and being glorified, should see his Lord.[4]

---

[1] *A.H.* III. xx. 2.   [2] *A.H.* IV. xxxvii. 5.

[3] *A.H.* IV. xxxviii. 2. Cf. IV. xxxviii. 1.

[4] *A.H.* IV. xxxviii. 3. Cf. IV. xxxvii. 7. Irenaeus traces the course of God's education of man first through the Law and the prophets of the Old Testament and then through the Incarnation. See IV. xiv–xvi.

Within God's providence man is being taught by his contrasting experience of good and evil to value the one for himself and to shun the other. Hence the mixture of good and evil in our world. For

> how, if we had no knowledge of the contrary, could he have had instruction in that which is good? . . . For just as the tongue receives experience of sweet and bitter by means of tasting, and the eye discriminates between black and white by means of vision, and the ear recognises the distinctions of sounds by hearing; so also does the mind, receiving through the experience of both the knowledge of what is good, become more tenacious of its preservation, by acting in obedience to God. . . . But if any one do shun the knowledge of both kinds of things, and the twofold perception of knowledge, he unawares divests himself of the character of a human being.[1]

We must, then, accept in trustful gratitude all that comes to us from God's hand.

> If, then, thou art God's workmanship, await the hand of thy Maker which creates everything in due time; in due time as far as thou art concerned, whose creation is being carried out [*efficeris*]. Offer to Him thy heart in a soft and tractable state, and preserve the form in which the Creator has fashioned thee, having moisture in thyself, lest, by becoming hardened, thou lose the impressions of His fingers. But by preserving the framework thou shalt ascend to that which is perfect. . . .[2]

There is thus to be found in Irenaeus the outline of an approach to the problem of evil which stands in important respects in contrast to the Augustinian type of theodicy. Instead of the doctrine that man was created finitely perfect and then incomprehensibly destroyed his own perfection and plunged into sin and misery, Irenaeus suggests that man was created as an imperfect, immature creature who was to undergo moral development and growth and finally be brought to the perfection intended for him by his Maker. Instead of the fall of Adam being presented, as in the Augustinian tradition, as an utterly malignant and catastrophic event, completely disrupting God's plan, Irenaeus pictures it as something that occurred in the childhood of the

---

[1] *A.H.* IV. xxxix. 1.   [2] *A.H.* IV. xxxix. 2.

race, an understandable lapse due to weakness and immaturity rather than an adult crime full of malice and pregnant with perpetual guilt. And instead of the Augustinian view of life's trials as a divine punishment for Adam's sin, Irenaeus sees our world of mingled good and evil as a divinely appointed environment for man's development towards the perfection that represents the fulfilment of God's good purpose for him.

Irenaeus was the first great Christian theologian to think at all systematically along these lines, and although he was far from working out a comprehensive theodicy his hints are sufficiently explicit to justify his name being associated with the approach that we are studying in this part. It is true that Irenaeus' name does not belong to this type of theodicy as clearly and indisputably as Augustine's name belongs to the predominant theodicy of Western Christendom; it is also true that within Irenaeus' own writings there are cross-currents and alternative suggestions that I have left aside here.[1] Nevertheless, to speak of the Irenaean type of theodicy is both to name a tradition by its first great representative and at the same time to indicate the significant fact that this mode of responding to the problem of evil originated in the earliest and most ecumenical phase of Christian thought.

5. EASTERN CHRISTIANITY

After Irenaeus many of his themes were echoed in Clement of Alexandria (died *c.* 220). Clement gave only peripheral attention to the doctrines of sin and the fall, but his occasional references show that he shared the Irenaean point of view. For example, he confronts the Gnostics' challenge: if man was created good, how has he sinned; but if he was

---

[1] Harnack (*History of Dogma*, E.T. vol. ii, pp. 268 f.), following H. H. Wendt, points to two lines of thought in Irenaeus, describing the one outlined above as 'moralistic and apologetic' and saying that it is 'alone developed with systematic clearness'. For a fuller exposition of this latter Irenaean position, see, for example, F. R. M. Hitchcock, *Irenaeus of Lugdunum* (Cambridge University Press, 1914), chap. 4. On the other hand, Gustav Wingren (*Man and the Incarnation: A Study in the Biblical Theology of Irenaeus*, E.T., Edinburgh: Oliver & Boyd, 1959, pp. 26 f.) criticizes this distinction and defends the unity of Irenaeus' thought.

not, how can his Creator have been good? To this dilemma Clement offers Irenaeus' answer that man was created immature :

> Above all, this ought to be known, that by nature we are adapted for virtue ; not so as to be possessed of it from our birth, but so as to be adapted for acquiring it. By which consideration is solved the question propounded to us by the heretics, Whether Adam was created perfect or imperfect? For they shall hear from us that he was not perfect in his creation, but adapted to the reception of virtue. . . .[1]

With Irenaeus, Clement thinks of Adam as a child.[2] And again he uses the distinction between the 'image' and 'likeness' of God : 'For is it not thus that some of our writers have understood that man straightway on his creation received what is "according to the image", but what is according "to the likeness" he will receive afterwards on his perfection?'[3]

Clement contributes nothing new ; nevertheless his use of these ideas is of interest to us today as evidence that they were freely accepted within the Great or Catholic Church during the period of its long struggle against Gnosticism.

These are the principal representatives in the early Church of the Irenaean point of view in theodicy. However, Methodius (died *c.* 311)[4] and St. Gregory of Nazianus (*c.* 329–*c.* 389)[5] apparently accepted the picture of Adam as immature and infantile, which symbolizes in terms of the Adamic myth the insight that man was not created perfect but that his perfecting lay in the future.

N. P. Williams sums up this earliest type of systematic Christian thinking concerning sin and evil as follows :

> Our study of these primitive Greek-Christian writers has thus revealed, gradually taking shape within the Catholic Church of the late second century, an interpretation of the Fall doctrine which, whilst preserving the essential outlines of the Pauline teaching, wears a humane, reasonable, and curiously modern complexion. It does not, indeed, betray any suspicion that Adam and Eve may not have been historical personages. But it gives us a picture of primitive man as frail, imperfect, and child-like — a

---

[1] *Stromata*, bk. vi, chaps. 11 and 12. (Trans. in the Ante-Nicene Library.)
[2] *Protrepticus*, chap. 11.　　　　[3] *Stromata*, bk. ii, chap. 22.
[4] *Symposium*, iii. 5.　　　　[5] *Oration*, 45, 7, 8.

picture which is on the whole unaffected by the Rabbinical fig-
ment of Adam's 'Original Righteousness', and is by no means
incapable of harmonisation with the facts revealed by the science
of to-day. It exaggerates neither the height from which, nor the
depth to which, the first men are alleged to have fallen. It finds
in the inherited disorder of our nature rather a weakness to be
pitied than an offence to be condemned. . . .[1]

Here, then, in the thought of the early Hellenistic theologians of the Christian Church, is the historical origin of the
tradition of theodicy that I am describing as Irenaean. It
was not carried at that time beyond these relatively inchoate
beginnings, for the Eastern Church, centring on Constanti-
nople, did not continue to develop theologically as did the
Roman Church in the West. However, the basic Irenaean
conception of man as a creature made initially in the 'image'
of God and gradually being brought through his own free
responses into the divine 'likeness', this creative process
being interrupted by the fall and set right again by the
incarnation, has continued to operate in the minds of
theologians of the Orthodox Church down to the present
day. A developmental or teleological view of man is evident,
for example, in the work of the Orthodox thinker whose
writings are most familiar to the West, Nicholas Berdyaev.[2]
Another leading contemporary Orthodox theologian, Paul
Evdokimov, expounding the conception of man that prevails
within his tradition, writes:

Creation in the biblical sense is like a grain of wheat which
brings forth a hundred-fold and never stops developing: 'My
Father works until now, and I work also'. Creation is the Alpha
moving towards the Omega, and indeed the Omega is already
contained within it. This makes every moment of time very definitely eschatological; it opens out into and is judged by its ultimate fulfilment. The Messiah is called *tsemach* [seed], and the
very notion of the Messiah derives from the Pleroma. Creation
demands incarnation, which in turn finds its consummation in
the parousia of the Kingdom. Time is built into the structure of
the created world, which means that the world is 'unfinished',
'embryonic', so as to further and direct that synergy of divine

[1] Williams, op. cit., pp. 199–200.
[2] See, for example, *The Meaning of the Creative Act*, trans. Donald Lowrie
(London: Victor Gollancz Ltd., 1955), chap. 5.

power and human power until the Day of the Lord wherein the seed attains its final fruition.[1]

Again, another Orthodox exponent, P. Bratsiotis, expresses this developmental view when he says that 'The first man was endowed by the Creator with all spiritual and moral powers, which constituted him in the image of God and which were to lead him forward by God's help through their growth and development to the realization of his life's destiny, which is to come into the likeness of God'.[2]

It cannot, however, be said that there is an 'Eastern Orthodox theodicy' which is Irenaean as distinct from Augustinian. Although, as we have seen, there are the foundations for a theodicy in the work of the early Hellenistic Fathers, Orthodox thought has never built systematically upon these foundations, and indeed has always taken very seriously the quite different theme of the pre-mundane fall of the angels and their activity in resisting God and tempting mankind.

---

[1] Paul Evdokimov, *L'Orthodoxie* (Paris: Delachaux et Niestlé, 1959), p. 84.
[2] P. Bratsiotis, 'Das Menschenverständnis in der griechisch-orthodoxen Kirche', *Theologische Zeitschrift* (1950), p. 379. See also Vladimir Lossky, *The Mystical Theology of the Eastern Church* (Paris, 1944. E.T., London: James Clarke & Co. Ltd., 1957), chap. 6.

CHAPTER X

THE IRENAEAN TYPE OF THEODICY IN SCHLEIERMACHER

A N alternative to the Augustinian type of theodicy existed in germ, as we have seen, in the thought of some of
the early Hellenistic Fathers, and above all Irenaeus.
But in the East this alternative theodicy remained undeveloped (like Orthodox theology as a whole) whilst in the West
it was overborne by the immense contrary influence of
Augustine. Thus the Irenaean or eschatological approach
to the problem of evil lay virtually dormant within Christendom from the time of Augustine until it awoke again in the
mind of the great nineteenth-century Protestant theologian
Friedrich Schleiermacher (1768–1834). It does not appear
that Schleiermacher was influenced by Irenaeus,[1] or that he
was consciously renewing and continuing an approach to the
mystery of evil whose foundations had been laid in some
of the earliest thinking of the Christian Church. We are thus
dealing not so much with a continuous Irenaean *tradition* of
theodicy as with a *type* of theodicy, or a manner of theological
approach to the subject. It is permissible and convenient
to name this approach after Irenaeus, as its first great representative, in spite of the fact that it has not been maintained
and developed in a continuity of teaching linking its origins
with the present day. The picture is rather one of unfulfilled
beginnings in the first four centuries, and then an independent
new start in the nineteenth century, carrying some of the
same themes much further and in a more systematic form.
From Schleiermacher onwards this movement of thought is
clearly visible, the stream becoming wider but also shallower

---

[1] In his *Geschichte der christlichen Kirche* Schleiermacher treated Irenaeus
simply as an anti-Gnostic writer. (*Sämtliche Werke*, Berlin, 1840, vol. xi/vi,
pp. 142–3.)

(as we shall see in the next chapter) as it runs into the currents of nineteenth-century evolutionary thinking, and being deepened again under the impact of war in the present century.

1. SCHLEIERMACHER ON 'ORIGINAL PERFECTION'

As is well known, Schleiermacher's method in *Der christliche Glaube*[1] is to start from the religious self-consciousness or piety of the Christian and to consider what theological propositions are implied in it. The essence of that piety is the feeling of absolute dependence, which is our conscious relation to God, and the various phases of this feeling are specified as they correspond to the various aspects of the divine nature. This is the procedural framework within which Schleiermacher writes about the problem of evil.

Traditional Christian mythology speaks of the fall of man from a prior state of original righteousness, and of a consequent disruption of the original perfection of his world. Schleiermacher reinterprets these ideas from the standpoint of a theology based upon the Christian religious self-consciousness. For him, the 'original perfection' of the creation is its suitability for accomplishing the purpose for which God created it. This purpose centres upon man, and its feasibility accordingly has both an objective and a subjective aspect, traditionally termed the 'original perfection' of the world and the 'original righteousness' or original perfection of man.

In his discussion of the former idea Schleiermacher proceeds from a premise that he has previously established: namely, that God-consciousness occurs in man always in connection with his consciousness of an environing world. As a later theologian in the Schleiermacherian tradition has expressed it, the supernatural is known in and through the natural.[2] It is as embodied beings organic to nature that we become aware of our condition of absolute dependence,

---

[1] 1st ed., 1821, 2nd ed., 1831. The quotations in this chapter are from the Eng. trans. of the 2nd ed. *The Christian Faith*, ed. H. R. Mackintosh and J. S. Stewart (Edinburgh: T. & T. Clark, 1928).

[2] John Oman, *The Natural and the Supernatural* (Cambridge University Press, 1931).

which is our conscious relationship with God.[1]   The perfection of the world is accordingly its character as permitting
the 'co-existence of the God-consciousness with the consciousness of the world',[2] and the occasioning of the higher awareness by the lower.   The world's 'original perfection' is thus
simply its suitability as an environment for the emergence of
man's God-consciousness.   To quote a typically Schleiermacherian sentence :

Such perfection is affirmed in the above sense, i.e. it is laid down
that all finite being, so far as it co-determines our self-consciousness, is traceable back to the eternal omnipotent casuality, and all
the impressions of the world we receive . . . include the possibility
that the God-consciousness should combine with each impression
of the world [Welteindruck] in the unity of the moment.[3]

Thus the world's perfection does not refer to its state at some
favoured period in the past, but to the basic character
determining all its states.   This is accordingly not a doctrine
of the original perfection of the world in the sense of a harmonious primordial condition which has subsequently
vanished.   The perfection of the world, in virtue of which
the God-consciousness can occur within it, still exists ; it is
'original' in the non-temporal sense of being fundamental
and constitutive.[4]

Man's own original perfection (or 'original righteousness')
is likewise, for Schleiermacher, not a primordial and long-
since-forfeited condition of human virtue but the structure
of human nature whereby 'in our clear and waking life a
continuous God-consciousness as such is possible'.[5]   For the
consciousness of God can be stimulated in us, not only in
moments of 'life-enhancement' and pleasure, but also in
moments of 'life-hindrance' and pain.[6]   There is, indeed, no
aspect of our human experience that is not in principle
capable of serving to awaken the awareness of God ; and this
religious receptivity (together with the social nature in virtue
of which we can communicate it to one another) constitutes
the 'original' or basic perfection of our nature.

[1] *C.F.*, para. 5.   [2] *C.F.*, p. 233 (para. 57, 1).
[3] *C.F.*, p. 234 (para. 57, 1).   [4] *C.F.*, p. 235 (para. 57, 2).
[5] *C.F.*, p. 245 (para. 60, 1).   [6] Cf. *C.F.*, p. 240 (para. 59, 3).

We have already acknowledged (in Chapter VIII, Section 3) that the notions of an original righteousness of man and an original perfection of his environment, when taken as describing empirical states of affairs in the distant past, are untenable in the light of modern scientific knowledge. These traditional notions have to be merged in, or indeed replaced by, the conception of the goodness of the created universe as an environment within which the divine plan for man may be carried out, and of man's capacity to respond to God and to share in the fulfilment of His purpose. Schleiermacher was the first major theologian explicitly to make this transposition of themes. But having made it, there seems little point in retaining the term 'original' (*ursprünglich*), as he does. It tends to obscure rather than to emphasize the important transition that Schleiermacher is making from the mythological conception of the goodness of the created world, as having characterized a brief period in the past, to the theologically more realistic conception of it as characterizing the world throughout its history as a sphere of creaturely existence in which finite persons may emerge and may become 'children of God' and 'heirs of eternal life'.

2. Schleiermacher's Account of Sin

Schleiermacher offers (1) a phenomenological[1] and (2) a theological definition of sin.

1. Phenomenologically, sin is equated with 'all arrestments of the disposition to the God-consciousness'.[2] Such arrestment is always due to 'the independence of the sensuous functions'[3] whereby 'the spirit is obstructed in its action by the flesh'.[4] Phenomenologically, then, sin is a retardation or a diminution of the consciousness of God due to the interference of our bodily nature, with its involvement in the material world. This definition has been criticized (e.g. by Julius Müller)[5] on the ground that an exclusive

---

[1] This is not a term that Schleiermacher himself uses, although it seems to describe accurately what he is offering at this point.
[2] *C.F.*, p. 262 (para. 63, 1). Cf. p. 271 (para. 66, 1).
[3] *C.F.*, p. 273 (para. 66, 2.).   [4] *C.F.*, p. 274 (para. 67, 2).
[5] *The Christian Doctrine of Sin*, i. 341–4.

reference to our bodily nature as the source of sin is too narrow and fails to take account of the more inward and spiritual sins, such as pride, lovelessness, and envy. However, it is likely, as Karl Barth points out in the partial defence of Schleiermacher's doctrine of evil which characteristically precedes his own trenchant criticism and eventual emphatic rejection of it, that by the lower or sensual nature Schleiermacher means 'world-consciousness as opposed to God-consciousness. . . . It is world-consciousness evading or resisting the majesty and claim of God-consciousness or spirit, and thereby falling into disorder and becoming flesh.'[1] Interpreted in this way, Schleiermacher's position will be adopted in a modified form in Part IV. For it will be argued there that the basic cause of sin, which renders it virtually inevitable, is man's situation as a creature who is organic to the animal kingdom and whose consciousness is accordingly focused upon the natural environment within which he must struggle to maintain himself.

2. Schleiermacher's theological account of sin is supported by, but not dependent upon, his phenomenology. He says: 'we may regard sin on the one hand as simply that which would not be unless redemption was to be; or on the other hand as that which, as it is to disappear, can disappear only through redemption'.[2] The first half of this statement, Schleiermacher claims, excludes the dualist mistake, whilst the second half safeguards against the contrary error of Pelagianism. The fuller implications of his formula are unfolded by Schleiermacher when he considers the relation between sin and the wider purpose of God, which we shall examine below.[3]

Schleiermacher, thinking as he does in terms of the *development* of God-consciousness in man from a mere potentiality towards 'a state of absolute strength',[4] notes that there is an early stage in the life both of the race and of the individual when the bodily nature is already fully operative but when God-consciousness has as yet hardly begun to develop. For 'in each individual the flesh manifests itself as a reality before

---

[1] *C.D.* III/3, p. 324. [2] *C.F.*, p. 270 (para. 65, 2).
[3] See pp. 234 ff. below. [4] *C.F.*, p. 278 (para. 68, 3).

the spirit comes to be such',[1] and it is for this reason that the awakening God-consciousness, finding itself in tension with the demands of the flesh, creates in us the sense of sin. If the growth of the God-consciousness in us were a perfectly smooth development whereby the spirit gains a progressive ascendancy over the flesh the process would not be accompanied by any sense of sin. 'We actually find, however, that our development is always an irregular one, and also that the spirit is obstructed in its action by the flesh — the circumstance, indeed, to which our consciousness of sin is due.'[2] Accordingly the development of the God-consciousness in man is in fact accompanied by a consciousness of guilt and of the need for redemption. And indeed since our spiritual nature begins to develop after our bodily nature, it would seem that the sense of sin, expressing the fact that the higher in us does not instantaneously come to rule the lower, is virtually inevitable. However, Schleiermacher insists that sin is not inevitable in any sense that would cancel our personal responsibility for it. For we are aware that 'the God-consciousness could have developed progressively from the first man to the purity and holiness which it manifests in the Redeemer', and accordingly we know that 'sin is a derangement of human nature'.[3]

Schleiermacher thus affirms the paradoxical conjunction of 'original sin' and personal responsibility. He recognizes a racial solidarity in sin, whereby each new individual, independently of any choice of his own, is born into and becomes part of a corrupt society; but he recognizes at the same time that this common sinfulness, embodied in all the structures of our common life, is built up out of the innumerable wrong volitions of individuals. And so, in a famous phrase Schleiermacher says that sin is 'in each the work of all, and in all the work of each'.[4] On the one hand, 'The sinfulness that is present in an individual prior to any action of his own, and has its ground outside his own being, is in every case a complete incapacity for good'.[5] But, on the other

---

[1] *C.F.*, p. 274 (para. 67, 2).   [2] *C.F.*, p. 274 (para. 67, 2).
[3] *C.F.*, p. 279 (para. 68, 3).   [4] *C.F.*, p. 288 (para. 71, 2).
[5] *C.F.*, p. 282 (para. 70). However, this incapacity to originate good is not an incapacity to respond to the goodness of God in His redeeming activity

hand, this involuntary incapacity for good issues in personal and therefore responsible actions, so that original sin becomes 'the personal guilt of every individual who shares in it'.[1] However, having affirmed this personal guilt, Schleiermacher at once relates it again to the common guilt. For 'if the sinfulness which is prior to all action operates in every individual through the sin and sinfulness of others, and if, again, it is transmitted by the voluntary actions of every individual to others and implanted within them, it must be something genuinely common to all'.[2] Again, 'What appears as the congenital sinfulness of one generation is conditioned by the sinfulness of the previous one, and in turn conditions that of the later; and only in the whole series of forms thus assumed, as all connected with the progressive development of man, do we find the whole aspect of things denoted by the term "original sin"'.[3] It is out of the recognition of this corporate character of sin that there arises the sense of a universal need for redemption.

Schleiermacher summarizes very clearly some of the main amendments which he is proposing in the Augustinian conception of sin and the fall:

for the contrast between an original nature and a changed nature we substitute the idea of a human nature universally and without exception — apart from redemption — the same; and ... for the contrast between an original righteousness that filled up a period of the first human lives and a sinfulness that emerged in time (an event along with which and in consequence of which that righteousness disappeared), we substitute a timeless original sinfulness always and everywhere inhering in human nature and co-existing with the original perfection given along with it ... finally, for the antithesis between an original guilt and a transmitted guilt we substitute the simple idea of an absolutely common guilt identical for all. . . .[4]

It is clear that Schleiermacher's teaching is at this point much closer to the thought of Irenaeus and the Greek Fathers of the Christian Church than to that of Augustine

towards ourselves; for 'were we to affirm that the capacity for redemption has been lost, we should come into conflict with our very belief in redemption' (ibid., p. 283, para. 70, 2).
[1] *C.F.*, p. 285 (para. 71). [2] *C.F.*, pp. 287–8.
[3] *C.F.*, p. 288 (para. 71, 2). [4] *C.F.*, p. 303 (para. 72, 6).

and the Latins. According to Schleiermacher man was created in an undeveloped state; and sin (which as a free personal act entails guilt) was virtually inevitable in finite creatures in which the consciousness of God had not yet developed. But this 'original' sinfulness, which takes the form of a racial solidarity in sin, is nevertheless compatible with the original perfection of man and his world in the sense that this entire situation — humanity inhabiting a material environment and developing in slow and halting stages towards a full consciousness of God — represents God's positive will for man : it is *this* that God saw, and found it to be very good.[1]

3. The Relation between Sin and Suffering

By evil (*Übel*) Schleiermacher means those aspects of our material environment which are experienced as inimical to us and as obstructing our lives : death, pain, disease, etc. Many of the elements of the world which are experienced in this way are unavoidable : 'there must always have been a relative opposition, making itself felt with varying intensity, between the existent as externally given and the corporeal life of human individuals, otherwise human beings could not have been mortal'.[2] But in itself this is not evil. On the contrary,

care for the preservation of life and the avoidance of what would disturb it, which is conditioned by mortality, is among the most powerful motives of human development . . . [and] enduring sinlessness would have stood out far more strongly and conspicuously if man, unimpeded in the development and use of his powers, bore evil, and, combining God-consciousness with love of his race, overcame the impulse to cling to his own life and accepted death.[3]

But this is now an excluded possibility. Our human nature is distorted by sin and we act in and react to the world in sinful, i.e. non-God-conscious, ways. Accordingly evil exists, both as social evil (*geselliges Übel*) resulting from human behaviour, and as natural evil (*natürliches Übel*). Schleiermacher

[1] Genesis i. 31. See *C.F.*, p. 235 (para. 57, 2).
[2] *C.F.*, p. 315 (para. 71, 1).
[3] *C.F.*, pp. 243–4 (para. 59, postscript).

states clearly the pregnant thought — which we have also found within the Augustinian camp in Karl Barth[1] — that our finite and transient mode of existence as human animals, with the pain and eventual dissolution which it entails, is not in itself evil, being simply the mode of creatureliness which God in His love has appointed for us, but that it becomes evil through the sinful fear and anxiety of our self-centred reactions to it. Schleiermacher summarizes the relation between sin and evil as follows :

As summarising the foregoing, our proposition implies, first, that without sin there would be nothing in the world that could properly be considered an evil (Übel), but that whatever is directly bound up with the transitoriness of human life would be apprehended as at most an unavoidable imperfection, and the operation of natural forces which impede the efforts of men as but incentives to bring these forces more fully under human control Secondly, it is implied that the measure in which sin is present is the measure in which evil (Übel) is present, so that, just as the human race is the proper sphere of sin, and sin the corporate act of the race, so the whole world in its relation to man is the proper sphere of evil (Übel), and evil the corporate suffering of the race.[2]

This situation can be expressed in more traditional language by saying that the evil suffered by mankind is a punishment for sin. This proposition, as Schleiermacher sponsors it, does not mean that God specifically sends suffering upon us, but that God's good world becomes evil to us as the result of our own sinful way of living in it and responding to it. Further, it does not mean that the evil suffered by each individual is proportionate to the degree of his own personal guilt. Evil is brought by the race as a whole upon itself, and the varying extents to which different individuals participate in the resulting evil is accidental.[3] In the case of social evils such as war, poverty, and political injustice, it is evident that the human race is directly responsible for inflicting suffering upon itself. In the case of natural evils the connection with man's sin is indirect, in that 'as man, were he without sin, would not feel what are merely hindrances of sensuous functions as evils, the very fact that he does so feel them is due to

---

[1] See above, pp. 134 f.   [2] *C.F.*, p. 317 (para. 75, 3).
[3] *C.F.*, para. 77.

sin, and hence that type of evil, subjectively considered, is a penalty of sin'.[1]

4. GOD AS ULTIMATELY ORDAINING SIN AND SUFFERING

We have at various points in Part II found it to be an inescapable conclusion that the *ultimate* responsibility for the existence of sinful creatures and of the evils which they both cause and suffer, rests upon God Himself.[2] For monotheistic faith there is no one else to share that final responsibility. The entire situation within which sin and suffering occur exists because God willed and continues to will its existence; and we must believe that from the first He has known the course that His creation would take. To say this is not to deny man's blameworthiness for his own sins, for our individual human responsibilities hold good on a different plane from that of the ultimate divine responsibility, in such a way that the one does not lessen the other.[3] We have seen that the Augustinian theological tradition, although unwilling openly to acknowledge this, nevertheless cannot help doing so implicitly in its doctrine of divine predestination. But the first great Christian theologian to affirm it openly was Schleiermacher.

Negatively, Schleiermacher argues that the distinction between divine causing and divine permitting, behind which Christian thought has traditionally sheltered itself from this aspect of logic of monotheism, is untenable.[4] But, more positively, he links sin with redemption. For 'everywhere human evil exists only as attached to good, and sin only as attached to grace'.[5] And looking at sin through grace, we see that redemption presupposes that from which we are redeemed. Thus, starting from the Christian experience of redemption, which expresses the immediate activity of God, and seeing this as occurring within the universal divine sovereignty, we can only conclude that the total event that culminates in man's redemption represents the outworking of

---

[1] *C.F.*, p. 319 (para. 76, 2).
[2] See pp. 69 f., 75, 119 f., 179 f., 197 above.
[3] For further elaboration of this, see below, p. 327.
[4] See *C.F.*, p. 338 (para. 81, 4).   [5] *C.F.*, p. 327 (para. 80, 2).

God's purpose. Accordingly Schleiermacher is not suggesting that 'sin could have a place in the creative dispensation of God apart from redemption, since in the divine will bearing upon the existence of the whole human race the two are ordained to stand in relationship to each other'.

For [he continues] the mere fact that the emergence of sin preceded the advent of redemption in no sense implies that sin was ordained and willed purely for itself; on the contrary, the very statement that the Redeemer appeared when the fullness of time was come makes it quite clear that from the beginning everything had been set in relation to His appearing. And if we add the fact that the sin which persists outside direct connexion with redemption never ceases to generate more sin, and that redemption often begins to operate only after sin has attained to a certain degree, we need have no misgivings in saying that God is also the Author [der Urheber] of sin—of sin, however, only as related to redemption.[1]

Schleiermacher sees his teaching as simply a systematic acknowledgement of the facts of Christian experience. We are conscious of God acting redemptively; and this activity presupposes the need for redemption. The entire situation — the polarity of sin and salvation — thus exists within the sovereign purpose of God. To deny God's omni-sovereignty, and therefore His omni-responsibility, would be to take the decisive first step towards Manichaean dualism. For 'if sin is in no sense grounded in a divine volition and is nevertheless held to be a real act, we must assume another will so far completely independent of the divine will as to be itself the ultimate ground of all sin as such'.[2]

On the other hand, Schleiermacher insists, again upon the basis of Christian experience, that God's activity in relation to sin is of a different order from His activity in redemption. For sin and grace stand within our experience in opposition to one another. 'The universal co-operation of God is the same in either domain, but in the case of sin there is lacking that specific divine impartation which gives to every approach to salvation the character of grace.'[3]

However, the developmental view of man, which holds that 'the merely gradual and imperfect unfolding of the

---

[1] *C.F.*, p. 328 (para. 80, 2).   [2] *C.F.*, p. 329 (para. 80, 4).
[3] *C.F.*, p. 327 (para. 80, 1).

power of the God-consciousness is one of the necessary conditions of the human stage of existence ',[1] and which entails that man is still in process of being created, with redemption as the climax of this process, leads Schleiermacher to see sin and evil as occurring within the total sphere of the divine purpose.[2] For until this purpose is fulfilled there must in the nature of the case be spiritual defect in the creature, and a corresponding need for redemption, which we know subjectively as sin. 'God has ordained that the earlier insuperable impotence of the God-consciousnes shall become for us, as our own act, the consciousness of sin, in order to intensify that longing without which even the endowment possessed by Jesus would have met with no living faculty for the reception of what He communicates.'[3] Accordingly :

What is here asserted is that in the whole range of life between the inward state of the Redeemer, in whom no break in the supremacy of the God-consciousness could issue from His highest spiritual activity, and those states of human disorder in which the spiritual functions are brought under the power of disease, and responsibility ceases owing to a lack of freedom, free self-development is always attended by sin. Hence if this whole form of existence — the life of the natural man — subsists in virtue of divine appointment [Anordnung], sin, as proceeding from human freedom, has also a place in that appointment.[4]

And yet at the same time it must be insisted, as Schleiermacher himself repeatedly insists, that the human creature who is thus being gradually brought by divine grace towards the full corporate consciousness of God in His Kingdom, is a free individual, bearing a personal moral responsibility and obligation. Indeed,

The very phrase 'freedom of the will' conveys a denial of all external necessity, and indicates the very essence of conscious life — the fact, namely, that no external influence determines our total condition in such a way that the reaction too is determined and given, but every excitation really receives its determinate quality from the inmost core of our own life, from which quality, again, proceeds the reaction, so that the sin proceeding from that core is in every case the act of the sinner and of no other.[5]

[1] *C.F.*, p. 338 (para. 81, 4).   [2] *C.F.*, para. 81, 1.
[3] *C.F.*, p. 366 (para. 89, 1).   [4] *C.F.*, p. 334 (para. 81, 2).
[5] *C.F.*, p. 334 (para. 81, 2).

But in setting over us His call towards higher states of God-consciousness, and thereby constituting our present situation as one of sin and guilt and of the need for salvation, God is ultimately responsible for sin as the precondition or our reception of His redeeming grace. Thus 'sin has been ordained [geordnet] by God, not indeed sin in and of itself, but sin merely in relation to redemption; for otherwise redemption itself could not have been ordained'.[1]

5. SCHLEIERMACHER AND THE INSTRUMENTAL VIEW OF EVIL

In his teaching that sin and evil are ordained by God as the preconditions of redemption Schleiermacher has sponsored the thesis that evil ultimately serves the good purpose of God. This idea is of such central importance that we must pause here to define it further.

We may begin by noting an obvious objection to Schleiermacher's teaching. Schleiermacher, it might be complained, is arguing that since God wills man's redemption, and since redemption presupposes sin, God wills man's sin! But, it may be said, sin and redemption are, on the contrary, related more like a culpable motor accident and the surgical operation by which life is saved and the damage repaired. Surgery is good; but we do not arrange accidents in order that there may be opportunities for surgery! Likewise, from the fact that divine redemption is a great good we must not infer that God has ordained the sin that calls for it. We should instead regard sin as being utterly contrary to God's desire and plan, and see redemption as His way of dealing with the unintended disaster of man's fall.

However, these considerations only serve to emphasize the difference between the premises of the Augustinian doctrine of sin and those from which Schleiermacher's thought proceeds. The surgery analogy presupposes the Augustinian picture of man as created in a state of finite perfection and wantonly falling from this into sin and death and woe. This conception of the economy of salvation has already been

[1] *C.F.*, p. 337 (para. 81, 3).

criticized in Part II, on the grounds that the notion of men (or angels) rebelling against their Maker from a state of 'original righteousness' amounts to postulating the self-creation of evil *ex nihilo*, and also that historically the paradise-fall story is incredible in the light of modern knowledge. Schleiermacher, however, and the Irenaean tradition generally, see man as having been initially created as an imperfect being, but a being who may in interaction with divine grace eventually be brought to the perfection that God desires for him. To use the distinction that Irenaeus and others of the Greek Fathers used, man has been created in the 'image' of God but has yet to be brought into the divine 'likeness' revealed in Christ.[1] Now if man has been so created that his perfection lies before him in the future rather than behind him in the past, his present imperfection belongs to his God-given nature. His imperfection (which issues directly in sin) and his redemption both have their place in the divine plan, and belong together, so that the latter does indeed, as Schleiermacher contended, presuppose the former. For 'the merely gradual and imperfect unfolding of the power of the God-consciousness is one of the necessary conditions of the human stage of existence'.[2] Thus the doctrine that God has ordained sin as the precondition of redemption has ample grounds within Schleiermacher's own Irenaean framework of thought, and is by no means the wanton paradox that it must seem from an Augustinian standpoint.

For Schleiermacher, then, man has been created at a distance from God, both as regards his own imperfect nature and as regards the circumstances of his life, precisely in order that he may freely come to God, drawn by redemptive grace. Accordingly, sin has occurred as a preparation for grace rather than, as the Augustinian tradition teaches, grace occurring to repair the damage of sin. This fundamental reversal on Schleiermacher's part has naturally drawn a vehement repudiation from Augustinian theologians. For if sin ultimately contributes to the fulfilment of God's purpose, they have demanded, does it not cease to be *sin*? According to H. R. Mackintosh, 'what Schleiermacher totally ignores

---

[1] See above, pp. 217 ff.   [2] *C.F.*, p. 338 (para. 81, 4).

is the fact that sin is rebellion against the Divine will'.[1] Is not God's irreconcilable enemy being falsely depicted as His faithful servant? And so Karl Barth says in his discussion of this aspect of Schleiermacher's theology:

> Sin is now understood positively. Without sin grace could not exist. . . . At this point we can only protest. When sin is understood positively, when it is esteemed and justified and established, when it counterbalances grace and is indispensable to it, it is not real sin. For real sin cannot be vindicated in this way. We cannot say of it that it is in any sense necessary to a stage of human existence and therefore willed and posited by God. . . .[2]

We are undoubtedly here at one of the major watersheds of theodicy. On the one hand, we must treat sin as *wrong* and evil as *bad*. We want to see them, as they are seen in the New Testament, as manifestations of a principle that is irreconcilably at enmity with both God and man. But, on the other hand, we must not so magnify this contrary power as to create a final dualism. *Ultimately* God alone is sovereign, and evil can exist only by his permission. This means that God has willed to create a universe in which it is better for Him to permit sin and evil than not to permit them. And this brings us back, however reluctantly, to some kind of instrumental view of evil.

Indeed, although the protests against Schleiermacher's way of relating evil to the divine purpose come from within the broadly Augustinian tradition, we have already seen (in Chapter IV) that within that tradition itself there is a metaphysical form of instrumentalism in the aesthetic view of evil as contributing by inner contrasts to the complex perfection of the universe. It is in fact impossible to repudiate entirely the view of evil as serving a good purpose without falling into the Manichaean heresy of an ultimate cosmic duality. But, on the other hand, it is very difficult, and some would say impossible, to admit any kind of service of evil to the divine will without in effect denying the 'exceeding sinfulness of sin' and the appalling reality of suffering. In Part IV an attempt will be made to follow a way of theodicy that avoids being

[1] *Types of Modern Theology* (London: Nisbet & Co. Ltd., 1937), p. 84.
[2] *C.D.* III/3, p. 333.

overwhelmed by either of these opposite dangers; but at the moment we simply note the form which the problem takes at this crucial point.

6. Man's Beginning and End

This view of sin and evil, unfolding the implications of Irenaeus' starting-point much further than he himself had done, is surrounded in Schleiermacher's *The Christian Faith* with other characteristic features of the Irenaean type of theodicy.

Like Irenaeus, Schleiermacher has a two-stage conception of the creation of man. But instead of the terminology of the 'image' and 'likeness' of God, he uses that of the first and the second Adam.[1] The first Adam possessed the potentiality for the full and perfect God-consciousness that has, however, become actual only in the second Adam, Christ, who is now drawing all men into a community of God-consciousness with himself. Thus Schleiermacher sees man being gradually perfected, and sin as virtually inevitable at the stages of his existence prior to full redemption.

In his eschatology, dealing with the other end of the creative process, Schleiermacher rejects the gratuitous increment to the problem of evil that is proposed by the Augustinian and Calvinist doctrine of double predestination — the predestinating of some to a joyous fellowship with God and others to eternal torment. Schleiermacher chooses to understand the biblical and confessional statements indicating a radical division among mankind — the sheep and the goats, the elect and the reprobate — in the universalist sense that at any given historical moment some have and others have not yet been brought to Christ. 'Only in this limited sense, therefore — that is, only at each point where we can make a comparison between those laid hold of by sanctification and those not yet laid hold of — ought we to say that God omits or passes over some, and that He rejects those He passes over, and hence that election always and only appears with reprobation as its foil.'[2] Indeed, 'no divine fore-ordination can

[1] *C.F.*, para. 89.   [2] *C.F.*, p. 548 (para. 119, 2).

240

be admitted as a result of which the individual would be lost to fellowship with Christ. Thus we may reasonably persist in holding this single divine fore-ordination to blessedness. . . .'[1] And so at the end of a long and complex discussion of the notion of a double destiny (in heaven and hell) he rejects the Augustinian and Calvinist doctrine of eternal damnation and concludes, undogmatically, in favour of the eventual universal efficacy of Christ's redeeming work.[2] He points out that perpetual torment could not coexist with the bliss of heaven. For,

> Even if externally the two realms were quite separate, yet so high a degree of bliss is not as such compatible with entire ignorance of others' misery, the more so if the separation itself is the result purely of a general judgment, at which both sides were present, which means conscious each of the other. Now if we attribute to the blessed a knowledge of the state of the damned, it cannot be a knowledge unmixed with sympathy. If the perfecting of our nature is not to move backwards, sympathy must be such as to embrace the whole human race, and when extended to the damned must of necessity be a disturbing element in bliss, all the more that, unlike similar feelings in this life, it is untouched by hope.[3]

And Schleiermacher concludes that,

> From whichever side we view it, then, there are great difficulties in thinking that the finite issue of redemption is such that some thereby obtain highest bliss, while others (on the ordinary view, indeed, the majority of the human race) are lost in irrevocable misery. We ought not to retain such an idea without decisive testimony to the fact that it was to this that Christ Himself looked forward; and such testimony is wholly lacking. Hence we ought at least to admit the equal rights of the milder view, of which likewise there are traces in Scripture [e.g. I Corinthians xv. 26, 55]; the view, namely, that through the power of redemption there will one day be a universal restoration of all souls.[4]

[1] *C.F.*, pp. 548–9 (para. 119, 2).
[2] *C.F.*, para. 120. Cf. para. 162; appendix to para. 163; and para. 169.
[3] *C.F.*, p. 721 (para. 163, appendix).
[4] *C.F.*, p. 722 (para. 163, appendix).

CHAPTER XI

RECENT TELEOLOGICAL THEODICIES

1. INTRODUCTORY

DURING the closing years of the nineteenth century, and during the twentieth century as it has thus far elapsed, there has been an abundant stream of thought in the Irenaean tradition. In this most recent period the problem of evil has held a prominent place in the attention of the more philosophically inclined Protestant theologians of the English-speaking world, especially in Britain. A recognizable pattern of ideas recurs in their writings, and themes such as the following are common to a large number of authors : a teleological approach to theodicy which looks forward to an eventual all-justifying bringing of good out of evil ; the two-stage conception of God's creation of man (generally without explicit reference to Irenaeus[1]) ; the free-will defence of God in respect of the existence of sin ; a view of nature, with all its hard and painful aspects, as an environment suited to the development in man of moral personality ; and a questioning of the notion of eternal hell in favour of belief in an ultimate universal salvation.

Within the literature in which these ideas constitute a common currency there are of course many differences of emphasis ; and there is also a development of thought which has manifestly been influenced by the historical events of the time. When we go back to the earlier works of this category we see very clearly the influence of the dominant and pervasive idea of the nineteenth century, the concept of evolution. We also feel, in many of the British works of the later years of the reign of Queen Victoria and the expansive Edwardian period, an air of satisfaction and optimism which contrasts sharply with the more disturbed and apocalyptic outlook of

---

[1] J. K. Mozley, in *Essays Catholic and Critical*, ed. by E. G. Selwyn (London : S.P.C.K., 1926), p. 240 n., is an exception.

so many of the writings produced during and after the First World War. There is a difference in this respect between nineteenth-century theodicies (regarding the nineteenth century as a cultural epoch that ended in 1914) and their twentieth-century successors. For in the lurid light of the holocaust of 1914–18, in which more than thirty-seven million human beings were slaughtered or maimed, the moral self-confidence of the West was profoundly shaken. Thereafter a new note of religious criticism was heard, on the Continent above all from Karl Barth, and in the United States from Reinhold Niebuhr. In Britain, too, a more realistic correction of evolutionary and idealist thought became general, and has issued in a significant reconstruction of teleological theodicy.

2. EVOLUTIONARY INFLUENCES FROM THE NINETEENTH CENTURY, AND THE SHOCK OF THE FIRST WORLD WAR

The leading theme of nineteenth-century, and especially later nineteenth-century, thought was the conception of an evolution or development taking place in all life. Although the Darwinian theory of the gradual evolution of the forms of life by means of natural selection was long and fiercely resisted by many churchmen as incompatible with the biblical account of creation, the developmental idea had by the close of the century become widely influential within Christian thought. It is reflected in the theodicies of that period, in the conception of the created order, centring upon man, as moving towards a divinely appointed end. Sin is seen as belonging to a stage of man's slow growth from primitive beginnings to the future earthly kingdom of God; and suffering is seen as a disciplinary and pedagogic experience required by the developing race.

The Gifford Lectures of the period, delivered annually since 1885 in one or other of the Scottish universities, provide a mirror in which to observe major trends in the more philosophical type of theology. Alexander Campbell Fraser of Edinburgh[1] gave the Gifford Lectures in 1894–6 under the

---

[1] As the first holder of the Campbell–Fraser Scholarship, instituted in 1948 to take an Edinburgh University graduate to Oriel College, Oxford, I gratefully salute his memory.

title *Philosophy of Theism*. He presented a 'conception of man as at present in physical progress towards a happy millennium',[1] and argued that 'Physical and intellectual evil — pain as well as ignorance and error — seem to be means of advancement towards the imperfectly comprehensible end to which the universe is moving'.[2] As regards sin, he wrote :

That mankind should be only in progress, not actually perfect from the first, may be implied in the idea of moral personality, A world of persons could not be a world of always perfect persons. Providential progress, not perfection from the beginning, appears as the condition suited to moral agents, distinguished from natural things.[3]

A. B. Bruce, in his own Gifford Lectures, delivered in 1897–8, expressed the general developmental conception in its application to religion :

Jesus [he said] taught a providence that works and achieves its ends through the processes of nature, and that reaches the accomplishment of its purpose *gradually*, not *per saltum*. In His conception of Divine Providence Jesus gave no undue prominence to the *unusual* and the *catastrophic*. His watchwords were : Nature God's instrument ; and, Growth the law of the moral as of the physical world.[4]

The spacious and optimistic spirit of upper-class life in Victorian England pervades Robert Flint's Baird Lectures for 1876, in which he says :

Due weight ought also to be given to the circumstance that the system of God's moral government of our race is only in course of development. We can see but a small part of it, for the rest is as yet unevolved. History is not a whole, but the initial or preliminary portion of a process which may be of vast duration, and the sequel of which may be far grander than the past has been. That portion of the process which has been already accomplished, small though it be, indicates the direction which is being taken ; it is, on the whole, a progressive movement ; a movement bearing humanity towards truth, freedom, and justice. Is it scientific,

[1] Alexander Campbell Fraser, *Philosophy of Theism* (Edinburgh : William Blackwood & Sons, 2nd ed., 1899), p. 283.
[2] Ibid., p. 286.   [3] Ibid.
[4] A. B. Bruce, *The Moral Order of the World* (London : Hodder & Stoughton, 1899), p. 409.

or in any wise reasonable, to believe that the process will not advance to its legitimate goal? Surely not.[1]

But the greatest popularizer of evolutionary Christianity was Henry Drummond, whose Lowell Lectures on *The Ascent of Man* exulted in the cosmic development which is carrying us ever upwards:

Evolution is . . . a light revealing in the chaos of the past a perfect and growing order, giving meaning even to the confusions of the present, discovering through all the deviousness around us the paths of progress, and flashing its rays upon a coming goal. Men begin to see an undeviating ethical purpose in this material world, a tide, that from eternity has never turned, making for perfectness.[2]

All this connects, as we have seen, with an ancient strand of Christian thought, stressing the idea that the divine creativity is still at work in relation to man and drawing him towards a perfection not yet realized.[3] But as it was coloured by Darwinian evolutionary theory in the expansive atmosphere of England during the proud imperial years of Queen Victoria, this strand of thought passed through a phase of easy optimism which was to be rudely shattered in 1914 by the most fearful outbreak of hatred and violence that the world had ever known.[4] This dire event was to nineteenth-century evolutionary optimism what the Lisbon earthquake had been to the eighteenth-century doctrine of the best possible world. Otto Piper, a theologian who experienced the moral and spiritual, as well as physical, shock of the First World War in the German trenches, has graphically described the transformation which it wrought in the thinking of his own generation:

Before the War we believed that the world in which we live was the result of a creative intelligent force. So the world appeared rational, and man seemed able to change and transform

[1] Robert Flint, *Theism* (Edinburgh: William Blackwood & Sons, 1877, 10th ed., 1902), pp. 258–9.
[2] Henry Drummond, *The Ascent of Man* (London: Hodder & Stoughton, 1894), pp. 436–7.   [3] See above, pp. 217 f.
[4] Other works of the pre-war period expressing an evolutionary type of theodicy include James Martineau, *A Study of Religion* (Oxford: Clarendon Press, 2nd ed., 1889), bk. ii, chap. 3; J. Iverach, *Evolution and Christianity* (London: Hodder & Stoughton, 1894); A. S. Pringle-Pattison, *The Idea of God in the Light of Recent Philosophy*, Gifford Lectures for 1912–13 (Oxford: Clarendon Press, 1917).

it according to his rational purposes. This interpretation was a result of an increasing intellectualism which had lost touch with the concrete world. The War gave us quite a new feeling of what reality is. . . . In the war the soldiers discovered that there is no natural harmony of ends in this world. Teleology and dysteleology are linked with one another. What a wonderful achievement of the human mind are modern weapons and battleships! But at the same time, do they not show that technical discoveries are turning against their inventors? They had their origin in man's wish to dominate Nature, yet man is not able to control them, and finally they became a means of his destruction.

Another discovery was that history develops to a large extent independently of personal will, and according to its own laws and inherent forces. The longer the War lasted, the more urgently were the soldiers oppressed by such problems as : How was it possible that such a dreadful war could happen in a world of order and highly-developed civilisation, and that there was no hope of finishing it? Since the soldiers in the conflicting armies had not wanted the War, and had no personal grudge against each other, why were they driven to fight each other? What was this unknown and illusive power of war? They all felt their lives dominated by a frightful necessity. . . . In pre-war times man felt secure, because he had infinite confidence in the powers of his mind. During the War life was in constant danger. This special experience became, by its striking nature, a decisive knowledge for the whole of life. For the War showed that the circumstances which caused the danger were accidental, and that in every situation life is dangerous. Because of the unknowable and destructive nature of the world, life is a struggle against an invisible but powerful enemy, who, in spite of all our courage and ability, finally overcomes us in death. . . .

The deepest reason for the mysterious character of the world is seen in the existence of evil in it. The world in which we live not only resists our attempts at rational transformation, there is even evidence that it has in itself a tendency to hinder and counteract the moral good. A happier time believed that this world, though imperfect in some parts, was good. But if, indeed, there has been a perfect stage, it no longer exists. . . . In this world there are tendencies which work against its order and harmony, and for its destruction.[1]

The unanswerable refutation of comfortable theory by grim events was expressed above all by P. T. Forsyth in *The*

[1] Otto Piper, *Recent Developments in German Protestantism* (London : S.C.M. Press Ltd., 1934), pp. 40–47.

Justification of God: Lectures for War-Time on a Christian Theodicy, published in 1917. Forsyth pointed out that the war, so far from being a cruel blow directed against mankind from some malign external power, was an irruption out of the depths of human nature itself and a valid conclusion to the logic of man's own history. 'War', he said, 'makes at least one contribution to human salvation — it is sin's apocalypse.'[1] The war came as such a shocking blow to faith in God only because men had forgotten both the evil in their own hearts and the reality of God as the ultimate of righteousness and truth as well as of tenderness and love:

> The whole habit of leisurely apologetic has had in view an evil too remote, passionless, and unrealised, and a God who, if He is not kindness to man, was no God for him. . . . Truly we cannot exaggerate the love of God, if we will take pains to first understand it. But we have been taught to believe only in a beneficent and not in a sovereign God, in a tender God in no sense judge, in an attractive God, more kindly than holy, more lovely than good. . . . Such a habit of mind, now that the lid is off hell, is suddenly struck from its perch, feels taken in, and asks if such a world as we see can be the means to a loving end, if it could ever be made to contribute to a Divine Kingdom.[2]

In the light of history, a history which had then culminated in the most bloody and destructive conflict of all time, Forsyth was constrained to reject outright the picture of mankind as advancing on an upward way towards ever higher levels of civilization and of moral and spiritual attainment: 'shall we look for a plan of beneficent progress looming up through man's career? History shows no such plan, especially in the moral region where we need it most.'[3] And Forsyth addressed his contemporaries, distressed by a great war's revelation of demonic evil within and without, with the eloquence and spiritual vision of a prophet:

> What is it that would justify God to you? You have grown up in an age that has not yet got over the delight of having discovered in evolution the key to creation. You saw the long expanding series broadening to the perfect day. You saw it foreshortened in

[1] P. T. Forsyth, *The Justification of God* (London : Latimer House Ltd., 1948), p. 19.   [2] *J.G.*, pp. 35-36.   [3] *J.G.*, p. 17.

the long perspective, peak rising on peak, each successively catching the ascending sun. The dark valley, antres vast, and deserts horrible, you did not see. They were crumpled in the tract of time, and folded away from sight. The roaring rivers and thunders, the convulsions and voices, the awful conflicts latent in nature's ascent and man's — you could pass these over in the sweep of your glance. . . . But now you have been flung into one of these awful valleys. You taste what it has cost, thousands of times over, to pass from range to range of those illuminated heights. You are in bloody, monstrous, and deadly dark. . . . Every aesthetic view of the world is blotted out by human wickedness and suffering. The air is as red as the rains of hell. The rocks you stood on fall on you . . .[1]

There is no upward curve linking the centuries and the belief in such a curve was only the brief illusion of a favoured class in a sheltered age. But Christian faith in its more healthy and realistic moments has sprung from a source quite different from the expansive optimism of a special historical epoch. 'Our faith did not arise from the order of the world; the world's convulsions, therefore, need not destroy it. Rather it rose from the sharpest crisis, the greatest war, the deadliest death, and the deepest grave the world ever knew — in Christ's Cross.'[2] This is the note that was altogether absent from the evolutionary theodicies of the nineteenth century. Forsyth's thought, like that of Karl Barth a little later in the new century, was Christocentric. He saw sin as that which Christ has combated and overcome, as the abyss from which the Redeemer has rescued mankind. This act constitutes God's great work of theodicy, for His justification of man by His saving participation in human evil is at the same time His justification of Himself in respect of that evil; 'the only possible theodicy is an adequate atonement'.[3] Again, 'The final theodicy is in no discovered system, no revealed plan, but in an effected redemption. It is not in the grasp of ideas, nor in the adjustment of events, but in the destruction of guilt and the taking away of the sin of the world.'[4] There is thus for Christian faith as it exists in actual Christian living a solution to the problem of evil; but it is a practical rather than a theoretical solution. 'It is not really

[1] *J.G.*, p 159. [2] *J.G.*, p. 57. [3] *J.G.*, p. 167. [4] *J.G.*, p. 53.

an answer to a riddle but a victory in a battle. . . . We do not
see the answer; we trust the Answerer, and measure by
Him. We do not gain the victory; we are united with the
Victor.'[1]

So Forsyth brought teleological theodicy back to reality,
both divine and human; and, as we shall see, some who have
continued the Irenaean tradition since Forsyth have learned
the lesson that he himself learned and taught, whilst others
have still largely ignored or been comparatively unaffected
by it.

As we turn to these more recent thinkers it may be well to
note again the broad Irenaean framework of thought which
is common to Forsyth and so many other twentieth-century
British theodicists, and within which Forsyth's sharp correc-
tive was set. Forsyth thought in terms of God's continuing
creation of man and of this process as having its climax in the
redemptive work of Christ. Seen from this Christological
centre, 'the economy of salvation becomes the principle of
the movement of the universe. Nature is but a draft scheme
of salvation with the key on another sheet, where the eternal
act of redemption is found to carry and crown the long
process of creation.'[2] And the justification of the process lies
in its end, already accomplished in Christ though not yet
fully manifested in the world. For 'by redemption what do
we mean? We mean that the last things shall crown the first
things, and that the end will justify the means, and the goal
glorify a Holy God.'[3] That final and all-justifying end lies
beyond history. 'The lines of life's moral movement and of
thought's *nisus* converge in a point beyond life and history.
This world is only complete in another; it is part and parcel
of another, and runs up into it, and comes home in it as body
does in soul.'[4] But even amid the clashes and tragedies and
failures and frustrations of this world it is still true that 'All
life has God and His vast providence and purpose in it.'[5]
The glorious end was already in view from the beginning:
'Man is born to be redeemed. The final key to the first
creation is the second; and the first was done with the second

[1] *J.G.*, pp. 211 and 220–1.   [2] *J.G.*, p. 48.   [3] *J.G.*, p. 69.
[4] *J.G.*, p. 212.   [5] *J.G.*, p. 212.

in view.'[1]   And so Forsyth's contribution to the development of this ancient Irenaean theme is to recall it to its original awareness that God's end for man is not to be the outcome of a natural evolution, but of a supernatural redemption, and that man's final perfection is already a reality in Christ.   For 'We possess, in a living and present Christ, God's goal and destiny of the soul and of the world'.[2]

3. Unreconstructed Teleological Theodicy: F. R. Tennant

F. R. Tennant (1866–1957) was perhaps the last of the natural theologians to defend theism as an explanatory hypothesis.   Today this seems to many to be a fundamentally misconceived programme, because it substitutes philosophical inference for religious experience.[3]   However, so far as the theodicy-problem is concerned Tennant's programme did not hinder him from seeing very clearly the value of the Irenaean approach, of which he must accordingly rank as a leading twentieth-century exponent.   At the same time it is consonant with his rationalist approach that he shows comparatively little sense of the depth of human sinfulness or of the crushing character of human suffering.   On the one hand, then, Tennant's treatment of the problem of evil is an important example of the Irenaean type of theodicy; but, on the other hand, he offers an essentially nineteenth-century version of it, not transformed in the light of either the revelation of the terrifying power of evil in the twentieth century or of a matching faith in the conclusive act of God in Christ in the first century.[4]

---

[1] *J.G.*, pp. 123–4. Cf.: 'The world was made for grace, made in the first creation by One who had in reserve all the resources of the second. Man was made at first to be redeemed at last' (pp. 155–6).   [2] *J.G.*, p. 48.

[3] To quote one of Tennant's statements of his basic position: Theism is 'the only sufficient interpretation of the intelligibility, orderliness, and progressiveness of the world as a whole'. (*Philosophical Theology*, vol. ii, Cambridge University Press, 1930, p. 206.)

[4] Other 'unreconstructed' teleological theodicies include: D. S. Cairns, *The Reasonableness of Christianity* (London: Hodder and Stoughton, 1918), chap. 5 (but see p. 261 below for later writings by the same author which come in the 'reconstructed' category); B. H. Streeter, *Reality* (London: Macmillan & Co., 1927), chap. 8; W. R. Matthews, *The Purpose of God* (London: Nisbet

Tennant begins from the Irenaean doctrine that man is still in process of being created as a free moral being. He rejects the Augustinian belief that man was made finitely perfect and then incomprehensibly rebelled against God and against his own paradisal condition.[1] On the contrary, 'moral goodness cannot be created as such; it cannot be implanted in any moral agent by an "almighty" Other. It is the outcome of freedom, and has to be acquired or achieved by creatures. We cannot imagine a living world, in which truly ethical values are to be actualized, save as an evolutionary cosmos in which free agents live and learn, make choices and build characters.'[2] And given that in God's providence the function of our world is to nurture moral personality, the theist can claim that this is the best possible world for that purpose.

What the theist means by 'best', in this connexion, is best in respect of moral worth, or of instrumentality thereto. But those who have allowed themselves to 'charge God foolishly' have substituted for this meaning that of happiness, or sensuously pleasantest. Certainly our world is not, in this sense, the best that we can imagine. Equally certainly, the theist maintains, it was not meant to be. If it were, it would not be truly the best; for we cannot go behind our judgement, rational or non-rational, that the highest value in the hierarchy of values is moral worth, or — what is the ultimate essence of all morality — personal love.[3]

In a world in which a process of ethical creation is taking place, moral evil or sin is an unavoidable possibility. God must allow His children to choose wrongly as well as to choose rightly, for 'Without freedom to choose the evil, or the lower good, a man might be a well-behaved puppet or a sentient automaton, but not a moral agent'.[4] Indeed, 'To preclude moral evil would be to preclude moral goodness, to do evil, to prefer a worse to a better world'.[5]

Tennant regarded these considerations as constituting in principle a theodicy in respect of moral evil. But he was also

& Co. Ltd., 1935), chap. 5; D. E. Trueblood, *Philosophy of Religion* (London: Rockliff, 1957), chap. 17.
  [1] Cf. Tennant's *The Origin and Propagation of Sin* (Cambridge: The University Press, 1902), pp. 10 f.   [2] F. R. Tennant, *P.T.*, vol. ii, p. 185.
  [3] *P.T.* ii, p. 186.   [4] *P.T.* ii, p. 188.   [5] *P.T.* ii, p. 191.

sympathetic to the hope that 'moral evil is not ultimately insuperable, that it will eventually give place to the good, or at least that it is not destined to become supreme over the good so that the world, though a moral order in one sense, would mock our highest moral aspirations'.[1] And he offers a general probability argument to support this hope :

> Moral advance, in spite of relapses, has undoubtedly marked mankind's history, on the whole, hitherto ; and though it cannot be argued from that fact that progress will continue throughout future ages, neither are we at liberty to regard humanity's past progress as a mere accident, or as a state of things which is likely to be permanently reversed. . . .
> [Hence] it is no flimsy and sentimental optimism, but a reasoned and reasonable expectation, that, as history establishes the fact of moral progress up to date, that progress is not an accident, but will maintain itself.[2]

Nevertheless it is not entailed by Tennant's theodicy that 'moral evil is destined, here or hereafter, to become extinct through the response of free will to fuller light : conceivably, evil may continue everlastingly while unable to become universal and supreme'.[3] According to him, however, this question is not vital to theodicy ; indeed, he goes so far as to say that 'the question of the continuance or the self-extinction of moral evil is irrelevant to theodicy'.[4] Yet surely the exclusion of such an all-justifying end to God's creative process gravely weakens Tennant's position. The basic claim of a teleological theodicy is that good will finally be brought out of evil ; and therefore, to the extent that such final good is qualified or restricted, the theodicy fails. Is it not self-evident that a world in which moral evil is eventually converted to good is better than one in which moral evil continues as a permanent sinful rejection of the good? But in that case it must be critical for the idea of 'the "best possible" evolutionary world'[5] whether sinners are finally redeemed or not.

At this point, then, we must charge Tennant with a lack of thoroughness and consistency in working out the basic

[1] *P.T.* ii, p. 193.   [2] *P.T.* ii, pp. 193 and 195.
[3] *P.T.* ii, p. 195.   [4] *P.T.* ii, p. 197.   [5] Ibid.

principle of teleological theodicy. We must also note that although Tennant rests much of his case upon the traditional 'free-will defence', the kind of challenge to it that has recently been formulated by such secular philosophers as J. L. Mackie and Antony Flew did not occur to him and accordingly finds no answer in his pages.[1]

Tennant's discussion of the problem of natural evil — pain, in distinction from sin — has virtues and defects similar to those of his treatment of moral evil. That is to say, it is a classic statement of certain Irenaean themes as these appear in modern teleological thought; yet Tennant feels the problem in a relatively untroubled way, and is therefore not driven, as was P. T. Forsyth, to take seriously the distinctively Christian revelation of God's dramatic sharing of human suffering and His bearing of human sin in Jesus Christ. Tennant offers a teleological theodicy that has not allowed itself to be threatened by the agony of the twentieth century and is accordingly not transformed by those deeper resources of the Christian faith which that agony has again thrown into vivid relief.

Tennant's programme is to show that 'pain is either a necessary by-product of an order of things requisite for the emergence of the higher goods, or an essential instrument to organic evolution, or both'.[2] The first of these possibilities is to be shown to apply to human suffering and the second to animal pain.

His basic contention, so far as human suffering is concerned, is that 'a world which is to be a moral order must be a physical order characterised by law or regularity'.[3]

Without such regularity in physical phenomena there could be no probability to guide us : no prediction, no prudence, no accumulation of ordered experience, no pursuit of premeditated ends, no formation of habit, no possibility of character or of culture. . . . And without rationality, morality is impossible : so, if the moral status of man be the goal of the evolutionary process, the reign of law is a *sine qua non*. It is a condition of the forthcomingness of the highest good, in spite of the fact that it is not an unmixed good but a source of suffering.[4]

[1] See below, pp. 302 f.   [2] *P.T.* ii, p. 198.
[3] *P.T.* ii, p. 199.   [4] *P.T.* ii, p. 199–200.

Thus 'the disadvantages which accrue from the determinateness and regularity of the physical world cannot be regarded either as absolute or as superfluous evils. . . . They are not good, when good is hedonically defined; but they are good for good, when good is otherwise defined, rather than good for nothing.'[1]

And so Tennant concludes that physical evil is inevitable in a world governed by general laws and inhabited by embodied creatures. Since it is only in such a world that moral personality can develop at all, the Creator's goodness is vindicated. Further, we must not think that God specifically dispenses to each individual the troubles that he meets in the course of his life. 'They are rather inevitable, if incidental, accompaniments or by-products of the world-order which, as a whole, and by means of its uniformity, is a pre-requisite of the actualization of the highest good that we can conceive a world as embodying.'[2]

Finally Tennant, without positively asserting a life after death, points out that theism renders this probable, since 'it would not be a perfectly reasonable world which produced free beings, with Godward aspirations and illimitable ideals, only to cut them off in everlasting death, mocking their hopes and frustrating their purposes'.[3] And a continuation of life's pilgrimage beyond the grave to a final homecoming in the divine Kingdom would transform the quality of our experience of evil, because this would then be seen as not only necessary to our moral growth but also as ' "but for a moment" in the time-span of just men made perfect'.[4]

However, Tennant's treatment of immortality is extremely tentative. It indicates a likelihood precariously poised upon a probability. He uses such locutions as, 'it is a question whether theism . . . can stop short of adding the doctrine of a future life to its fundamental articles of belief . . .',[5] and concludes with the purely historical remark, 'Hence theists generally regard the Supreme Being as a God, not of the dead, but of the living'.[6] However, in an Appendix — and its location in an Appendix indicates its peripheral status in Tennant's argument — he concludes

[1] *P.T.* ii, pp. 201–2.   [2] *P.T.* ii, p. 204.   [3] *P.T.* ii, p. 205.
[4] Ibid.   [5] Ibid.   [6] Ibid.

more positively that the world 'cannot safely be regarded as realising a *divine* purpose unless man's life continues after death'.[1] Nevertheless he does not see that what is for him a tentative and secondary addendum to his position is absolutely essential to the theodicy that he has offered. For (as Tennant was well aware)[2] this world is in fact only very partially a scene of successful soul-making. Although there are many striking instances of good being brought triumphantly out of evil through a man's or a woman's reaction to it, there are many other cases in which the opposite has happened. Sometimes obstacles breed strength of character, dangers evoke courage and unselfishness, and calamities produce patience and moral steadfastness. But sometimes they lead instead to resentment, fear, grasping selfishness, and disintegration of character. Therefore, it would seem that any divine purpose of soul-making which is at work in earthly history must continue beyond this life if it is ever to achieve more than a very partial and fragmentary success. And only if there is such a continuation beyond death, and within it either a completion or at least an ever-increasing approach to a completion of the soul-making process, can our human situation be reconciled with the existence of an all-powerful and infinitely good and loving Creator. In laying such little stress upon eschatology, and also (as we have seen) in allowing the possibly permanent existence of moral evil, Tennant fails to present the teleological type of theodicy at its strongest. And in omitting altogether the significance of the life and death of Christ in relation to the problem of evil he still remains, with the nineteenth-century evolutionary thinkers, at a distance from a fully and distinctively *Christian* theodicy.[3]

4. RECONSTRUCTED TELEOLOGICAL THEODICY

By the 'reconstruction' of teleological theodicy I mean its coming to take seriously, and to see as crucially relevant to

[1] *P.T.* ii, p. 272.   [2] See *P.T.* ii, p. 203.
[3] For a sympathetic and constructive use of Tennant's discussion of the problem of evil, improving upon him at the points criticized above, see Ninian Smart, *Philosophers and Religious Truth* (London: S.C.M. Press Ltd., 1964), chap. 6.

the problem of evil, the central fact of the Christian religion
— the life, death, and resurrection of Jesus, responded to in
human faith as God's decisive act of redemption. In traditional label-language the reconstruction consists in an advance of theodicy beyond natural to revealed theology, and
especially to Christology. This involves taking evil far more
seriously than it was generally taken in the idealist and
evolutionary apologetics of the late nineteenth century; for
it involves seeing sin and suffering as they are depicted in
the New Testament, as deadly enemies of God and man. It
also involves seeing God Himself as intimately involved and
self-implicated in the reality of evil : personally combating
human suffering in Christ's healing acts, personally accepting and absorbing the onslaught of moral evil in his death,
and personally proclaiming in his resurrection the coming
triumphant completion of His creative work, despite the
powers of sin and death.

This reconstructive task has been carried forward by a
number of recent writers operating within the broadly
Irenaean framework. Their reconstructive efforts are, however, for the most part not very extensively developed. They
hint and point, some more explicitly than others, but
they do not present a fully articulated theodicy such as
might result from a union of the data of Christian faith
received from the New Testament with the Irenaean interpretative framework. Nevertheless, various aspects of such
a theodicy are suggested by a number of contemporary
thinkers, and I shall assemble points of value from three
such, namely Leonard Hodgson, J. S. Whale, and H. H.
Farmer.

Leonard Hodgson (1889–) must be counted among those
who have tried to go beyond the considerations of natural
theology, as formulated, for example, by F. R. Tennant,
to reconsider the problem of evil in the light of distinctively Christian faith. This intention appears in the structure of his treatment of the topic in his Gifford Lectures. In
the first series he devotes a chapter to the problem of evil,
dealing with it solely from the point of view of natural
theology, and then in the second series he again devotes a

chapter to it, now enlarging the scope of his thought to include the data of Christian revelation.

Hodgson's earlier and deliberately restricted discussion presents the standard 'unreconstructed' teleological theodicy based on the principle that the universe is 'the creation of a Creator whose aim is the production of a community of finite persons characterized by the goodness which is the expression of perfect freedom'.[1] Evil in its various forms is then presented as incidental to the free growth of moral personality. As regards moral evil,

if we can believe that the aim of creation is the production of persons whose goodness is the perfection of freedom, it makes sense to regard their permission to be wicked in the course of their making as incidental to the achievement of that purpose. And if at times we are appalled by the depths to which, in the history of this world, wickedness has been allowed to descend, and the extent to which it has been allowed to prevail, I can see no light in the darkness except by taking these depths and this extent as the measure of the value set by God upon the created freedom being genuinely free.[2]

And human suffering, other than that which is due to human sin, is seen as incidental to a stable world-order such as can alone sustain the development of moral personality.[3]

Throughout this discussion there is, deliberately, no attempt to see the problem of evil in the light of Christ — no attempt to find its practical solution in His redeeming work, and its theoretical solution in reflection upon the wider significance of that work. However, in the corresponding chapter of the second series of lectures, in which he surveys the same ground again from an explicitly theological standpoint, Hodgson makes a beginning at least to the re-thinking of the problem of evil from a Christological point of view.

[1] Leonard Hodgson, *For Faith and Freedom* (Oxford : Basil Blackwell, 1956–7), i, 192. Cf. *Towards a Christian Philosophy* (London : Nisbet & Co. Ltd., 1942), p. 88.   [2] *F.F.F.*, i, 204.

[3] In an attempt to bring animal pain also within the scope of this explanation Hodgson suggests a further and more speculative epicycle to his theory, according to which fallen spiritual beings with some degree of control over our world have corrupted its development prior to the appearance of man. *F.F.F.*, i. 213–14. On this theory, see below, pp. 367 f.

He insists that 'evil, and other irrationalities, are obstacles to thought which cannot be removed by thinking, which have to be changed in deed in order to become transparent to thought'.[1] Hence the irrational fact of evil is at least partly transparent to Christian thought; for it is already in principle abolished, and is in process of being abolished in fact, by God's redeeming act in Christ.[2] Here, then, is an indication of the Christological transformation of the older teleological idea. Good is to come out of the world's evil and is to come in such rich and enduring abundance as to justify the whole creative process. But good is not evolving naturally out of sin and suffering; it has been violently wrung from it by the hard and bitter pain of Christ's cross and is now being brought to fruition through the continued redemptive sufferings of Christ's disciples. All this is pointed at from a distance, however, rather than being fully developed in Hodgson's pages.

A further important hint of Hodgson's, again only sketched in brief, is that the traditional Augustinian assumption that theodicy requires that God must somehow be relieved of responsibility for the existence of evil, is a fundamental misconception. For 'in His actions God has revealed Himself as claiming this responsibility'.[3] It is indeed 'the gist of the biblical revelation, that it tells of God claiming responsibility for having allowed the existence of evil and taking upon Himself to overcome it'.[4] Presumably Hodgson's meaning is that in undertaking in Christ the work of man's salvation God has shouldered responsibility for the total situation of the existence and fall of frail mortals, and of their consequent need for divine redemption. On this view Christ's atoning act endorses the thought, expressed above whilst criticizing the Augustinian theodicy-tradtion, that the *ultimate* responsibility for the existence of sinful creatures can rest nowhere else than with the eternal Being who has created them.[5] In Christ, God accepts this responsibility, and as He accepts it we see that His world is not after all out of control but that the process of creating 'children of

[1] *F.F.F.* ii. 57. [2] *F.F.F.* ii. 56. [3] *F.F.F.* ii. 57.
[4] *F.F.F.* ii. 66. [5] See above, pp. 69 f., 75, 119 f., 179 f., 197.

God' is far more costly to God and man, but also far more precious in its outcome, than natural reason would have dared to guess.

Another writer who begins with a rational theodicy of the teleological kind but who uses the resources of distinctively Christian faith to deepen it, is J. S. Whale (1896–), in his widely circulated booklet *The Christian Answer to the Problem of Evil.*[1] What he there offers as 'the answer of theism' is the type of theodicy formulated by F. R. Tennant. But he goes on to present also 'the Christian answer', centring upon the death and resurrection of Christ. Speaking of that death he says, 'Here is a crime if there ever was one; no more flagrant example of injustice can be adduced than this; and if divine providence seems flatly contradicted anywhere in human history, it is here. At the Cross the whole human problem of suffering and sin comes to a burning focus.'[2] So far as the enigma of man's pain is concerned, the light that comes from the cross is not an intellectual illumination but the startling discovery that God Himself bears our pain with us. 'The pains He had to bear may not give you and me a theory about pain, but they help us to bear pain. Men and women have not been mistaken in their conviction that He who triumphed through pain is with them in all their darkness and suffering.'[3] And likewise the answer of the cross to the problem of sin is that in his suffering, caused by man and yet endured on man's behalf, Christ has initiated a healing process that will ultimately overcome all sin. 'At the Cross we see God using our sin as the instrument of our redemption; His best is given in terms of our worst.'[4] And that paradoxical divine act has in principle and in prolepsis transformed and redeemed the entire human situation. 'We know well enough that we still sin, suffer and die; the fragmentariness and pain of life are not taken away; we do not yet see these things put under our feet. But we see Jesus crowned, with a victory in which we already share and which Death itself cannot touch.'[5]

[1] London : S.C.M. Press Ltd., 1936, 4th ed., 1957.
[2] Whale, op. cit., p. 39.
[3] Ibid., p. 43.   [4] Ibid., p. 46.   [5] Ibid., p. 49.

Two further points may be added from another contemporary theologian whose thought concerning the problem of evil is likewise teleological and eschatological. Among the products of recent British theology the writings of H. H. Farmer (1892–) are perhaps the most faithfully and consistently close to the religious realities with which all theology seeks to deal. Characteristically, he insists that the all-important, though not the sole, source of Christian faith in the good purpose of God which securely grasps our human lives is 'the experience of being reconciled to God through Christ in the personal life'.[1] Like J. S. Whale, Farmer holds that a Christian theodicy can arise only in thought which is focused upon the cross of Christ. For,

It seems clear that there is only one way in which faith in the overshadowing wisdom and love of God can be truly succoured and that is for it to be able to grasp its object, or be grasped by it, out of the heart of those historical happenings which otherwise give it the lie. ... In the Christian experience such a revelation is given in the life of Jesus, and supremely in the culmination of it in His Cross.[2]

Farmer goes on to consider the implications of this for the Christian life. Speaking of the committed Christian, he says that,

in respect of such suffering and deprivation as may visit his own life, he finds ... that if he accepts it, not as a meaningless stroke of fate, but as an opportunity to share in the vast fellowship of human pain and to make some contribution to its redemption through patience and self-forgetfulness and love, so the victory over it is achieved; it ceases to be sterile and becomes a sacrament of higher things both to himself and to others.[3]

The other important insight that we may gain from Farmer for our constructive task concerns the positive significance of unresolved mystery in the Christian life.

Nothing more plainly marks the difference between the religious and the philosophic approaches to the problem of evil than this. To the philosophic mind the evils of life, in so far as they

[1] *The World and God* (London: Nisbet & Co. Ltd., 2nd ed., 1936), p. 231.
[2] Ibid., p. 243.
[3] Ibid., p. 246. Cf. Emil Brunner, *The Christian Doctrine of Creation and Redemption*, trans. by Olive Wyon (London: Lutterworth Press, 1952), pp. 182 ff.

remain unexplained, represent so many gaps, irreducible dark spots, over which perhaps at best a flimsy bridge of speculative possibility may be built. But to the religious mind, so far from being mere gaps, they become sacramental of deeper trust in, and therefore deeper knowledge of, God. . . . [The] darkness and perplexities of life are to the reconciled life a continuous and indispensable opportunity for that attitude of trust in God which is both the source and the consummation of a truly personal relationship to Him. A relationship of genuine sonship to God must have this element of sheer confidence in it, without a perpetual demanding of precise explanations and written guarantees against all risks; wherefore a world whose purpose is to fashion into sonship must leave room for such confidence.[1]

In these and other writings,[2] in which the attempt is made to develop a teleological theodicy that takes with full seriousness both the depth and hideousness of evil as this has been unmasked in the history of our own time, and God's practical grappling with evil in Christ, there are valuable guide-lines which will be followed in Part IV. What is to be presented there is essentially a continuation of the tradition studied in this part, though a continuation of it which will also draw upon certain elements in the Augustinian tradition.

[1] Ibid., pp. 237–9.
[2] Other reconstructed or partially reconstructed attempts include : William Temple, *Mens Creatrix* (London : Macmillan & Co. Ltd., 1917), chap. 20, and *Nature, Man and God* (London : Macmillan & Co. Ltd., 1934), chap. 14; D. S. Cairns, *The Faith That Rebels* (London : S.C.M. Press Ltd., 1937), chaps. 9 and 10 (see p. 250 above for an earlier work by Cairns of the unreconstructed type) ; O. C. Quick, *Doctrines of the Creed* (London : Nisbet & Co. Ltd., 1938), especially pp. 205–15 ; Nels F. S. Ferré, *Evil and the Christian Faith* (New York : Harper & Row, 1947) — but not the chapter on the problem of evil in his later *Reason in Religion* (Edinburgh : Thomas Nelson & Sons, Ltd., 1963) ; C. E. Raven, *Natural Religion and Christian Theology*, vol. ii (Cambridge University Press, 1953), pp. 121 f. ; Geddes Macgregor, *Introduction to Religious Philosophy* (Boston : Houghton Mifflin Co., 1959, and London : Macmillan & Co. Ltd., 1960), pt. vii; and Hugh Montefiore, *Awkward Questions on Christian Love* (London : Collins, 1964), chap. 1. See also H. H. Farmer's *Towards Belief in God* (London : S.C.M. Press Ltd., 1942), chap. 13.

CHAPTER XII

THE TWO THEODICIES — CONTRASTS AND AGREEMENTS

1. THE CONTRAST BETWEEN THE TWO TYPES OF THEODICY

WE now have before us the two historical types of Christian theodicy. I shall summarize the main points of contrast between them, and then take notice of certain convergent tendencies, working for the most part beneath the surface, which qualify these contrasts and even suggest the possibility of a common future development. But first the contrasts:

1. The main motivating interest of the Augustinian tradition is to relieve the Creator of responsibility for the existence of evil by placing that responsibility upon dependent beings who have wilfully misused their God-given freedom. In contrast the Irenaean type of theodicy in its developed form, as we find it in Schleiermacher and later thinkers, accepts God's ultimate omni-responsibility and seeks to show for what good and justifying reason He has created a universe in which evil was inevitable.

2. The Augustinian tradition embodies the philosophy of evil as non-being, with its Neo-Platonic accompaniments of the principle of plenitude, the conception of the great chain of being, and the aesthetic vision of the perfection of the universe as a complex harmony. In contrast, the Irenaean type of theodicy is more purely theological in character and is not committed to the Platonic or to any other philosophical framework.

3. The Augustinian theodicy, especially in Thomist thought and in the Protestantism of the eighteenth-century 'optimists' (as distinct from that of the Reformers and of twentieth-century neo-Reformation theologians), sees God's

relation to His creation in predominantly non-personal terms. God's goodness is His overflowing plenitude of being bestowing existence upon a dependent realm; man has accordingly been created as part of a hierarchy of forms of existence which would be incomplete without him ; evil is traceable to the necessary finitude and contingency of a dependent world which however exhibits an aesthetic perfection when seen from the divine standpoint ; and the existence of moral evil is harmonized within this perfect whole by the balancing effect of just punishment. These are all ideas to which the category of the personal is peripheral. According to the Irenaean type of theodicy, on the other hand, man has been created for fellowship with his Maker and is valued by the personal divine love as an end in himself. The world exists to be an environment for man's life, and its imperfections are integral to its fitness as a place of soul-making.

4. The Augustinian type of theodicy looks to the past, to a primal catastrophe in the fall of angels and/or men, for the explanation of the existence of evil in God's universe. In contrast, the Irenaean type of theodicy is eschatological, and finds the justification for the existence of evil in an infinite (because eternal) good which God is bringing out of the temporal process.

5. Accordingly, in the Augustinian tradition the doctrine of the fall plays a central role, whereas in the Irenaean type of theodicy, whilst it is not necessarily denied in all its forms, the doctrine becomes much less important. The accompanying notions of an original but lost righteousness of man and perfection of his world, and of inherited sinfulness as a universal consequence of the fall, which jointly render that event so catastrophic and therefore so crucial for the Augustinian theodicy, are both rejected.

6. The Augustinian tradition points, at the other end of history, to a final division of mankind into the saved and the damned, whereas Irenaean thinkers (at any rate since Schleiermacher) have been inclined to see the doctrine of eternal hell, with its implicates of permanently unexpiated sin and unending suffering, as rendering a Christian theodicy impossible.

2. Points of Hidden Agreement

Despite these very large differences there are also points of agreement between the two types of theodicy. These do not lie fully open to view on the surface, but are perhaps all the more significant for that, as unintended witnesses to certain basic necessities of Christian thought concerning the problem of evil.

1. The aesthetic conception of the perfection of the universe in the Augustinian tradition has its equivalent in the Irenaean type of theodicy in the thought of the eschatological perfection of the creation. This is the belief that the Kingdom of God, as the end and completion of the temporal process, will be a good so great as to justify all that has occurred on the way to it, so that we may affirm the unqualified goodness of the totality which consists of history and its end. Augustine's 'To thee there is no such thing as evil'[1] is matched by Mother Julian's eschatological 'But all shall be well, and all shall be well, and all manner of thing shall be well'.[2] Thus, despite the large difference that the Augustinian tradition attributes to the world the goodness of a balanced harmony of values in space and time, whilst the Irenaean type of theodicy sees it as a process leading to an infinitely (because eternally) good end, each proclaims the unqualified and unlimited goodness of God's creation as a whole.

2. Both alternatives acknowledge explicitly or implicitly God's ultimate responsibility for the existence of evil. Theodicies of the Irenaean type, from Schleiermacher onwards, do this explicitly. They hold that God did not make the world as a paradise for perfect beings but rather as a sphere in which creatures made as personal in the 'image' of God may be brought through their own free responses towards the finite 'likeness' of God. From this point of view sin and natural evil are both inevitable aspects of the creative process. The Augustinian tradition, on the other hand, implicitly teaches an ultimate divine responsibility for the

[1] *Conf.* vii. 13, 19.
[2] *The Revelations of Divine Love*, trans. by James Walsh, S.J. (London: Burns & Oates, 1961), chap. 27, p. 92.

existence of evil by bringing the free and culpable rebellion of men (and angels) within the scope of divine predestination. Augustine and Calvin both see the fall as part of the eternal plan which God has ordained in His sovereign freedom. The real issue between the two theodicies at this point is not so much as to the fact of the ultimate divine omni-responsibility, as the proper attitude of a theologian to that fact. The Augustinian thinks it is impious to state explicitly what his doctrine covertly implies; the Irenaean, in a more rationalist vein, is willing to follow the argument to its conclusion.

3. The 'O felix culpa' theme is common to both types of theodicy. The profound paradox expressed in the famous words (of unknown authorship)[1] of the ancient Eastern liturgy — 'O fortunate crime, which merited such and so great a redeemer'— are quoted with approval by theologians in both traditions. In agreement with this paradox Augustine explicitly affirms that 'God judged it better to bring good out of evil than not to permit any evil to exist',[2] and Aquinas that 'God allows evils to happen in order to bring a greater good therefrom';[3] though neither of them permits this insight to affect his theodicy as a whole, as it does those of the Irenaean type. Nevertheless, the recognition in both kinds of theodicy that the final end-product of the human story will justify the evil within that story points to an eschatological understanding of the divine purpose which gives meaning to human life.

4. Both types of theodicy acknowledge logical limitations upon divine omnipotence, though neither regards these as constituting a real restriction upon God's power; for the inability to do the self-contradictory does not reflect an impotence in the agent but a logical incoherence in the task proposed. This principle was misused, because over-extended, by Leibniz, who regarded all empirical relationships as logical and accordingly saw the characters of all possible worlds as determined by logical necessity rather than by the divine will. The principle is invoked in a more modest way by the developed Irenaean type of theodicy when it claims

---

[1] See below, p. 280, n. 1.   [2] *Ench.* viii. 27.
[3] *S.T.*, pt. III, Q. i, art. 3.

that there is a logical impossibility in the idea of free persons being ready made in the state (which is to constitute the end-product of the creative process) of having learned and grown spiritually through conflict, suffering, and redemption.

5. The Augustinian tradition affirms, whilst theodicies of the Irenaean type need not deny, the reality of a personal devil and of a community of evil powers. The traditional notion of Satan as the prince of darkness can have permanent value for Irenaean thought, at least as a vivid symbol of 'the demonic' in the sense of evil solely for the sake of evil, as we meet it, for example, in sheerly gratuitous cruelty. On the other hand, the notion is misused if it is absolutized so as to provide a solution — but necessarily a dualistic solution — to the problem of evil. A permissible doctrine of Satan must be logically peripheral to Christian theodicy in both its Augustinian and Irenaean forms.

6. The Augustinian tradition affirms a positive divine valuation of the world independently of its fitness as an environment for human life. The Irenaean way of thinking is not concerned to deny this, although it is inclined to stress that we can know God's purposes and evaluations only in so far as He has revealed them to us in their relation to mankind. Again, neither type of theodicy has any interest in denying the possibility, or indeed the probability, that there are divine purposes at work within the created universe other than and in addition to that of providing a sphere for man's existence. This thought has, however, been more cherished in Augustinian and Catholic than in Irenaean and Protestant thought, and probably represents a point at which the latter should be willing to learn from the former.

These are significant agreements, upon which we must attempt to build in Part IV.

3. NEW MOVEMENTS WITHIN THE AUGUSTINIAN TRADITION

It is, I think, clear that on the points of difference between the Augustinian and Ireanean types of theodicy the latter is in general more consonant with modern knowledge, freer from mythology, and more fully cleansed from morally repugnant ideas. We might therefore expect that in response to the attraction of insights that are consonant with the modern temper a tendency would develop for the Augustinian tradition to move towards the Irenaean. Such a tendency has in fact been evident during the decades of the present century, and may be expected to grow in strength and scope as the century proceeds. In this section some of these new developments will be noted.

Fr. Joseph Rickaby, S.J. (1845–1932), a prolific and courageous Catholic writer, recommended the use of hypotheses in theology; and in an interesting early article[1] and in two subsequent books, *In an Indian Abbey*[2] and *Studies on God and His Creatures*,[3] he presents his hypotheses concerning the problem of evil.

One of the hypotheses is that 'to intelligent and rational creatures final and perfect good accrues not otherwise than as the meed of victorious struggle',[4] and that in providing the conditions of this God has made sin among His human creatures virtually inevitable :

I cannot tell whether it be a law natural and necessary, — a law which must obtain, if intelligent creatures are to be at all, — or a positive law set up by God's free will and wisdom, but a law it certainly is, that the happiness of intelligent creatures . . . is not obtainable otherwise than as the meed of victory. Now victory means conflict, and conflict is a trial of courage and fidelity, and wherever there is a real (not a sham, and fantastic) trial, there must be a real incidence of prevarication and failure; and where such incidence is real over a large area of conflict, some are bound to prevaricate, some are bound to fail, some are bound to sin. . . . Where many are severely tempted, and all and each are likely to fall, we are sure that some will fall, though in no one case will the fall be a necessity. Much temptation then, over a wide area of many persons, cannot be without some sin; and God has chosen

---

[1] 'The Greek Doctrine of Necessity : a Speculation on the Origin of evil', *The Month*, vol. xcii, no. 413 (November 1898). See also criticisms of the article, from an orthodox Thomist point of view, by Dom Bruno Webb in 'God and the Mystery of Evil' (*The Downside Review*, vol. 75, no. 242, autumn 1957).

[2] London : Burns & Oates Ltd., 1919. See the preface for his conception of theological hypotheses.

[3] London : Longmans, Green & Co., 1924.   [4] *Indian Abbey*, p. 125.

to make a wide world of much temptation, because he wishes to be glorified by men's fidelity in conflict.[1]

In short, 'If it be asked how there can be a necessity [of sin occurring] which is not a necessity in any individual, let it be observed that where there is real trial of free will there is real likelihood of sin : but a multitude of such likelihoods mean a necessity of sin somewhere'.[2] This is an approach, other than that indirectly implied in the doctrine of predestination, to the affirmation of an ultimate divine responsibility for the occurrence of sin, together with an appeal to the Irenaean insight that sin is bound up with the conditions required for man's moral growth and his eventual attainment of genuine goodness.

The other hypothesis which Rickaby offers — likewise in a tentative and relatively undeveloped form — is that the problem of evil requires an eschatological solution. The goodness of the universe is its fitness for the fulfilment of the purpose that God had in view in creating it. 'On my present hypothesis, the parts of the universe are interlocked, like railway signals — interlocked with one another and with the end in God's view. Another universe might have been created for another end ; but this universe, and no other, can possibly realise, as assuredly it will realise, the one mysterious end and purpose, selected by God out of many, when He resolved upon the present creation.'[3] Apart from that future fulfilment to which history is leading there is no meaning in men's sufferings. But 'This world is the next world a-building. Stop the building, and how can you possibly expect the construction left on your hands ever to be satisfactory?'[4]

Rickaby does not, however, as one might have hoped, carry his eschatological resolution to the point of seeing that it must involve the final redemption of all God's human creatures ; for otherwise there will be a permanent stain of evil in the universe. He does not venture beyond the point at which his tradition has officially stopped : 'Then evil and good shall be no longer blended on earth : but there shall be

---

[1] *Studies*, pp. 121–2.
[2] 'The Greek Doctrine of Necessity', *The Month*, p. 510.
[3] *Indian Abbey*, pp. 132–3.   [4] *Studies*, p. 131.

a region of pure good, and a region of evil, heaven and over
against it hell : and the men who are in one or other of those
receptacles shall be there according to their deserts. Thus
a world of perfect goodness will be at last reached, but over
against there will stand evil, evil now cast out and cut off
from good.'¹

We may turn now to the late Pierre Teilhard de Chardin
(1881–1955), a distinguished Roman Catholic palaeon-
tologist, professionally respected as such, who also wrote
books of fascinating speculative interest dealing with the
borderland between science and theology. Père Teilhard's
activities in this area were restricted in various ways by the
Jesuit Order of which he remained a loyal member, and he
was unable to obtain an official *Imprimatur* for the publica-
tion of his philosophical writings, which have appeared
since his death under the auspices of an international com-
mittee of scientists and scholars.²

Père Teilhard set aside, gently but firmly, the notion of a
'special creation' of man, in favour of the contemporary
understanding of *homo sapiens* as the product of a long evolutionary process. He saw evil, in the forms not only of
disorder and biological failure, decomposition and death,
solitude, and anxiety and the pains of growth, but also of
moral evil, as experiences which are inevitable for man as a
being who is still in process towards his spiritual goal ; for we
are creatures 'who already exist, but are not yet complete'.³
Père Teilhard says that 'a world, assumed to be progressing
towards perfection, or "rising upward", is of its nature precisely still partially disorganized. A world without a trace
or a threat of evil would be a world already consummated.'⁴
And looking towards a future definitive divine drawing of
good out of evil he says :

But God will make it good — He will take His revenge, if one
may use the expression — by making evil itself serve a higher good

¹ 'The Greek Doctrine of Necessity', *The Month*, p. 511.
² On these ecclesiastical restrictions and hindrances see the Preface by
Martin Jarrett-Kerr to the English edition of Nicholas Corte's *Pierre Teilhard de
Chardin: His life and Spirit* (London : Barrie and Rockliff, 1960).
³ Pierre Teilhard de Chardin, *The Divine Milieu*, trans. by Bernard Wall
et al., p. 64.   ⁴ Ibid., p. 65 n.

of His faithful, the very evil which the present state of creation does not allow Him to suppress immediately. Like an artist who is able to make use of a fault or an impurity in the stone he is sculpting or the bronze he is casting so as to produce more exquisite lines or a more beautiful tone, God, without sparing us the partial deaths, nor the final death, which form an essential part of our lives, transfigures them by integrating them into a better plan — *providing we lovingly trust in Him*. Not only our unavoidable ills but our faults, even our most deliberate ones, can be embraced in that transformation, provided always we repent of them. Not everything is immediately good to those who seek God; but everything is capable of becoming good: *Omnia convertuntur in bonum*.[1]

We thus have in the thought of Teilhard de Chardin the elements of a theodicy which instead of finding its solution in a past fall of man — although Père Teilhard does refer to an unspecified 'original fall' as having complicated the difficulties of man's development — finds it instead in the supreme future good which God is bringing out of our sin and suffering.[2] One would have expected that this way of thinking would have led him, by its own inherent logic, to at least a tacit abandonment of the doctrine of hell and of the eternal frustration of God's good purpose which this entails. But Père Teilhard retains this doctrine, even if with ambiguous comments that may well have seemed dangerously speculative to his ecclesiastical superiors.[3]

The chief significance of Teilhard de Chardin's work, in its relation to the problem of evil, is that he is a committed participant in the Catholic theological tradition who accepts the modern scientific picture of man as organic to nature and as having emerged from the gradual evolution of the forms of life during the course of millions of years. To acknowledge man's organic involvement in the natural order is to be set on the way to seeing not only his mortality, his

---

[1] Pierre Teilhard de Chardin, *The Divine Milieu*, trans. by Bernard Wall *et al.*, pp. 65–66.
[2] Ibid., p. 68. See also Teilhard de Chardin's central philosophical work, *The Phenomenon of Man* (London: Collins, 1959), Appendix. Cf. Charles E. Raven, *Teilhard de Chardin: Scientist and Seer* (London: Collins, 1962), chap. 9.
[3] Ibid., pp. 140–3.

liability to disease, discomfort, danger, and all the other 'thousand natural shocks that flesh is heir to' but also the survival impulses, the competitive attitude to his fellows, and the self-regarding tendencies of thought and action which are evoked by his very physical vulnerability, and the deep anxiety caused by his consciousness of mortality, as belonging to the conditions in which God has chosen to place His human children. Thus the besetting evils of our lot and weaknesses of our nature are not consequences of some primeval sin, but are factors in a divinely appointed situation whose meaning and value lie in its potentiality for a good which God is producing in and out of it, greater than could be attained by personal beings existing without inner stress or strain and living in an environment devoid of dangers and uncertainties, baffling mystery, and the inevitable blank wall of death.

Here again we see a distinct leaning towards a more Irenaean restatement of the Augustinian theodicy-tradition. The same is true, though to a lesser extent, of the work of the late Père Antonin Delmace Sertillanges, O.P., a Roman Catholic theologian whose good standing was never questioned. In the posthumously published second volume of his *Le Problème du Mal* (a work both volumes of which, however, appeared without the ecclesiastical *Imprimatur*), in which he offers his own constructive views, he rejects the idea that all the afflictions of human life are a result of the fall. He insists, surely with greater realism than the official Augustinian-Thomist tradition, that even apart from sin mankind would be involved in pain :

can one suppose that even in the absence of all sin evil under all its forms could be spared to humans? We may say that if we are speaking in general terms ; but the least analysis contradicts the suggestion if it claims a universal validity. One can dream of an Eden as rosy as the fingers of Aurora, but one will never make it favourable enough to avoid for its inhabitants all, even slight, pain. . . . By what permanent miracle could it be avoided that a human being of flesh and bone, in a nature which is not a nature of dreams, should be exposed to a fatal accident, an unexpected fall, an unwelcome encounter, a lesion of tissues which are more delicate the more perfect they are, or to an accident of climate producing suffering . . .?[1]

In short, 'a complete absence of evil in a real universe, for a humanity constructed like ours, is not thinkable'.[2] Père Sertillanges acknowledges that St. Thomas argued for man's actual fall from a prior state of original righteousness,[3] rather than for the alternative possibility that man 'is placed at an inferior level in the chain of being, and this level explains his defects as a material being, in spite of his aspirations as spirit'.[4] However, Père Sertillanges regards the argument used by St. Thomas as philosophically and theologically without weight, and claims that Thomas himself was aware of this.[5] Man, Sertillanges insists, 'is an unstable composite, hardly more constant, in the ever moving ocean of cosmic being, than the crest of a wave on the sea. . . . In sum, [the idea of a primitive perfection] is a matter of changing the universe, and the Fathers of the Church always understood this when they speculated on the state called that of original righteousness.'[6] The implication of all this is that man's mortality, frailty, and fallibility belong to the situation in which God created him and are not the result of a disastrous moral fall near the distant beginning of man's history. Sertillanges is here adopting into his thought a major position of the Irenaean theodicy.

Still more recently Dom Mark Pontifex of Downside Abbey, writing in the composite volume *Prospect for Metaphysics*, edited by Ian Ramsey,[7] has outlined a theodicy that looks towards an eschatological resolution of the problem of evil. Dom Mark's general suggestion is that man is in process towards an all-justifying future good : 'A passing evil at the earlier stages of the creature's progress may be of small consequence in comparison with the good eventually to be gained, and the evil which we experience in this life may be of this kind.'[8] Within this process, both pain and the likelihood of sin are inevitable. Referring to the first

---

[1] *Le Problème du Mal*, vol. ii, p. 26.   [2] Ibid.   [3] *S.c.G.* iv, chap. 52.
[4] Sertillanges, op. cit., p. 27.   [5] Ibid., p. 28.   [6] Ibid., pp. 28–29.
[7] London : George Allen & Unwin Ltd., 1961. Cf. Dom Mark Pontifex, *Providence and Freedom* (London : Burns & Oates, 1960), chaps. 4–8.
[8] Ibid., p. 124.

appearance of sin, Dom Mark says, 'even then a conflict of desires and the possibility of choice may have been inevitable. Unless indeed it was inevitable why should God have ever allowed the first evil to be felt?'[1] And turning to the eventual future fulfilment of the divine purpose, in which God will have brought good out of evil, he appears even to hint at the prospect of universal salvation : 'If in some circumstances God can, without spoiling his plan, give enough help to the creature to ensure right conduct, then we may suppose he can do so often enough to bring good out of evil for creation as a whole.'[2]

4. Austin Farrer

One of the best recent books on the problem of evil — perhaps the best — is Austin Farrer's *Love Almighty and Ills Unlimited*.[3] In general, Farrer, an Anglican theologian, stands within the Catholic tradition ; and his theodicy represents a basically Augustinian approach which nevertheless freely criticizes certain aspects of the Catholic thought-world (for example, belief in a personal devil),[4] and which also emphasizes the teleological and eschatological character of God's creative work in a way that is Irenaean rather than traditionally Augustinian. Farrer takes up most of the aspects of our problem and discusses none without illuminating it by the distinctive freshness and vividness of his thought and writing. I shall, however, refer only to one section of his book, in which he sketches the Irenaean alternative to the Augustinian picture of the Creator's plan. Farrer considers the inevitable question : Why has the omnipotent creator made so imperfect a world as this? And in response he paints two alternative pictures of the divine intention.

The first is a dramatization of the ancient principle of plenitude.

A gardener may have filled the best beds he has, where the aspect is fair, and the soil deep. He may still wish to plant other

[1] Ibid., p. 136.
[2] Ibid., p. 135. Cf. *Providence and Freedom*, pp. 104–5.
[3] New York : Doubleday & Co. Inc., 1961, and London : Collins, 1962.
[4] Farrer, *Love Almighty and Ills Unlimited*, chap. 7.

grounds, where much beauty, though not the highest, can be brought to flourish. So God plants over the next best soil available to him; he extends existence to archangels, pure spirits though finite; each a limited mirror of his own perfection, each viewing him from a distinct point of vantage; each answering the vision with a unique obedience. These beds being planted, the divine gardener takes pity on the next best, and after those, on the next best again; and so down through many ranks and hierarchies of angels, as far as the humblest sort of pure spirits. These having been created, a fresh choice has to be made. All the possibility of spiritual nature has been realised — everything you could call garden soil has been brought under cultivation. Only dry walls and rocks remain. What can the gardener do with stones? He can slip little plants into the crannies; and he may reckon it the furthest stretch of his art, to have made such barrenness bloom. So, beyond the spiritual there lies the possibility of the material. The creator does not hold his hand. It is better a physical universe, with its inevitable flaws, should be, than not be; and from the stony soil of matter he raises first living, and then reasonable creatures.[1]

As Farrer points out, we must think away the temporal dimension of this story.

Let God see, or invent, at a single glance the whole cascade of creatures, falling from the frontier of eternal light to the brink of blank darkness. The implied justification of our own material universe is unaffected. Whether proceeding step by step, or in a single sweep, God makes the material realm as a creation worth having in addition to many hierarchies of splendid spirits, and not as constituting in or by itself the best of possible worlds.[2]

Farrer then points out the flaw embodied in his myth. This is the flaw that was noted in our discussion of Augustine's use of the principle of plenitude, namely that it ultimately implies a limited God. The myth assumes that the top-grade soil has all been used up and that therefore the gardener must, if he is to go on planting, turn to the second-grade soil, and so on downwards. But no such necessity can apply to the divine creative activity.

It is a baseless suggestion, then, that the Creator, wishing to go outside what he has done already, must go downwards. He might go sideways for ever, for anything we can tell. And so there

---

[1] *L.A.I.U.*, pp. 65–66.   [2] *L.A.I.U.*, p. 67.

is no reason why he should ever, in the invention of new creatures, reach the material level. And seeing it was precisely this that the gardener-parable set out to explain, we must judge it to have failed entirely.[1]

Farrer now spins his other myth, which turns out to represent the Irenaean alternative to the Augustinian conception.

God's desire was to create beings able to know and to love him. Yet, in the nature of the case, there lay a dilemma. In proportion to their capacity for such love or knowledge, the created minds or wills would be dominated by the object of their knowledge or their love; they would lose the personal initiative which could alone give reality to their knowing or their loving. The divine glory would draw them into itself, as the candle draws the moth. You might say, 'Why should he not shade the light? Could not God put a screen between himself and his creatures?' But of what would the screen consist? A screen, literally understood, is a physical barrier; and it screens a physical object from an organ of physical vision. God is a spirit; and the hypothesis we are examining is of purely spiritual creatures also. What sort of screen could God interpose between himself and them?[2]

The answer of Farrer's myth is to 'suppose he creates a whole physical world, and places creaturely minds in it; suppose he so attaches them to it, that they are initially turned towards it, and find in it their natural concern. . . . Might we not perhaps say [Farrer elaborates] that the first requirement is to have a created world which is quite other than God? Then, by identification with such a world, godlike creatures may keep their distinctness from God, and not fall straight back into the lap of creating power.'[3] Such a story, as Farrer says, 'has the advantage not only of explaining why God should create at so low a level as the physical, but also of squaring with what we know about the world we live in. For, to all evidence, the world-process begins with the most elementary organisation of energy, and builds gradually up, level by level.'[4]

However, having told his two stories (or fables, parables, or myths, as he also calls them), Farrer apparently

[1] *L.A.I.U.*, p. 68.   [2] *L.A.I.U.*, pp. 69–70.
[3] *L.A.I.U.*, pp. 70–71.   [4] *L.A.I.U.*, p. 73.

repudiates them both equally. 'There must', he says, 'be something radically false about a line of speculation which reduces the most august of mysteries to the triviality of a nursery tale.'[1] We can contemplate with awe the actual creation: 'But speculate on the reason why such-and-such existences have been appointed rather than others, and you fall into a silly, heathenish mythology, with no savour of godhead to it.'[2]

It seems to me that Farrer is unduly displeased with his own stories. They suffer, of course, from the limitation of all theological theories, whether pictorially expressed or not, in that they deal by thin human analogies with infinitely rich and concrete divine realities. But if we are not to eschew theology altogether, we must accept its inherently abstract and inadequate nature. Waiving, then, this general misgiving about theological speculation as such, Farrer's two myths are both worthy of consideration. As we saw when we met it within the Augustinian theodicy tradition, the first has a long and honourable history. It is not mere rootless fancifulness but an expression of that Neo-Platonic vision of the universe which entered into Christian thought through Augustine. And the theological criticism of it is not that it is mythological but that by implication it limits the power of the Creator. Farrer's alternative myth likewise need not be rejected merely on account of its speculative character or picturesque form. It represents a permissible attempt to understand from a Christian point of view the existence of so unideal a creature as man and of the material world that he inhabits. In Part IV this same speculation will be developed as one of the bases of a non-Augustinian theodicy.

[1] *L.A.I.U.*, p. 75.   [2] *L.A.I.U.*, p. 69.

PART IV

A THEODICY FOR TODAY

CHAPTER XIII

THE STARTING-POINT

1. The Negative Task of Theodicy

At the outset of an attempt to present a Christian theodicy — a defence of the goodness of God in face of the evil in His world — we should recognize that, whether or not we can succeed in formulating its basis, an implicit theodicy is at work in the Bible, at least in the sense of an effective reconciliation of profound faith in God with a deep involvement in the realities of sin and suffering. The Scriptures reflect the characteristic mixture of good and evil in human experience. They record every kind of sorrow and suffering from the terrors of childhood to the 'stony griefs of age': cruelty, torture, violence, and agony; poverty, hunger, calamitous accident; disease, insanity, folly; every mode of man's inhumanity to man and of his painfully insecure existence in the world. In these writings there is no attempt to evade the clear verdict of human experience that evil is dark, menacingly ugly, heart-rending, crushing. And the climax of this biblical history of evil was the execution of Jesus of Nazareth. Here were pain and violent destruction, gross injustice, the apparent defeat of the righteous, and the premature death of a still-young man. But further, for Christian faith, this death was the slaying of God's Messiah, the one in whom mankind was to see the mind and heart of God made flesh. Here, then, the problem of evil rises to its ultimate maximum; for in its quality this was an evil than which no greater can be conceived. And yet throughout the biblical history of evil, including even this darkest point, God's purpose of good was moving visibly or invisibly towards its far-distant fulfilment. In this faith the prophets saw both personal and national tragedy as God's austere but gracious disciplining of His people. And even the greatest

279

A Theodicy for Today

evil of all, the murder of the son of God, has been found by
subsequent Christian faith to be also, in an astounding
paradox, the greatest good of all, so that through the
centuries the Church could dare to sing on the eve of its
triumphant Easter celebrations, 'O felix culpa, quae talem
ac tantum meruit habere redemptorem'.[1] For this reason
there is no room within the Christian thought-world for the
idea of tragedy in any sense that includes the idea of finally
wasted suffering and goodness.[2]

In all this a Christian theodicy is latent; and our aim
must be to try to draw it out explicitly. The task, like that
of theology in general, is one of 'faith seeking understanding',
seeking in this case an understanding of the grounds of its own
practical victory in the face of the harsh facts of evil. Accordingly, from the point of view of apologetics, theodicy has a
negative rather than a positive function. It cannot profess
to create faith, but only to preserve an already existing faith
from being overcome by this dark mystery. For we cannot
share the hope of the older schools of natural theology of
inferring the existence of God from the evidences of nature;
the one main reason for this, as David Hume made clear in
his *Dialogues*, is precisely the fact of evil in its many forms.
For us today the live question is whether this renders impossible a rational belief in God: meaning by this, not a belief
in God that has been arrived at by rational argument (for it
is doubtful whether a religious faith is ever attained in this
way), but one that has arisen in a rational individual in
response to some compelling element in his experience, and
decisively illuminates and is illuminated by his experience as

---

[1] 'O certe necessarium Adae peccatum, quod Christi morte deletum est!
O felix culpa, quae talem ac tantum meruit habere redemptorem!' (O truly
necessary sin of Adam, which is cancelled by Christ's death! O fortunate
crime (*or*, O happy fault), which merited [to have] such and so great a redeemer!) These famous phrases occur in the Roman Missal in the *Exultet* for
the evening before Easter Day. The date and authorship of this *Exultet* are
uncertain. It has been attributed, but without adequate evidence, to St.
Augustine, to St. Ambrose, and to Gregory the Great. As part of the Easter
liturgy it goes back at least to the seventh century and possibly to the beginnings
of the fifth century. On its history see Arthur O. Lovejoy, *Essays in the History
of Ideas*, 1948 (New York: Capricorn Books, 1960), pp. 286–7.

[2] Cf. D. D. Raphael, *The Paradox of Tragedy* (London: George Allen &
Unwin Ltd., 1960), pp. 43 f.

a whole. The aim of a Christian theodicy must thus be the relatively modest and defensive one of showing that the mystery of evil, largely incomprehensible though it remains, does not render irrational a faith that has arisen, not from the inferences of natural theology, but from participation in a stream of religious experience which is continuous with that recorded in the Bible.

2. The Traditional Theodicy based upon Christian Myth

We can distinguish, though we cannot always separate, three relevant facets of the Christian religion : Christian experience, Christian mythology, and Christian theology.

Religious experience is 'the whole experience of religious persons',[1] constituting an awareness of God acting towards them in and through the events of their lives and of world history, the interpretative element within which awareness is the cognitive aspect of faith. And distinctively *Christian experience*, as a form of this, is the Christian's seeing of Christ as his 'Lord and Saviour', together with the pervasive recreative effects of this throughout his life, transforming the quality of his experience and determining his responses to other people. Christian faith is thus a distinctive consciousness of the world and of one's existence within it, radiating from and illuminated by a consciousness of God in Christ. It is because there are often a successful facing and overcoming of the challenge of evil at this level that there can, in principle at least, be an honest and serious — even though tentative and incomplete — Christian theodicy.

By *Christian mythology* I mean the great persisting imaginative pictures by means of which the corporate mind of the Church has expressed to itself the significance of the historical events upon which its faith is based, above all the life, death, and resurrection of Jesus who was the Christ. The function of these myths is to convey in universally understandable

---

[1] William Temple, *Nature, Man and God* (London : Macmillan & Co. Ltd., 1934), p. 334.

ways the special importance and meaning of certain items of mundane experience.

By *Christian theology* I mean the attempts by Christian thinkers to speak systematically about God on the basis of the data provided by Christian experience. Thus it is a fact of the Christian faith-experience that 'God was in Christ';[1] and the various Christological theories are attempts to understand this by seeing it in the context of other facts both of faith and of nature. Again, it is another facet of this basic fact of faith that in Christ God was 'reconciling the world unto Himself';[2] and the various atonement theories are accordingly attempts to understand this further aspect of the experience. The other departments of Christian doctrine stand in a similar relationship to the primary data of Christian experience.

In the past, theology and myth have been closely twined together. For the less men knew about the character of the physical universe the harder it was for them to identify myth as myth, as distinct from history or science. This fact has profoundly affected the development of the dominant tradition of Christian theodicy. Until comparatively recent times the ancient myth of the origin of evil in the fall of man was quite reasonably assumed to be history. The theologian accordingly accepted it as providing 'hard' data, and proceeded to build his theodicy upon it. This mythological theodicy was first comprehensively developed by Augustine, and has continued substantially unchanged within the Roman Catholic Church to the present day. It was likewise adopted by the Reformers of the sixteenth century and has been virtually unquestioned as Protestant doctrine until within approximately the last hundred years. Only during this latest period has it been possible to identify as such its mythological basis, to apply a theological criticism to it, and then to go back to the data of Christian experience and build afresh, seeking a theodicy that can hope to make sense to Christians in our own and succeeding centuries.

But first, in order to see how the hitherto dominant theodicy has arisen, and why it is now utterly unacceptable,

[1] II Corinthians v. 19. [2] Ibid.

we must trace the outline of the mythology that underlies it. The story of the fall of man is part of a more comprehensive cosmic story. In this great amalgam of Jewish and Christian themes, God created spiritual beings, the angels and archangels, to be His subjects and to love and serve Him in the heavenly spheres. But a minority of them revolted against God in envy of His supremacy, and were defeated and cast into an abode suited to their now irreconcilably evil natures. Either to replenish the citizenry of heaven thus depleted by the expulsion of Satan and his followers, or as an independent venture of creation, God made our world, and mankind within it consisting initially of a single human pair. This first man and woman, living in the direct knowledge of God, were good, happy, and immortal, and would in due course have populated the earth with descendants like themselves. But Satan, in wicked spite, successfully tempted them to disobey their Creator, who then expelled them from this paradisal existence into a new situation of hardship, danger, disease, and inevitable death. This was the fall of man, and as a result of it the succeeding members of the human race have been born as fallen creatures in a fallen world, participating in the effects of their first parents' rebellion against their Maker. But God in Christ has made the atonement for man's sin that His own eternal justice required and has offered free forgiveness to as many as will commit themselves to Christ as their Saviour. At the last judgement, when faith and life alike will be tested, many will enter into eternal life whilst others, preferring their own darkness to God's light, will linger in a perpetual living death.

This great cosmic drama is the official Christian myth. With only minor variations it has constituted the accepted framework of thought of the great majority of Christians in the past, and still fulfils this role for the great majority today. By means of it Christian faith, which began as a crucial response of trust towards one in whom the disciples had experienced God directly at work on earth, broadened out into a comprehensive vision of the universe. The great creation–fall–redemption myth has thus brought within the scope of the simplest human soul a pictorial grasp of the universal significance of the life and death of Jesus. Jesus himself was not a mythological figure ; he lived in Palestine and his life and death and resurrection made their impact upon living people, and through them upon others in a long succession of faith down to ourselves today. But the cosmic picture, sketched by St. Paul and completed by St. Augustine, of the beginning of our present human situation in the fall of humanity from a condition of paradisal perfection into one of sin and pain and death, and of its end in the separation of mankind into those destined for the eternal bliss or torment of heaven or hell, is a product of the religious imagination. It expresses the significance of the present reality of sin and sorrow by seeing them as flowing from a first dramatic act of rebellion ; and the significance of the experience of reconciliation with God by means of the picture of a juridical arrangement taking place within the councils of the Trinity and being transacted in time on the cross of Christ ; and the significance of man's inalienable personal responsibility by the picture of a divine administration directing souls to their appropriate final destinations.

This great cosmic drama in three acts has constituted a valid myth in the sense that it has successfully fulfilled the conserving and communicating function of a myth in the minds of countless people. By means of natural images it has vividly brought home to the simplest understandings the claim that Christ stands at the centre of the universe and is of crucial importance for all men. And when religious myths thus work effectively it is as absurd to criticize them for being myths rather than science or history as it would be for us today to insist that they *are* science or history and to proceed to draw scientific or historical conclusions from them.

Because we can no longer share the assumption, upon which traditional Christian theodicy has been built, that the creation–fall myth is basically authentic history,[1] we inevitably look at that theodicy critically and see in it inadequacies

---

[1] One of the most eloquent recent presentations of the traditional conception of a temporal fall of man is that of C. S. Lewis in *The Problem of Pain* (London : The Centenary Press, 1940), pp. 65 f.

to which in the past piety has tended to blind the eyes of faith.

For, in general, religious myths are not adapted to the solving of problems. Their function is to illumine by means of unforgettable imagery the religious significance of some present or remembered fact of experience. But the experience which myth thus emphasizes and illumines is itself the locus of mystery. Hence it is not surprising that Christian mythology mirrors Christian experience in presenting but not resolving the profound mystery of evil. Nor is it surprising that when this pictorial presentation of the problem has mistakenly been treated as a solution to it, the 'solution' has suffered from profound incoherences and contradictions.

This traditional solution (representing the theological, in distinction from the philosophical, side of Augustine's thought on the theodicy problem) finds the origin of evil, as we have seen, in the fall, which was the beginning both of sin and, as its punishment, of man's sorrows and sufferings.[1] But this theory, so simple and mythologically satisfying, is open to insuperable scientific, moral, and logical objections. To begin with less fundamental aspects of the traditional solution, we know today that the conditions that were to cause human disease and mortality and the necessity for man to undertake the perils of hunting and the labours of agriculture and building, were already part of the natural order prior to the emergence of man and prior therefore to any first human sin, as were also the conditions causing such further 'evils' as earthquake, storm, flood, drought, and pest. And, second, the policy of punishing the whole succeeding human race for the sin of the first pair is, by the best human moral standards, unjust and does not provide anything that can be recognized by these standards as a theodicy. Third, there is a basic and fatal incoherence at the heart of the mythically based 'solution'. The Creator is preserved from any responsibility for the existence of evil by the claim that He made men (or angels) as free and finitely perfect creatures, happy in the knowledge of Himself, and subject to no strains or temptations, but that they themselves inexplicably and

[1] See Chapter III above.

inexcusably rebelled against Him. But this suggestion amounts to a sheer self-contradiction. It is impossible to conceive of wholly good beings in a wholly good world becoming sinful. To say that they do is to postulate the self-creation of evil *ex nihilo* ! There must have been some moral flaw in the creature or in his situation to set up the tension of temptation ; for creaturely freedom in itself and in the absence of any temptation cannot lead to sin. Thus the very fact that the creature sins refutes the suggestion that until that moment he was a finitely perfect being living in an ideal creaturely relationship to God. And indeed (as we have already seen) the two greatest upholders of this solution implicitly admit the contradiction. Augustine, who treats of evil at its first occurrence in the fall of Satan and his followers, has to explain the eruption of sin in supposedly perfect angels by holding that God had in effect predestined their revolt by withholding from them the assurance of eternal bliss with which, in contrast, He had furnished the angels who remained steadfast.[1] And Calvin, who treats the subject primarily at the point of the fall of man, holds that 'all are not created in equal condition ; rather, eternal life is foreordained for some, eternal damnation for others'.[2] Thus the myth, when mistakenly pressed to serve as a theodicy, can be saved only by adding to it the new and questionable doctrine of an absolute divine predestination. And this in turn only leads the theodicy to contradict itself. For its original intention was to blame evil upon the misuse of creaturely free will. But now this misuse is itself said to fall under the divine predestinating decrees. Thus the theodicy collapses into radical incoherence, and its more persistent defenders have become involved in ever more desperate and implausible epicycles of theory to save it. For example, to salvage the view of the fall of man as a temporal event that took place on this earth some definite (if unknown) number of years ago, it has been suggested that after emerging from his subhuman precursors man lived in the paradisal state for only a very brief period, lasting

[1] *C.G.*, bk. xi, chaps. 11 and 13 ; bk. xii, chap. 9. Cf. above, Chapter III, Sect. 9.
[2] *Inst.*, bk. iii, chap. xxi, para. 5. Cf. above, Chapter VI, Sects. 2 and 3.

perhaps no more than a matter of hours. Again, attempts have been made to protect the fall doctrine from the encroachments of scientific research by locating the primal calamity in a pre-mundane sphere. In the third century Origen had taught that some of the spirits whom God created rebelled against the divine majesty and were cast down into the material world to constitute our human race;[1] and in the nineteenth century the German Protestant theologian Julius Müller, impressed by the overwhelming difficulties of affirming an historical fall, in effect revived Origen's theory as an explanation of the apparently universal evil propensities of man. All men are sinful, he suggested, because in another existence prior to the present life they have individually turned away from God.[2]

The difficulties and disadvantages of such a view are, I think, not far to seek. The theory is without grounds in Scripture or in science, and it would have claim to consideration only if it could provide a solution, even if a speculative one, to the question of the origin of moral evil. But in fact it is not able to do this. It merely pushes back into an unknown and unknowable realm the wanton paradox of finitely perfect creatures, dwelling happily and untempted in the presence of God, turning to sin. Whether on earth or in heaven, this still amounts to the impossible self-creation of evil *ex nihilo*. If evil could thus create itself out of nothing in the midst of a wholly good universe, it could do so in a mundane Garden of Eden as easily as, or perhaps more easily than, in the highest heaven. Nothing, then, is gained for theodicy by postulating a pre-mundane fall of human souls.

As a variation which he regarded as superior to the notion of a pre-mundane fall of individuals, N. P. Williams proposed the idea of 'a collective fall of the race-soul of humanity at an indefinitely remote past'.[3] This collective fall occurred, according to Williams, during the long period between the first emergence of man as a biological species and

---

[1] *De Principiis*, bk. ii, chap. i, para. 1. Cf. ibid., chap. ix, para. 6.
[2] *The Christian Doctrine of Sin*, bk. iv, chap. 4. Cf. bk. iii, pt. i, chap. 3, sect. 1, and chap. 4, sect. 3.
[3] N. P. Williams, *The Ideas of the Fall and of Original Sin*, p. 513.

his subsequent development to the point at which there were primitive societies, and therefore moral laws which could be transgressed. 'We must', he says, 'postulate some unknown factor or agency which interfered to arrest the development of corporate feeling, just when man was becoming man, some mysterious and maleficent influence which cut into the stream of the genetic evolution of our race at some point during the twilit age which separates pre-human from human history.'[1] This evil influence which attacked and corrupted mankind is also 'the mysterious power which vitiates the whole of sub-human life with cruelty and selfishness',[2] and thus accounts not only for moral evil but also for the disorder, waste, and pain in nature.[3] Accordingly the original calamity was not merely a fall of man but of the Life-Force itself, which we must conceive 'as having been at the beginning, when it first sprang forth from the creative fecundity of the Divine Being, free, personal, and self-conscious'.[4] This World-Soul was created good, but 'at the beginning of Time, and in some transcendental and incomprehensible manner, it turned away from God and in the direction of Self, thus shattering its own interior being, which depended upon God for its stability and coherence, and thereby forfeiting its unitary self-consciousness, which it has only regained, after aeons of myopic striving, in sporadic fragments which are the separate minds of men and perhaps of superhuman spirits'.[5]

Williams is, I think, justified in claiming that such a speculation cannot be excluded *ab initio* as impermissible to a responsible Christian theologian. As he points out,

Such a substitution of the idea of a corruption of the whole cosmic energy at some enormously remote date for the idea of a voluntary moral suicide of Man in comparatively recent times would be no greater a revolution than that which was effected by St Anselm, when he substituted a satisfactional theory of the Atonement for the view which regarded the death of Christ as a

[1] N. P. Williams, *The Ideas of the Fall and of Original Sin*, pp. 518–19.
[2] Ibid., p. 520.
[3] The application of the notion of a pre-mundane fall to evil in nature will be discussed below, pp. 367 f.
[4] Williams, op. cit., p. 525.   [5] Ibid., p. 526.

ransom paid to the Devil — a view which had behind it the vener-
able authority of a thousand years of Christian history.[1]

Williams' suggestion preserves the central thought of the
Augustinian fall doctrine that the ultimate source of evil lies
in an original conscious turning away from God on the part
of created personal life. But precisely because of its faithfulness to that tradition his theory fails to throw any new
light upon the problem of evil. Whether the self-creation of
evil *ex nihilo* be located in an historical Adam and Eve, or in
a multitude of souls in a pre-mundane realm, or in a single
world-soul at the beginning of time, it is equally valueless
from the point of view of theodicy. In order for a soul or
souls to fall there must be, either in them or in their environment, some flaw which produces temptation and leads to sin ;
and this flaw in the creation cannot be traced back to any
other ultimate source than the Creator of all that is. Thus
Williams' theory is open to the same objection as Müller's :
namely, that it is a speculation whose only point would be to
solve or lighten the problem of evil, but that it fails to do this.[2]

3. The 'Vale of Soul-making' Theodicy

Fortunately there is another and better way. As well as
the 'majority report' of the Augustinian tradition, which has
dominated Western Christendom, both Catholic and Protestant, since the time of Augustine himself, there is the 'minority report' of the Irenaean tradition. This latter is both
older and newer than the other, for it goes back to St.
Irenaeus and others of the early Hellenistic Fathers of the
Church in the two centuries prior to St. Augustine, and it has
flourished again in more developed forms during the last
hundred years.

Instead of regarding man as having been created by God
in a finished state, as a finitely perfect being fulfilling the

---

[1] Ibid., p. 524.
[2] A pre-mundane fall has been propounded by Canon Peter Green in
The Problem of Evil (London : Longmans, Green & Co., 1920), chap. 7, and in
The Pre-Mundane Fall (London : A. R. Mowbray & Co., 1944) ; and by C. W.
Formby in *The Unveiling of the Fall* (London : Williams & Norgate, 1923).

divine intention for our human level of existence, and then falling disastrously away from this, the minority report sees man as still in process of creation. Irenaeus himself expressed the point in terms of the (exegetically dubious) distinction between the 'image' and the 'likeness' of God referred to in Genesis i. 26 : 'Then God said, Let us make man in our image, after our likeness.'[1] His view was that man as a personal and moral being already exists in the image, but has not yet been formed into the finite likeness of God. By this 'likeness' Irenaeus means something more than personal existence as such; he means a certain valuable quality of personal life which reflects finitely the divine life. This represents the perfecting of man, the fulfilment of God's purpose for humanity, the 'bringing of many sons to glory',[2] the creating of 'children of God' who are 'fellow heirs with Christ' of his glory.[3]

And so man, created as a personal being in the image of God, is only the raw material for a further and more difficult stage of God's creative work. This is the leading of men as relatively free and autonomous persons, through their own dealings with life in the world in which He has placed them, towards that quality of personal existence that is the finite likeness of God. The features of this likeness are revealed in the person of Christ, and the process of man's creation into it is the work of the Holy Spirit. In St. Paul's words, 'And we all, with unveiled faces, beholding the glory of the Lord, are being changed into his likeness (εἰκών) from one degree of glory to another; for this comes from the Lord who is the Spirit';[4] or again, 'For God knew his own before ever they were, and also ordained that they should be shaped to the likeness (εἰκών) of his Son.'[5] In Johannine terms, the movement from the image to the likeness is a transition from one level of existence, that of animal life (*Bios*), to another and higher level, that of eternal life (*Zoe*), which includes but transcends the first. And the fall of man was seen by Irenaeus

---

[1] *A.H.* v. vi. 1. Cf. pp. 217 f. above. [2] Hebrews ii. 10.
[3] Romans viii. 17. [4] II Corinthians iii. 18.
[5] Romans viii. 29. Other New Testament passages expressing a view of man as undergoing a process of spiritual growth within God's purpose, are: Ephesians ii. 21; iii. 16; Colossians ii. 19; I John iii. 2; II Corinthians iv. 16.

as a failure within the second phase of this creative process, a failure that has multiplied the perils and complicated the route of the journey in which God is seeking to lead mankind.

In the light of modern anthropological knowledge some form of two-stage conception of the creation of man has become an almost unavoidable Christian tenet. At the very least we must acknowledge as two distinguishable stages the fashioning of *homo sapiens* as a product of the long evolutionary process, and his sudden or gradual spiritualization as a child of God. But we may well extend the first stage to include the development of man as a rational and responsible person capable of personal relationship with the personal Infinite who has created him. This first stage of the creative process was, to our anthropomorphic imaginations, easy for divine omnipotence. By an exercise of creative power God caused the physical universe to exist, and in the course of countless ages to bring forth within it organic life, and finally to produce out of organic life personal life ; and when man had thus emerged out of the evolution of the forms of organic life, a creature had been made who has the possibility of existing in conscious fellowship with God. But the second stage of the creative process is of a different kind altogether. It cannot be performed by omnipotent power as such. For personal life is essentially free and self-directing. It cannot be perfected by divine fiat, but only through the uncompelled responses and willing co-operation of human individuals in their actions and reactions in the world in which God has placed them. Men may eventually become the perfected persons whom the New Testament calls 'children of God', but they cannot be created ready-made as this.

The value-judgement that is implicitly being invoked here is that one who has attained to goodness by meeting and eventually mastering temptations, and thus by rightly making responsible choices in concrete situations, is good in a richer and more valuable sense than would be one created *ab initio* in a state either of innocence or of virtue. In the former case, which is that of the actual moral achievements of mankind, the individual's goodness has within it the strength of temptations overcome, a stability based upon an accumulation of right choices, and a positive and responsible character that comes from the investment of costly personal effort. I suggest, then, that it is an ethically reasonable judgement, even though in the nature of the case not one that is capable of demonstrative proof, that human goodness slowly built up through personal histories of moral effort has a value in the eyes of the Creator which justifies even the long travail of the soul-making process.

The picture with which we are working is thus developmental and teleological. Man is in process of becoming the perfected being whom God is seeking to create. However, this is not taking place — it is important to add — by a natural and inevitable evolution, but through a hazardous adventure in individual freedom. Because this is a pilgrimage within the life of each individual, rather than a racial evolution, the progressive fulfilment of God's purpose does not entail any corresponding progressive improvement in the moral state of the world. There is no doubt a development in man's ethical situation from generation to generation through the building of individual choices into public institutions, but this involves an accumulation of evil as well as of good.[1] It is thus probable that human life was lived on much the same moral plane two thousand years ago or four thousand years ago as it is today. But nevertheless during this period uncounted millions of souls have been through the experience of earthly life, and God's purpose has gradually moved towards its fulfilment within each one of them, rather than within a human aggregate composed of different units in different generations.

If, then, God's aim in making the world is 'the bringing of many sons to glory',[2] that aim will naturally determine the kind of world that He has created. Antitheistic writers almost invariably assume a conception of the divine purpose which is contrary to the Christian conception. They assume that the purpose of a loving God must be to create a hedon-

[1] This fact is symbolized in early Christian literature both by the figure of the Antichrist, who continually opposes God's purposes in history, and by the expectation of cataclysmic calamity and strife in the last days before the end of the present world order.
[2] Hebrews ii. 10.

istic paradise; and therefore to the extent that the world is other than this, it proves to them that God is either not loving enough or not powerful enough to create such a world. They think of God's relation to the earth on the model of a human being building a cage for a pet animal to dwell in. If he is humane he will naturally make his pet's quarters as pleasant and healthful as he can. Any respect in which the cage falls short of the veterinarian's ideal, and contains possibilities of accident or disease, is evidence of either limited benevolence or limited means, or both. Those who use the problem of evil as an argument against belief in God almost invariably think of the world in this kind of way. David Hume, for example, speaks of an architect who is trying to plan a house that is to be as comfortable and convenient as possible. If we find that 'the windows, doors, fires, passages, stairs, and the whole economy of the building were the source of noise, confusion, fatigue, darkness, and the extremes of heat and cold' we should have no hesitation in blaming the architect. It would be in vain for him to prove that if this or that defect were corrected greater ills would result: 'still you would assert in general, that, if the architect had had skill and good intentions, he might have formed such a plan of the whole, and might have adjusted the parts in such a manner, as would have remedied all or most of these inconveniences'.[1]

But if we are right in supposing that God's purpose for man is to lead him from human *Bios*, or the biological life of man, to that quality of *Zoe*, or the personal life of eternal worth, which we see in Christ, then the question that we have to ask is not, Is this the kind of world that an all-powerful and infinitely loving being would create as an environment for his human pets? or, Is the architecture of the world the most pleasant and convenient possible? The question that we have to ask is rather, Is this the kind of world that God might make as an environment in which moral beings may be fashioned, through their own free insights and responses, into 'children of God'?

Such critics as Hume are confusing what heaven ought to

---

[1] *Dialogues Concerning Natural Religion*, pt. xi. Kemp-Smith's ed. (Oxford: Clarendon Press, 1935), p. 251.

be, as an environment for perfected finite beings, with what this world ought to be, as an environment for beings who are in process of becoming perfected. For if our general conception of God's purpose is correct the world is not intended to be a paradise, but rather the scene of a history in which human personality may be formed towards the pattern of Christ. Men are not to be thought of on the analogy of animal pets, whose life is to be made as agreeable as possible, but rather on the analogy of human children, who are to grow to adulthood in an environment whose primary and overriding purpose is not immediate pleasure but the realizing of the most valuable potentialities of human personality.

Needless to say, this characterization of God as the heavenly Father is not a merely random illustration but an analogy that lies at the heart of the Christian faith. Jesus treated the likeness between the attitude of God to man, and the attitude of human parents at their best towards their children, as providing the most adequate way for us to think about God. And so it is altogether relevant to a Christian understanding of this world to ask, How does the best parental love express itself in its influence upon the environment in which children are to grow up? I think it is clear that a parent who loves his children, and wants them to become the best human beings that they are capable of becoming, does not treat pleasure as the sole and supreme value. Certainly we seek pleasure for our children, and take great delight in obtaining it for them ; but we do not desire for them unalloyed pleasure at the expense of their growth in such even greater values as moral integrity, unselfishness, compassion, courage, humour, reverence for the truth, and perhaps above all the capacity for love. We do not act on the premise that pleasure is the supreme end of life ; and if the development of these other values sometimes clashes with the provision of pleasure, then we are willing to have our children miss a certain amount of this, rather than fail to come to possess and to be possessed by the finer and more precious qualities that are possible to the human personality. A child brought up on the principle that the only or the supreme value is pleasure would not be likely to become an ethically mature adult or an attractive or happy personality. And to most parents it seems more important to try to foster quality and strength of character in their children than to fill their lives at all times with the utmost possible degree of pleasure. If, then, there is any true analogy between God's purpose for his human creatures, and the purpose of loving and wise parents for their children, we have to recognize that the presence of pleasure and the absence of pain cannot be the supreme and overriding end for which the world exists. Rather, this world must be a place of soul-making. And its value is to be judged, not primarily by the quantity of pleasure and pain occurring in it at any particular moment, but by its fitness for its primary purpose, the purpose of soul-making.[1]

In all this we have been speaking about the nature of the world considered simply as the God-given environment of man's life. For it is mainly in this connection that the world has been regarded in Irenaean and in Protestant thought.[2] But such a way of thinking involves a danger of anthropocentrism from which the Augustinian and Catholic tradition has generally been protected by its sense of the relative insignificance of man within the totality of the created universe. Man was dwarfed within the medieval world-view by the innumerable hosts of angels and archangels above him — unfallen rational natures which rejoice in the immediate presence of God, reflecting His glory in the untarnished mirror of their worship. However, this higher creation has in our modern world lost its hold upon the imagination. Its place has been taken, as the minimizer of men, by the

[1] The phrase 'the vale of Soul-making' was coined by the poet John Keats in a letter written to his brother and sister in April 1819. He says, 'The common cognomen of this world among the misguided and superstitious is "a vale of tears" from which we are to be redeemed by a certain arbitrary interposition of God and taken to Heaven — What a little circumscribed straightened notion! Call the world if you Please "The vale of Soul-making".' In this letter he sketches a teleological theodicy. 'Do you not see', he asks, 'how necessary a World of Pains and troubles is to school an Intelligence and make it a Soul?' (*The Letters of John Keats*, ed. by M. B. Forman. London: Oxford University Press, 4th ed., 1952, pp. 334–5.)

[2] Thus Irenaeus said that 'the creation is suited to [the wants of] man; for man was not made for its sake, but creation for the sake of man' (*A.H.* v. xxix. 1), and Calvin said that 'because we know that the universe was established especially for the sake of mankind, we ought to look for this purpose in his governance also'. (*Inst.* i. xvi. 6.)

immensities of outer space and by the material universe's unlimited complexity transcending our present knowledge. As the spiritual environment envisaged by Western man has shrunk, his physical horizons have correspondingly expanded. Where the human creature was formerly seen as an insignificant appendage to the angelic world, he is now seen as an equally insignificant organic excrescence, enjoying a fleeting moment of consciousness on the surface of one of the planets of a minor star. Thus the truth that was symbolized for former ages by the existence of the angelic hosts is today impressed upon us by the vastness of the physical universe, countering the egoism of our species by making us feel that this immense prodigality of existence can hardly all exist for the sake of man — though, on the other hand, the very realization that it is not all for the sake of man may itself be salutary and beneficial to man!

However, instead of opposing man and nature as rival objects of God's interest, we should perhaps rather stress man's solidarity as an embodied being with the whole natural order in which he is embedded. For man is organic to the world; all his acts and thoughts and imaginations are conditioned by space and time; and in abstraction from nature he would cease to be human. We may, then, say that the beauties and sublimities and powers, the microscopic intricacies and macroscopic vastnesses, the wonders and the terrors of the natural world and of the life that pulses through it, are willed and valued by their Maker in a creative act that embraces man together with nature. By means of matter and living flesh God both builds a path and weaves a veil between Himself and the creature made in His image. Nature thus has permanent significance; for God has set man in a creaturely environment, and the final fulfilment of our nature in relation to God will accordingly take the form of an embodied life within 'a new heaven and a new earth'.[1] And as in the present age man moves slowly towards that fulfilment through the pilgrimage of his earthly life, so also 'the whole creation' is 'groaning in travail', waiting for the time when it will be 'set free from its bondage to decay'.[2]

[1] Revelation xxi. 1.   [2] Romans viii. 21–22.

And yet however fully we thus acknowledge the permanent significance and value of the natural order, we must still insist upon man's special character as a personal creature made in the image of God ; and our theodicy must still centre upon the soul-making process that we believe to be taking place within human life.

This, then, is the starting-point from which we propose to try to relate the realities of sin and suffering to the perfect love of an omnipotent Creator. And as will become increasingly apparent, a theodicy that starts in this way must be eschatological in its ultimate bearings. That is to say, instead of looking to the past for its clue to the mystery of evil, it looks to the future, and indeed to that ultimate future to which only faith can look. Given the conception of a divine intention working in and through human time towards a fulfilment that lies in its completeness beyond human time, our theodicy must find the meaning of evil in the part that it is made to play in the eventual outworking of that purpose ; and must find the justification of the whole process in the magnitude of the good to which it leads. The good that outshines all ill is not a paradise long since lost but a kingdom which is yet to come in its full glory and permanence.

From this point of view we must speak about moral evil ; about pain, including that of the lower animals ; about the higher and more distinctively human forms of suffering ; and about the relation between all this and the will of God as it has been revealed in Jesus Christ.

CHAPTER XIV

MORAL EVIL

1. THE SHAPE OF SIN

THE religious concept of sin covers the domains of two basic ethical ideas, those of wrong action and of bad moral character. It differs, however, from a combination of these purely ethical notions by setting them in a theological context and interpreting the human faults in question as expressions of a wrong relationship with God.

Since something defective is best understood by contrast with the norm from which it departs, the nature of a wrong relationship with God may be investigated by considering what a right relationship with Him would be, even though this latter represents an ideal that has not (with one exception) been exemplified in human history. The ideal relationship of a human person with God would consist in a vivid awareness of Him, at once joyous and awesome, and a consequent wholehearted worship of the infinite Goodness and Love by obedient service to His purposes within the creaturely realm. To know the creative centre of reality as active personal agape would be to accept gladly one's own status as a creature, utterly insignificant and yet loved and valued in God's free grace, within a universe that wholly depends upon His activity. Such a knowledge of God would make impossible the natural egoism in which we each treat ourselves as the centre of our own world, whilst the awareness of God's universal care and watchful love would render needless that protective self-concern by which we seek to safeguard our own interests in imagined competition with our neighbours. For if God were known as equally the Father of all men and the Lord of all history, there would be no need to fight against others for a share of His love. Accordingly a human being who lived in a wholly right relationship to God would not be defensive, fearful, or grasping. He would be saved from self-regarding anxieties by the knowledge that his own personal welfare forms part of the all-comprehensive good of God's Kingdom and is guaranteed for ever by the divine sovereignty. A sufficiently vivid awareness of our Creator and heavenly Father as the ultimate both of love and of power would thus exclude that anxious self-concern which expresses itself in greed, suspicion, cruelty, and the urge to gain power over others.

But who is portrayed by such a picture of man dwelling in a right relationship with God? Only, according to the claim of Christian faith, one man : Jesus, who was the Christ. Born into the stream of human life running in this wrong channel Jesus nevertheless lived in perfect obedience to his Father's will. He thus represented a higher purpose which was in conflict with the prevailing dynamics of human society, and as a vulnerable human agent of that higher purpose he met a death which has decisively illumined both the horror of man's life apart from God and the depth of the divine love which is so costingly at work to rescue us from that horror. But, apart from that one individual, all men everywhere (so far as we know) are and have always been in varying degrees self-centred rather than God-centred, concerned for their own private welfare rather than for the fulfilment of God's greater purposes for mankind.

It is, of course, also happily true that man's story is illumined by heroism, self-sacrifice, love, and compassion. But these gleams of light only throw into darker relief the surrounding, and chronic, human self-centredness from which have flowed so many forms of man-made evil. Cruelty, greed, lovelessness, ruthless ambition, narrow suspiciousness, with their immense production of human misery, are all expressions of the practical atheism that St. Augustine diagnosed as the heart curved in upon itself. Indeed, by far the greatest bulk of human suffering is due either wholly or in part to the actions or inactions of other human beings. Not only the hurts that weighed upon Hamlet's mind :

Th' oppressor's wrong, the proud man's contumely,
The pangs of dispriz'd love, the law's delay,

The insolence of office, and the spurns
That patient merit of th' unworthy takes . . .[1]

but also, in our own age of scientific achievement, the mass evils of undernourishment and poverty, as well as the various forms of social injustice and exploitation, and the ancient scourge of war and the grave distortions of man's common life due to preparations for war, are all man-made ills. This vast range and depth of sorrow spells out the sad theme of man's inhumanity to man. Further, the range of moral and personal failure may well be greater even than appears. We cannot at present set a limit to the extent to which mental and emotional factors, both individual and collective, enter into our liability to both disease and physical accident. It is a fact that more than half the hospital beds in the most 'advanced' countries are occupied by patients suffering from conditions to which the mind is a major contributing factor. And, beyond this, it may well be that accident-proneness and lack of resistance to disease are to a significant extent caused or promoted by psychological factors. To the extent to which this is so — an extent that cannot be precisely defined in the present state of psychosomatic medicine — we may say that the incidence of disease might be much lower or even non-existent if these psychic conditions were sufficiently favourable.

Sin, then, is a disorientation at the very centre of man's being where he stands in relationship with the Source and Lord of his life and the Determiner of his destiny. That vertical relationship affects all our horizontal relationships within the created realm, so that our sinfulness expresses itself in various kinds of broken, distorted, perverted, or destructive relationships to our fellows and to the natural world, and in correspondingly wrong attitudes to ourselves as beings organic both to nature and to human society. And because sin thus belongs to our own innermost nature and is at the same time the source of so many forms of evil it has usually, and surely rightly, been seen as constituting the heart of the problem of evil. Accordingly at this central

[1] *Hamlet* III. i. 71–73.

point the theodicy-problem takes the form : Why has an omnipotent, omniscient, and infinitely good and loving Creator permitted sin in His universe?

2. The Traditional Free-will Defence

To this question the Christian answer, both in the Augustinian and in the Irenaean types of theodicy, has always centred upon man's freedom and responsibility as a finite personal being. This answer has recently been critically discussed in the philosophical journals under the name of the free-will defence.[1] The discussion has been of great value in clarifying the issues involved, and as a result it is evident that, fully stated, the free-will defence falls into three stages.

The first stage establishes a conception of divine omnipotence. It is argued that God's all-power does not mean that He can do anything, if 'anything' is held to include self-contradictions such as making a round square, or a horse that has none of the characteristics of a horse, or an object whose surface both is and is not red all over at the same time. The self-contradictory, or logically absurd, does not fall within the scope of God's omnipotence ; for a self-contradiction, being a logically meaningless form of words, does not describe anything that might be either done or not done. As Aquinas comments, 'it is more appropriate to say that such things cannot be done, than that God cannot do them'.[2] Thus, for example, God will never make a four-sided triangle. However, this is not because He cannot make figures with four or any other number of sides, but merely because the meaning of the word 'triangle' is such that it would never be correct to call a four-sided figure a triangle. Clearly this

---

[1] J. L. Mackie, 'Evil and Omnipotence', *Mind* (April 1955) ; Antony Flew, 'Divine Omnipotence and Human Freedom', *N.E.* ; S. A. Grave, 'On Evil and Omnipotence', *Mind* (April 1956) ; P. M. Farrell, 'Evil and Omnipotence', *Mind* (July 1958) ; Ninian Smart, 'Omnipotence, Evil and Supermen', *Philosophy* (April/July 1961) ; and replies in this latter journal by Flew (January 1962) and Mackie (April 1962). Mackie's 'Evil and Omnipotence' and Smart's 'Omnipotence, Evil and Supermen' are reprinted in Nelson Pike, ed., *God and Evil* (Englewood Cliffs, N.J.: Prentice-Hall, Inc., 1964).

[2] *S.T.*, pt. i, Q.xxv, art. 3. Aquinas's entire discussion of this point is classic and definitive.

does not involve any limitation upon God's power such that if He had greater power He would be able to accomplish these logical absurdities. Not even infinite might can adopt a meaningless form of words as a programme for action.

The first phase of the argument is, I think, clearly sound, and since it is also accepted by the recent philosophical critics of the free-will defence[1] I shall not defend it further here.

The second phase of the argument claims that there is a necessary connection between personality and moral freedom such that the idea of the creation of personal beings who are not free to choose wrongly as well as to choose rightly is self-contradictory and therefore does not fall within the scope of the divine omnipotence. If man is to be a being capable of entering into personal relationship with his Maker, and not a mere puppet, he must be endowed with the uncontrollable gift of freedom. For freedom, including moral freedom, is an essential element in what we know as personal as distinct from non-personal life. In order to be a person man must be free to choose right or wrong. He must be a morally responsible agent with a real power of moral choice. No doubt God could instead have created some other kind of being, with no freedom of choice and therefore no possibility of making wrong choices. But in fact He has chosen to create persons, and we can only accept this decision as basic to our existence and treat it as a premise of our thinking.[2]

This second phase of the argument, like the first, seems to be clearly sound and is likewise not challenged by the contemporary philosophical critics of the free-will defence. I shall therefore not elaborate or defend it further.

3. The Recent Critique of the Free-will Defence

It is upon the third phase that discussion centres. Granted that God is going to make finite persons and not mere

---

[1] See J. L. Mackie, 'Evil and Omnipotence', *Mind*, p. 203, and Antony Flew, 'Divine Omnipotence and Human Freedom', *N.E.*, p. 145.

[2] Few would, after reflecting on the matter, be willing to follow T. H. Huxley: 'I protest that if some great Power would agree to make me always think what is true and do what is right, on condition of being turned into a sort of clock . . . I should instantly close with the offer.' *Collected Essays*, vol. i (London: Macmillan & Co., 1894), pp. 192–3.

puppets or automata; and granted that persons must be genuinely free; could not God nevertheless have so made men that they would always freely do what is right? For human persons, though all endowed with some degree of freedom and responsibility, nevertheless vary markedly in their liability to sin. The saint, at one end of the scale, of whom we can say that it is logically possible but morally impossible for him to sin, and the depraved and perverted human monster at the other extreme, of whom we can say that it is logically possible but morally impossible for him not to sin, are both persons. They both, we are supposing, possess the freedom which is the ground of moral responsibility and the basis of liability to praise or blame. And accordingly it would be true of a morally perfect person that it is logically possible for him to sin and yet that he will never in fact do so, either because he has no inclination to sin or because he is so strongly orientated towards the good that he always masters such temptations as he meets. His whole nature would be perfect (even though it might contain the tension of temptations overcome), and accordingly the actions flowing from that nature would constitute perfect responses to his environment. That moral perfection is compatible with liability to temptation is established, for Christian belief, by the fact of Christ, who 'in every respect has been tempted as we are, yet without sinning'.[1] It would therefore seem that an omnipotent deity, creating *ex nihilo*, and determining solely by His own sovereign will both the nature of the beings whom He creates and the character of the environment in which He places them, could if He wished produce perfect persons who, while free to sin and even perhaps tempted to sin, remain for ever sinless.

Antony Flew and J. L. Mackie, writing independently, have recently pressed this attack upon the free-will defence, arguing that God might have made His human creatures so that they would always in fact freely choose the right.

Flew begins by defining a free action as an action that is not externally compelled but flows from the nature of the agent.[2] This, he claims, is the way in which the word 'free'

---

[1] Hebrews iv. 15. [2] Flew, *N.E.*, pp. 149–51.

is normally used ; and he offers as a paradigm case of a free action a young man's choice of the girl whom he wishes to marry. This is a free decision, not in the sense that it is arbitrary or unrelated to his whole psychophysical make-up and to the course of his life up to this time, or unpredictable by those who know him best, but simply in the sense that it is *his* choice, arising out of his own nature, and arrived at without outside compulsion. (This corresponds exactly to the kind of freedom that Calvin attributes to man and on the basis of which he holds man responsible for his sins.[1])

Flew's next move is to point out that, so defined, acting freely is not incompatible with being caused to act in the way in which one does act. On the contrary, we are always caused to act as we do — we are 'caused' by our own nature considered in its entirety. If we were different people we would act differently ; but being precisely the people that we are, we act as we do. Hence, it would seem that God, in initially giving us the nature out of which we behave as we do, could have given us one that would always freely issue in right actions. So, Flew contends, 'Omnipotence might have, could without contradiction be said to have, created people who would always as a matter of fact freely have chosen to do the right thing.'[2] This is the essence of Flew's argument. It can be concentrated into a single question, If God made us, why did He not make us so that we should always want to do what is right?

The adequacy of Flew's definition of free will, as simply the absence of external constraint, can of course be questioned, and I shall in fact argue at a later stage that the Christian conception of the divine purpose for man requires as its postulate the stronger notion of free will as a capacity for choice whose outcome is in principle unpredictable.

---

[1] See, for example, Calvin's *Inst.* ii. iii. 5. (At another point Calvin grants that it would have been possible for God so to have made man that he 'either could not or would not sin at all'. 'Such a nature', he continues, 'would, indeed, have been more excellent. But to quarrel with God on this precise point, as if he ought to have conferred this upon man, is more than iniquitous, inasmuch as it was in his own choice to give whatever he pleased.' i. xv. 8.)

[2] Flew, *N.E.*, p. 152.

I shall, however, endeavour to show by internal criticism the insufficiency of the Flew–Mackie argument and the way in which it points to the need for such a larger conception of freedom. But let us first have before us another statement of the challenge that is to be met. J. L. Mackie formulates it as follows:

> If there is no logical impossibility in a man's freely choosing the good on one, or on several, occasions, there cannot be a logical impossibility in his freely choosing the good on every occasion. God was not, then, faced with a choice between making innocent automata and making beings who, in acting freely, would some-times go wrong: there was open to him the obviously better possibility of making beings who would act freely but always go right. Clearly, his failure to avail himself of this possibility is inconsistent with his being both omnipotent and wholly good.[1]

In a subsequent reply to critics[2] Mackie notes a progression of three questions that arise; and it will be useful to discuss the problem in the same three stages.

The first question is this: Granted that it is logically possible that one man should on one occasion freely choose the good, is it also logically possible that *all* men should *always* do so? Is there, in other words, any logical contra-diction in the idea of all men always acting rightly, and doing so of their own free choice, without external compulsion? Clearly the answer of the Christian theologian must be that this *is* logically possible, since it belongs to the expected fulfilment of God's purpose for human life.

A second question now arises: Granted that it is logically possible that men should always freely choose the good, is it also logically possible that they should be *so constituted* that they always freely choose the good? In other words, is there any logical contradiction in the idea of men being by nature such that they always spontaneously want to do the right thing, so that of their own free desire they live morally flaw-less lives?

One of the most interesting items in the recent discussion

[1] 'Evil and Omnipotence', *Mind*, p. 209.
[2] 'Theism and Utopia', *Philosophy*, vol. xxxvii, no. 140 (April 1962).

in the journals has focused upon this question. Ninian Smart[1] has characterized the claim that men could have been created wholly good as 'the Utopia thesis', and has attacked this on the ground that 'The concept *good* as applied to humans connects with other concepts such as *temptation, courage, generosity*, etc. These concepts have no clear application if men are built wholly good.'[2] He claims that the notion of goodness would be emptied of content if there were no such experience as temptation and therefore no occasion to choose good as distinct from evil. Imagining such a situation, he says:

> I think that none of the usual reasons for calling men good would apply in such a Utopia. Consider one of these harmless beings. He is wholly good, you say? Really? Has he been courageous? No, you reply, not exactly, for such creatures do not feel fear. Then he is generous to his friends perhaps? Not precisely, you respond, for there is no question of his being ungenerous. Has he resisted temptations? No, not really, for there are no temptations (nothing you could really call temptations). . . .[3]

Smart's conclusion is:

> that the concept *goodness* is applied to beings of a certain sort, beings who are liable to temptations, possess inclinations, have fears, tend to assert themselves and so forth; and that if they were immunized from evil they would have to be built in a different way. But it soon becomes apparent that to rebuild them would mean that the ascription of goodness would become unintelligible, for the reasons why men are called good and bad have a connection with human nature as it is empirically discovered to be. Moral utterance is embedded in the cosmic status quo.[4]

This is a persuasive argument; nevertheless the aid which it offers to the free-will defence must not be overstated. Smart has shown that a morally untemptable being could not properly be described as good as this term is normally used in ethical discussion. A creature not subject to temptation, or to fear, lust, envy, panic, anxiety, or any other demoralizing condition, would no doubt be innocent but could not

---

[1] 'Omnipotence, Evil and Supermen.'   [2] Ibid., p. 188.
[3] Ibid., p. 192.   [4] Ibid., pp. 190–1.

justifiably be praised as being morally good. In order to possess positive goodness men must be mutable creatures, subject to at least some forms of temptation. This is the valid conclusion of Smart's reasoning. But as both Flew and Mackie have pointed out in their replies to Smart,[1] they can admit this without their main position being affected. For their contention is that men might have been so constituted as to have been more resistant to temptation than they are — sufficiently more resistant, in fact, not to fall. And Smart's reasoning does not exclude this idea. It remains as a possibility; and the Flew–Mackie challenge recurs : Why did God not realize this possibility in His initial creation of mankind? Why did He not make men so that they would, out of their own God-given moral resources, always overcome temptation and freely act rightly?

4. Divine–Human Personal Relationship

Is it possible to develop an argument that will exclude this as impossible? Can we, for example, draw a distinction, and claim that whilst there is no contradiction in the idea of men being so constituted that they always freely act rightly, there is nevertheless a contradiction in the idea of God so forming them that they can be *guaranteed in advance* always freely to act rightly? At this point the issue moves into that raised by the third in Mackie's progression of questions : Granted that it is logically possible that men should be so constituted that they always freely choose the good, is it also logically possible that God should make them so? Here Mackie's argument is very simple : 'If their being of this sort is logically possible, then God's making them of this sort is logically possible.'[2]

Mackie's conclusion here seems undeniable. But having accepted it there is a fourth question to be asked. This arises from an important aspect of Christian belief that

[1] Antony Flew, 'Are Ninian Smart's Temptations Irresistible?', *Philosophy*, vol. xxxvii, no. 139 (January 1962), p. 58 ; J. L. Mackie, 'Theism and Utopia', *Philosophy* (April 1962), p. 155.
[2] 'Theism and Utopia', p. 157.

Mackie (unlike Flew) fails to take into account. According to Christianity, the divine purpose for men is not only that they shall freely act rightly towards one another but that they shall also freely enter into a filial personal relationship with God Himself. There is, in other words, a religious as well as an ethical dimension to this purpose. And therefore, having granted that it would be logically possible for God so to make men that they will always freely act rightly towards each other, we must go on to ask the further question : Is it logically possible for God so to make men that they will freely respond to Himself in love and trust and faith?

I believe that the answer is no. The grounds for this answer may be presented by means of an analogy with post-hypnotic suggestion, which Flew uses in this connection.[1] A patient can, under hypnosis, be given a series of instructions, which he is to carry out after waking — say, to go at a certain time to a certain library and borrow a certain book — and he may at the same time be told that he will forget having received these instructions. On coming out of the hypnotic trance he will then be obediently unaware of what transpired in it, but will nevertheless at the prescribed time feel an imperious desire to go to the library and borrow the book, a desire that the ordinary resources of the educated intellect will find no difficulty in rationalizing. The patient will thus carry out the hypnotist's commands whilst seeming both to himself and to others to be doing so of his own free will and for his own sufficient reasons. In terms of the definition of a free act as one that is not externally compelled but flows from the character of the agent, the actions of one carrying out post-hypnotic suggestions are free actions and the patient is a free agent in his performance of them. Nevertheless, taking account of the wider situation, including the previous hypnotic trance, we must say that the patient is not free as far as these particular actions are concerned *in relation to the hypnotist*. In relation to the hypnotist he is a kind of puppet or tool. And if the hypnotist's suggestion had been that the patient would agree with him about some

[1] Flew, *N.E.*, pp. 161 f.

controversial matter or, coming closer to an analogy with our relationship with God, trust the hypnotist, or love him, or devotedly serve him, there would be something inauthentic about the resulting trust, love, or service. They would be inauthentic in the sense that to the hypnotist, who knows that he has himself directly planted these personal attitudes by his professional techniques, there would be an all-important difference between the good opinion and trust and friendship of the patient and that of someone else whose mind had not been conditioned by hypnotic suggestion. He would regard and value the two attitudes in quite different ways. His patient's post-hypnotic friendship and trust would represent a purely technical achievement, whereas the friendship and trust of the other would represent a response to his own personal qualities and merits. The difference would be that between genuine and spurious personal attitudes — genuine and spurious, not in respect of their present observed and felt characters but in respect of the ways in which they have come about. For it is of the essential nature of 'fiduciary' personal attitudes such as trust, respect, and affection to arise in a free being as an uncompelled response to the personal qualities of others. If trust, love, admiration, respect, affection, are produced by some kind of psychological manipulation which by-passes the conscious responsible centre of the personality, then they are not real trust and love, etc., but something else of an entirely different nature and quality which does not have at all the same value in the contexts of personal life and personal relationship. The authentic fiduciary attitudes are thus such that it is impossible — logically impossible — for them to be produced by miraculous manipulation: 'it is logically impossible for God to obtain your love-unforced-by-anything-outside-you and yet himself force it'.[1]

For if God had done what Mackie and Flew claim that He ought to have done, namely so fashioned men's natures that

[1] John Wisdom, 'God and Evil', *Mind*, vol. xliv (Jan. 1935), p. 10. Wisdom's article is an important defence of the position that 'It is possible that there is or will be in this world something, say a kingdom of heaven, of so great value that any world without it would be worse than this one and that further the present evil is a logically necessary condition of it.' (p. 4.)

they always freely act rightly, He would be in a relationship to His human creatures comparable with that of the hypnotist to his patient. That is to say, He would have pre-selected our responses to our environment, to one another, and to Himself in such a way that although these responses would from our own point of view be free and spontaneous, they would from God's point of view be unfree. He alone would know that our actions and attitudes, whilst flowing from our own nature, have in fact been determined by His initial fashioning of that nature and its environment. So long as we think of God's purpose for man, as Mackie does, exclusively in terms of man's performance in relation to his fellows, as a moral agent within human society, there is no contradiction in the idea of God's so making human beings that they will always freely act rightly. But if we proceed instead from the Christian view that God is seeking man's free response to Himself in faith, trust, and obedience, we see the necessity for our fourth question and for a negative answer to it. It would not be logically possible for God so to make men that they could be guaranteed freely to respond to Himself in genuine trust and love. The nature of these personal attitudes precludes their being caused in such a way. Just as the patient's trust in, and devotion to, the hypnotist would lack for the latter the value of a freely given trust and devotion, so our human worship and obedience to God would lack for Him the value of a freely offered worship and obedience. We should, in relation to God, be mere puppets, precluded from entering into any truly personal relationship with Him.

There might, indeed, be very great value in a universe of created beings who respond to God in a freely given love and trust and worship which He has Himself caused to occur by His initial formation of their nature. But if human analogies entitle us to speak about God at all, we must insist that such a universe could be only a poor second-best to one in which created beings, whose responses to Himself God has not thus 'fixed' in advance, come freely to love, trust, and worship Him. And if we attribute the latter and higher aim to God, we must declare to be self-contradictory the idea of God's so creating men that they will inevitably respond positively to Him.[1]

To summarize this proposed rebuttal of the Flew–Mackie challenge : God can without contradiction be conceived to have so constituted men that they could be guaranteed always freely to act rightly in relation to one another. But He cannot without contradiction be conceived to have so constituted men that they could be guaranteed freely to respond to Himself in authentic faith and love and worship. The contradiction involved here would be a contradiction between the idea of A loving and devoting him/herself to B, and of B valuing this love as a genuine and free response to himself whilst knowing that he has so constructed or manipulated A's mind as to produce it. The imagined hypnosis case reveals this contradiction as regards the relations between two human beings, and by analogy we apply the same logic of personal attitudes to the relation between God and man.

5. Freedom as Limited Creativity

The argument has pointed to the need for a stronger conception of man's freedom *vis-à-vis* God than that used by Mackie and Flew. It seems that there would be no point in the creation of finite persons unless they could be endowed with a degree of genuine freedom and independence over against their Maker. For only then could they be capable of authentic personal relationship with Him.

But how is such responsible creaturely freedom to be defined? It is not sufficient to say that a free action is undetermined even by the nature of the agent. For to divorce the action from the agent would be to equate freedom with randomness of behaviour.[2] This is not the conception of

[1] W. D. Hudson argues along essentially the same lines in 'An Attempt to Defend Theism', *Philosophy*, vol. xxxix, no. 147 (January 1964), p. 20. For a different response to the Flew–Mackie challenge, claiming that the notion of beings who are so created that they always freely act rightly is a logically incoherent notion, see Alvin Plantinga, 'The Free Will Defence', *Philosophy in America*, ed. by Max Black (London: Allen & Unwin Ltd., 1965).
[2] Cf. J. L. Mackie, 'Evil and Omnipotence', *Mind*, p. 209, and 'Theism and Utopia', *Philosophy*, pp. 156–7.

freedom generally used in ethical discussions. Morally responsible action is not normally regarded as being free in the sense that it is random and undetermined, but on the contrary as being determined by ethical principles or the demands of values. And personal action in general is usually regarded as expressing the individual character of the agent in interaction with his situation. It is, indeed, precisely the fact that his actions arise out of and express his character and personality that makes them *his* actions, for which he is morally accountable. It is very difficult to see how such concepts as responsibility and obligation could have any application if human volitions occurred at random instead of flowing from the individual nature of the agent. From the point of view of ethics the cost of equating freedom with volitional randomness would thus be so great as to be prohibitive.

Is there a third concept of freedom such that a free act is neither, on the one hand, the inevitable outworking of a man's character nor, on the other hand, a merely random occurrence? There is such a concept of freedom, and, indeed, it is the one that seems intuitively most adequate to our ordinary experience as moral agents. It is, however, not easily defined. It is roughly the notion of freedom as a limited creativity. This must be thought of as involving an element of unpredictability; for, whilst the action proceeds from the nature of the agent, the nature from which it proceeds is that of 'the actual self alive in the moment of decision'.[1] Thus, whilst a free action arises out of the agent's character it does not arise in a fully determined and predictable way. It is largely but not fully prefigured in the previous state of the agent. For the character is itself partially formed and sometimes partially re-formed in the very moment of free decision.

Some such concept of freedom seems to be a necessary postulate of the Christian view of the relation between man and God. The primary point at which it is required is that

[1] Charles Hartshorne, *The Logic of Perfection* (Lasalle, Illinois: Open Court, 1962), p. 20. The fullest and most adequate exposition and defence of this conception of freedom known to me is that of C. A. Campbell in *On Selfhood and Godhood* (London: George Allen & Unwin Ltd., 1957), chap. 9.

at which man in his freedom is willing or unwilling to become aware of God. For it is man's *cognitive* freedom in relation to his Creator that must be insisted upon. And the concept of freedom as creativity would make it possible to speak of God as endowing His creatures with a genuine though limited autonomy. We could think of Him as forming men through the long evolutionary process and leaving them free to respond or fail to respond to Himself in uncompelled faith. I do not know, however, how freedom at this point can be proved. Yet even in defending it in the philosophically questionable ways in which alone it can be defended, we are pointing to our own directly intuited status as responsible beings.[1] We know such creativity from within in our own moments of responsible moral choice ; and we can infer it as a presupposition of Christian theology. But it may be that it cannot be independently established by philosophical analysis. Perhaps, as Kierkegaard said, 'The fact that God could create free beings *vis-à-vis* of himself is the cross which philosophy could not carry, but remained hanging from.'[2]

But although the Flew–Mackie critique of the free-will defence fails, and reveals at the same time theology's need for this stronger conception of human liberty, yet the latter cannot in the end save the free-will defence. For although the traditional theodicy can withstand this external attack, it contains within itself tensions and pressures which it cannot withstand.

6. The Virtual Inevitability of the Fall

Let us consider an 'unfallen' being exercising this creative freedom in an 'unfallen' environment. The basic choice open to him is between God-centredness and self-centredness, obedience or disobedience. Let us further suppose that the very fact of his own finitude involves for him the possibility, and the awareness of the possibility, of being dissatisfied with

[1] Cf. D. M. Mackinnon, *A Study in Ethical Theory* (London : A. & C. Black, 1957), chap. 4.
[2] *The Journals of Søren Kierkegaard*, trans. and ed. Alexander Dru (London : Oxford University Press, 1938), p. 58. Journal for April 22, 1838.

his own inferior status *vis-à-vis* God, and hence a temptation
to renounce his free allegiance to his Maker and usurp to
himself an imagined autonomy. If he falls under this temptation his fall is an act of creative spontaneity which is permitted but in no sense ordained by God. He will thenceforth
be at enmity with his Maker, a sinner meriting punishment
or even annihilation, whose only hope is in God's mercy.
This picture amounts to the traditional free-will defence.
But now we must ask in what way and to what extent, in
his pre-fallen state, the free creature was immediately conscious of God? Two alternatives present themselves, both of
which, however, raise difficulties for the traditional free-will
defence.

1. We might suppose that he dwelt in the immediate
presence of God. The difficulty here is that when we think
of a created being thus living face to face with infinite plenitude of being, limitlessly dynamic life and power, and un-
fathomable goodness and love, there seems to be an absurdity
in the idea of his seeing rebellion as a possibility, and hence
in its even constituting a temptation to him. Surely his state
would be that defined by Augustine as one of *non posse peccare*
— not able to sin. For how could he plan to reject the
sovereignty of the divine omnipotence of which he is over-
whelmingly conscious; and why should he ever desire to
reject the Lordship of the infinite Love in which, above all
else, he rejoices? Surely, in order for there to be any impulse
or temptation in this direction he must either be stupid to
the point of being less than human, or else he must already
be possessed by a pride that draws him into enmity against
the Almighty? Thus, on the premise that the creature in his
unfallen state dwells consciously in the supremely glorious
and joyous presence of God, his fall is not sufficiently accounted for merely by attributing to him a pure freedom
of spontaneous creativity. If this is to be something other
than mere randomness the creature's defection requires some
motive or reason or apparent reason; there must be some
kind of temptation drawing him, however slightly, away from
God.

However, it may be objected at this point that it is a fatal

mistake to seek for *reasons* for a sinful choice. For it is of the
very essence of an evil volition to be irrational; devoid of
all reason, including even rational self-interest; without
intelligent motivation, and unguided by any positive aim.
It must, said Julius Müller, be 'inconceivable, *i.e.*, incomprehensible, seeing that it is realized by arbitrariness, and arbitrariness is a violation of rational reason and true sequence'.[1]
Accordingly it is by no means impossible, but on the contrary
all too easy, for a sinful deed to occur in the absence of any
motive or reason or apparent reason. We cannot, then,
exclude the possibility of finitely perfect and blessed angels,
living in a direct unhindered consciousness of God, falling
into sin; and indeed the greater the absurdity and rational
impossibility of such defection the greater the sin.

I do not think, however, that this argument is sound. To
say that a blessed angel, dwelling in the immediate presence
of God, suddenly becomes evil and commits the irrationality
of sin is to imply either that God had created him as an
irrational angel or that in his first sinful volition evil created
itself *ex nihilo*. The first of these possibilities is incompatible
with the perfect goodness and the second with the unlimited
sovereignty of God. It remains true, then, that an angelic
or a human fall presupposes some temptation such as is not
conceivable in finitely perfect creatures existing consciously
in the presence of God.

2. On the other hand, we might suppose that the unfallen
creature does *not* exist in such closeness to God, but rather in
a human (or angelic) world in which the divine reality is not
unambiguously manifest to him. Instead of the infinitely
loving and powerful Creator being continuously evident to
His creatures, it is, let us suppose, only by an act of faith
that they become aware of Him. The difficulty here for the
traditional view is that the situation is now weighted against
the creature. His fall is now rather more than a bare logical
possibility. For, instead of being situated so firmly within the
divine Kingdom, and in enjoyment of the *visio Dei*, that he
could fall away from it only by a titanic but unmotivated act

[1] Julius Müller, *The Christian Doctrine of Sin*, vol. ii, p. 173. Cf. Karl Barth,
C.D., vol. iv/1, p. 410; J. S. Whale, *Christian Doctrine*, pp. 49–50.

of insubordination, he is placed in a natural environment in which some positive effort on his own part is required if he is to be aware of God and rightly to relate himself to Him. In these circumstances, must there not be a sharing of responsibility for man's failure? If God has elected not to make Himself initially evident to His creature, can the latter be altogether to blame if he fails to worship his Maker with his whole being?

This, then, is the dilemma. The creature's fall is either impossible, or else so very possible as to be excusable. It does not seem feasible for the creature to be in a perfectly neutral position; either he is so vividly conscious of God as to be held in God's presence and service by the overwhelming immensity of the divine reality and goodness, or else he is not thus set consciously in God's presence, and self-centredness rather than God-centredness is a very natural possibility for the realization of which the creature can hardly bear the virtually unlimited guilt attributed to him in the traditional free-will defence. In short, then, men (or angels) cannot meaningfully be thought of as finitely perfect creatures who fall out of the full glory and blessedness of God's Kingdom. Sin — self-centredness rather than God-centredness — can only have come about in creatures placed in an environment other than the direct divine presence. Only in such an environment could they have the freedom in relation to God that is presupposed by that state of *posse peccare* (able to sin) that is evidenced by their actual fall.

7. MAN CREATED AS A FALLEN BEING

We may, I think, properly leave the angels out of account and pursue this thought in terms of our human knowledge of God and our human freedom and sinfulness. For I have already argued that the basic problems of finite personal life would not be altered by our transposing them into the heavens and speaking of angelic instead of human beings. Further, whilst we know a good deal about man and his history, we know nothing about angels, so that to derive our theodicy from a supposed angelic prehistory would be to build speculation upon speculation without the benefit of any assured data.

When we compare the way in which man has in fact, so far as we know, appeared on this earthly scene, emerging out of apedom with only dim and rudimentary notions of his Creator, with the epistemological conditions presupposed by man's status as a free and responsible agent in relation to his Maker, we find a remarkable agreement between them. For human freedom *vis-à-vis* God presupposes an initial separateness and a consequent degree of independence on man's part. In creating finite persons to love and be loved by Him God must endow them with a certain relative autonomy over against Himself. But how can a finite creature, dependent upon the infinite Creator for its very existence and for every power and quality of its being, possess any significant autonomy in relation to that Creator? The only way we can conceive is that suggested by our actual situation. God must set man at a distance from Himself, from which he can then voluntarily come to God. But how can anything be set at a distance from One who is infinite and omnipresent? Clearly spatial distance means nothing in this case. The kind of distance between God and man that would make room for a degree of human autonomy is epistemic distance. In other words, the reality and presence of God must not be borne in upon men in the coercive way in which their natural environment forces itself upon their attention. The world must be to man, to some extent at least, *etsi deus non daretur*, 'as if there were no God'. God must be a hidden deity, veiled by His creation. He must be knowable, but only by a mode of knowledge that involves a free personal response on man's part, this response consisting in an uncompelled interpretative activity whereby we experience the world as mediating the divine presence. Such a need for a human faith-response will secure for man the only kind of freedom that is possible for him in relation to God, namely cognitive freedom, carrying with it the momentous possibility of being either aware or unaware of his Maker.

A human environment in which these conditions are fulfilled may be expected by its apparently atheous character to require religious faith if man is ever to know God and yet also, by its fitness to mediate the divine presence and activity, to be such as to make faith possible. On the one hand, then, we should expect the reality of God to be other than automatically and undeniably evident to us; it will, on the contrary, be possible for our minds to rest in the world itself without passing beyond it to its Maker. But we should also expect the reality of God to become evident to men in so far as they are willing to live as creatures in the presence of an infinitely perfect Being whose very existence sets them under a sovereign claim of worship and obedience. We should expect the world to be such that, given this willingness (which is the volitional element in religious faith), we become able to recognize all around us the signs of a divine presence and activity. Men of faith will then see the heavens as declaring the glory of God and will discern His hand moving amidst the events of human history. Thus the world, as the environment of man's life, will be religiously ambiguous, both veiling God and revealing Him — veiling Him to ensure man's freedom and revealing Him to men as they rightly exercise that freedom.

As we thus deduce the basic character of a world in which man can exist as a free and responsible person in the presence of God, have we not found that the description fits the world in which we are living? For we are not set in the direct, unmistakable, divine presence, in which we should be unable to come to God freely, of our own creaturely choice, but instead we are set in a situation in which we may, by our own personal responses, either allow the knowledge of God to dawn upon us, or hold it at bay as a mere intellectual hypothesis. This basic human freedom is depicted in the biblical creation myth. For man is not there presented as an angelic being dwelling in the heavenly places and rejoicing in a continuous awareness of God's environing presence but as a frail, uncertain creature living in his own world, to which God is but an occasional visitor. We may express this today by saying that man was created at the epistemic distance from God at which the gospel has found him. That is to say, when God summoned man out of the dust of the evolutionary process He did not place him in the immediate consciousness of His own presence but in a situation from which man could, if he would, freely enter into the divine Kingdom and presence. And the creation of man in his own relatively autonomous world, in which the awareness of God is not forced upon him but in which he is cognitively free in relation to his Maker, is what mythological language calls the fall of man. Our present earthly existence is described in the myth as man's life after the fall. Man exists at a distance from God's goal for him, however, not because he has fallen from that goal but because he has yet to arrive at it.

All this is consonant with the biblical narrative when the latter is not interpreted as a literal account of events which occurred on earth thousands or tens of thousands of years ago. That story has more often been considered in the elaborated version of Pauline theology than in its original form in Genesis iii. From the point of view of theodicy there are, however, important clues to be found in the more primitive version.

Whereas in the theologically edited myth the serpent has become identified with Satan, the arch-enemy of God and man, there is no suggestion of this in the Genesis text.[1] On the contrary, there the serpent is a part of the animal creation, singled out only as being more subtle than the other beasts. He is not, as in Milton's epic, the titanic prince of darkness striking a blow in his warfare against the Almighty. Accordingly he does not, as in later tradition, represent the presence of explicit moral evil in paradise prior to man's first sin. Indeed he is not properly described as a tempter at all in the sense of one who deliberately solicits to evil. This is a later interpretation of the mythic material. The serpent is amoral, and the urge which he embodies is ethically neutral even though it leads to evil consequences.

The suggestion of the myth is that evil first arose through the interaction of certain factors inherent in man's God-given

[1] Cf. Gerhard von Rad, *Genesis*, trans. John H. Marks (Philadelphia: Westminster Press, 1961), p. 85.

situation, and not from the invasion of an already fully developed evil will from outside. This suggestion offers an account of the origin of moral evil, whereas the later satanic theory merely postpones the problem by implying that evil already existed in the person of the devil. When the latter theory goes on to explain that Satan had fallen through the sin of pride it comes no nearer to offering an explanation of evil's origin, since the question now arises as to the origin of this sinful pride. But the more primitive myth does indicate how moral evil arose. It suggests to the modern reader that in man's initial situation as a created being set in his own world, it was almost inevitable that he should direct his attention elsewhere than to God; and this not through some individual idiosyncrasy or culpable failing, but generically and unavoidably in all mankind.

What, we may now ask, is this fatal factor represented in the story by the serpent? It is agreed by the commentators that the 'knowledge of good and evil' that Adam and Eve attained by eating the forbidden fruit does not mean an awareness of *moral* good and evil.[1] It is rather knowledge in general, and especially knowledge of what is beneficial and harmful. It signifies technical rather than ethical information. It suggests, then, in the light of what has already been said, that the serpent, the most subtle of the beasts, represents the exploring experimental intelligence that is inherent in man's rational nature and that is evoked by an environment which challenges him to master it as a natural realm apparently existing in its own right. The serpent is the first scientist; his 'temptation' is the earliest hypothesis; and the fall is the first and most daring experiment. This primal experiment, in which the whole future scientific enterprise is implicit, is the experiment of regarding the world as an independent order with its own inherent structure and laws. The serpent rejects the religious significance affixed to the tree of knowledge by divine taboo. He declares that there is nothing to fear; the tree is just a tree like any other; the religious prohibition means nothing. It is this 'over-subtle' idea of the serpent's, treating the tree as a natural rather

[1] Cf. von Rad, *Genesis*, pp. 79 and 86–87.

than a religious object, and not the unintelligible defection of a perfect archangel, that brings evil into the world. Man's epistemic distance from God makes possible, and more than possible, man's own self-realization in relation to the world and in distinction from God. Man investigates nature, thereby treating it as a natural phenomenon, *etsi deus non daretur*, and in thus turning his primary attention upon the world as an environing order he forfeits the vision of God. He becomes immersed in nature and alienated from God. This is the basic separation, from which self-alienation and alienation from his fellows inevitably follow.

Thus the ancient myth seems to point, however obscurely and ambiguously, to a fall (or, as we must presently say, a 'fallenness') rendered virtually inevitable by the basic features of man's divinely appointed situation. When we detach the myth from its customary framework of Pauline theology we accordingly find that its atmosphere is more tragic than wicked and its standpoint more descriptive than condemnatory. Milton has well caught this spirit in the concluding lines of *Paradise Lost*:

> Some natural tears they dropped, but wiped them soon;
> The world was all before them, where to choose
> Their place of rest, and Providence their guide:
> They hand in hand with wandering steps and slow,
> Through Eden took their solitary way.[1]

The biblical myth pictures man's present alienation from God as resulting from a temporally prior fall. Man's state, which is as though he had lost an original perfection, is there pictured in terms of the drama of paradise lost. But the advancing knowledge which has identified the myth as myth rather than as history has taught us to see in it instead an analysis of man's state as it has always been. Accordingly when as Christians of the twentieth century we use the biblical myth, not to contradict what the sciences tell us about the circumstances of man's emergence on earth, but to express thereligious meaning of those circumstances, we arrive at the following picture. Animal life was first caused

[1] Bk. xii, 645–9. Cf. Roland M. Frye, *God, Man and Satan* (Princeton University Press, 1960), p. 91.

to develop into a primitive human consciousness which, so far from being qualified for citizenship in the Kingdom of God, was originally directed towards the absorbing task of mastering a largely hostile environment. Man lived in relation to the world rather than in relation to God. And in causing man to evolve in this way out of lower forms of life God has placed His human creature away from the immediate divine presence, in a world with its own structure and laws in which he has a certain relative but real autonomy and freedom over against his Creator. He exists in such a close organic relationship to the natural world that this is the first object of his knowledge and interest, and he can become conscious of God's presence in and beyond this only in so far as he is willing to know himself as subordinate to a personal Mind and Will infinitely superior to himself in worth as well as in power. Thus man's basic cognitive freedom in relation to God is established, and the drama of salvation, which is the story of the eighth day of the divine creative work, begins.

This arrangement has a dual effect. On the one hand, man's spiritual location at an epistemic distance from God makes it virtually inevitable that man will organize his life apart from God and in self-centred competitiveness with his fellows.[1] How can he be expected to centre his life upon a Creator who is as yet unknown to him? He must instead centre his life upon himself, even though he then immediately begins to feel, in a dim sense of moral demand and holy claim, the pressure upon his spirit of his unseen Creator. And this is what historical knowledge and reasonable inference reveal to us. However far back we go in human

---

[1] This 'virtual inevitability' of sin has been recognized by a number of theologians in the present century. For example, William Temple says that 'It is not utterly necessary that this [i.e. man's self-centredness] should be so; and therefore it is not true to say that God made man selfish, or predestined him to sin. But that it should be so was "too probable not to happen"; and it is true to say that God so made the world that man was likely to sin, and the dawn of moral self-consciousness was likely to be more of a "fall" than an ascent' (*Nature, Man and God*, p. 366); Reinhold Niebuhr says that 'man sins inevitably yet without escaping responsibility for his sin' (*The Nature and Destiny of Man*, vol. i (London: Nisbet & Co., 1941), p. 266); and Paul Tillich speaks of 'the point at which creation and the fall coincide' (*Systematic Theology*, vol. i, pp. 255–6.)

history and prehistory men have apparently lived in other than a perfect filial relationship to their Maker.[1] Our Judaic-Christian myth affirms this of everyone since Adam and Eve, and modern science has only completed the picture by adding that there never were an actual Adam and Eve. Man as he emerged from the evolutionary process already existed in the state of epistemic distance from God and of total involvement in the life of nature that constitutes his 'fallenness'. He did not fall into this from a prior state of holiness but was brought into being in this way as a creature capable of eventually attaining holiness. In Irenaeus' terminology, he was made in the image but had yet to be brought into the likeness of God.[2]

On the other hand, this divine economy makes it possible for man, thus reaching self-consciousness at an epistemic distance from God, freely to accept God's gracious invitation and to come to Him in uncompelled faith and love. For the absolute goodness of the Creator is such that there can be no neutrality in relation to Him. Those who are not for Him are against Him; and the paradox of creaturely freedom is that only those who are initially against Him can of their own free volition choose to be for Him. Man can be truly *for* God only if he is morally independent of Him, and he can be thus independent only by being first *against* Him! And because sin consists in self-centred alienation from God, only God can save us from it, thereby making us free for Himself. Thus man must come to heaven by the path of redemption from sin.

This is said in agreement both with the profound medieval insight of the 'O felix culpa'[3] and with the massive Protestant theology of Friedrich Schleiermacher in the nineteenth

[1] It is possible that weapons were used and murders committed even by our pre-human ancestors. Evidence for this is collected by Robert Ardrey in *African Genesis* (London: James Collins, 1861), pp. 186 f.
[2] Irenaeus, *A.H.* v. vi. 1. See above, pp. 217 ff.
[3] The 'O felix culpa . . .' theme is perennial in Christian thought and literature. For example, a fifteenth-century lyric has lately been revived as a Christmas carol:

> And all was for an appil,
>   An appil that he tok . . .
> Ne hadde the appil takê ben,
>   The appil taken ben,

century.[1] And yet having said it, and having acknowledged it to be valid Christian truth, we find — as so often in the course of this investigation — that the problem which has been thus encircled by the forces of reason breaks out unexpectedly at another point. Having recognized the inevitability of sin as a corollary of man's initial epistemic distance from his Creator we must now complicate our picture again in order to make it correspond more faithfully to the complex, messy, and often sickening realities of human life as we know it. For violence, treachery, and bestial cruelty have left their blood-stained marks on the pages of recent history and must be allowed to make their imprint upon our theology also. When we think, for example, of Belsen and Auschwitz and Buchenwald we are compelled to take seriously the idea of the demonic in the sense of evil which is utterly gratuitous — evil simply for evil's sake. We encounter this in sheer pointless sadistic cruelty; in the deliberate destruction or corruption of personality by the crushing of a human being into moral insensibility, or his perversion into an evil monster; in terrifying experiences that unhinge the mind; and in other shapes and forms. When we meet evil in such senselessly malevolent and malicious acts and attitudes we seem to be face to face with that which

> Ne haddê never our lady
> A bene hevene quene.
> Blessed be the time
> That appil takê was.
> Therefore we moun singen
> '*Deo gracias*'.

(*Early English Lyrics*, ed. E. K. Chambers and F. Sidgwick, London: A. H. Bullen, 1907, no. 50.)

And Milton makes his Adam, having been granted a vision of the future, say to the Archangel Michael:

> full of doubt I stand,
> Whether I should repent me now of sin
> By me done and occasioned, or rejoice
> Much more that much more good thereof shall spring,
> To God more glory, more good will to men
> From God — and over wrath grace shall abound.
> (*Paradise Lost*, xii, 473–8.)

On this theme both in Milton and more generally in literature, see Arthur O. Lovejoy, 'Milton and the Paradox of the Fortunate Fall' in *Essays in the History of Ideas*, 1948 (New York: Capricorn Books, 1960).

[1] See Chapter X above.

has been called the devil or, less anthropomorphically, the demonic. In the demonic, evil as a necessary element in a soul-making universe seems to have got out of hand and to have broken loose from God's control. It seems to have developed its own evil potentialities far beyond anything that can have been intended in any divine plan.

What are we to say about this development from the standpoint of the thesis of this chapter?

In the demonic we see laid bare the true nature of all evil in so far as it is purely and unambiguously evil. Evil *is* demonic! There is a continuity between mild evil and demonic evil, and a nisus in the one towards the other. But even the shock of this fact must not be allowed to force us out of monotheism into dualism. We must not, under the impact of our vision of the demonic, deify evil and dethrone God. Even in its most virulent forms evil is still not ultimate. It cannot be unforeseen by the Creator or beyond His control. We must not suppose that God intended evil as a small domestic animal, and was then taken aback to find it growing into a great ravening beast! The creator to whom this could happen is not God. Even as we recognize the demonic potentialities and actualities of evil we must still affirm the mystery that was revealed to Mother Julian of Norwich when she wondered 'why the beginning of sin was not prevented by the great foreseeing wisdom of God'. She heard the voice of Jesus say to her, 'Sin must needs be, but all shall be well. All shall be well; and all manner of thing shall be well.'[1] We have in the end, then, both to recognize the essentially demonic nature of evil, and to maintain the sole ultimate sovereignty and omni-responsibility of God. This final and most difficult problem of theodicy will be the subject of Chapter XVII.

However, before concluding the present chapter something more should be said about the idea of divine responsibility for evil.[2] This concept is generated by the belief that God

[1] *The Revelations of Divine Love of Julian of Norwich*, trans. by James Walsh, S.J. (London: Burns & Oates, 1961), chap. 27. Cf. O. C. Quick, *Doctrines of the Creed* (London: Nisbet & Co. Ltd., 1938), pp. 210–11.

[2] A contemporary theologian who directly uses this concept is Leonard Hodgson in *For Faith and Freedom*, ii. 57. See above, pp. 258–9.

has determined in His absolute freedom all the conditions of creaturely existence, including those out of which sin and suffering have arisen. Given this circumstance, how can God fail to have the final responsibility for the existence of His creation in its concrete actuality, including, as it does, evil as an element within it?

Whether or not we speak of God as being ultimately responsible for the existence of evil depends upon our definition of responsibility. For there are both differences and similarities between the sense in which men are responsible and any sense in which God could be said to be responsible. Human responsibility occurs within the context of an existing moral law and an existing society of moral beings. But God is Himself the source of the moral law and the Creator of all beings other than Himself. In His original decision to create He was accordingly not responsible *under* any moral law or *to* any existing person. Nevertheless, there is a technical sense of 'responsible' that can be applied to God as well as to man. One whose action, A, is the primary and necessary precondition for a certain occurrence, O, all other direct conditions for O being contingent upon A, may be said to be responsible for O, if he performs A in awareness of its relation to O and if he is also aware that, given A, the subordinate conditions will be fulfilled. Suppose, for example, that an alcoholic is in process of being cured, and alcohol is being carefully kept away from him. If I then deliberately contrive a situation in which alcohol is easily available to him, and he succumbs to an irresistible craving for it and harms both himself and others, I must then bear a responsibility for what happens, even though my accountability does not cancel any that is properly attributed to the alcoholic himself. In this case my responsibility is qualified by my ignorance as to whether the alcoholic will or will not succumb; I am guilty only of *risking* a disastrous outcome for him. But in the case of God no such qualification is possible. His decision to create the existing universe was the primary and necessary precondition for the occurrence of evil, all other conditions being contingent upon this, and He took His decision in awareness of all that would flow from it.

However, even this ultimate omni-responsibility of the Creator does not take away each human individual's accountability for his own deliberate actions. Our moral liability is not diminished by the fact that we have been created by a higher Being who bears the final responsibility for His creatures' existence, with all that this contingently encompasses. The divine and human responsibilities operate upon different levels and are not mutually incompatible. Man is responsible for his life within the creaturely world, whilst God is ultimately responsible for the existence of that creaturely world and for the fact that man lives responsibly within it.[1]

[1] See further on p. 396 below.

CHAPTER XV

PAIN

1. PAIN AND SUFFERING

THERE is a large physiological and medical literature
dealing with the subject of pain, and what follows,
representing a philosopher's attempt to relate the
scientific work in this field to the wider problem of evil, is
based primarily on the works mentioned in the footnote.[1]
Philosophers, prior to the modern scientific study of the
subject, linked pain with pleasure as opposite *qualia* of the
soul. These were regarded as psychic states rather than
physical sensations, and as such they were thought to stand
at opposite ends of what has sometimes been called the
hedonic scale. However, this polarization of the two terms
is now seen to be mistaken. For, whilst pleasure is a psychic
condition, pain is a physical sensation with its own nerve
structure. Pain in this sense has no opposite ; it is simply a
unique irreducible mode of sensation. On the other hand,
the opposite of the psychic state of pleasure is the psychic
state that is variously called unpleasure, suffering, discomfort, distress, anguish, negative hedonic tone, and which I
propose to call suffering. The relation between pain and
suffering is that the former normally gives rise to the latter.
However, suffering is not simply our reaction to physical
pain. It is our reaction also to many kinds of events and
circumstances, of which physical pain is only one and gener-

[1] Ferdinand Sauerbruch and Hans Wenke, *Pain: Its Meaning and Significance*,
trans. by Edward Fitzgerald (London: George Allen & Unwin Ltd., 1963) ;
C. A. Keele and Robert Smith, editors, *The Assessment of Pain in Man and
Animals*, The Proceedings of an International Symposium held under the
auspices of The Universities Federation for Animal Welfare (London: E. & S.
Livingstone Ltd., 1962) ; Harold G. Wolff and James D. Hardy, 'On the
Nature of Pain', *Physiological Reviews*, vol. xxvii, no. 2 (April 1947) (with full
bibliography) ; Karl M. Dallenbach, 'Pain: History and Present Status',
The American Journal of Psychology, vol. lii, no. 3 (July 1939) ; Thomas Lewis,
M.D., F.R.S., *Pain* (New York: The Macmillan Co., 1942).

ally not the most dreadful. We may suffer also, for example, from fear, anxiety, remorse, envy, humiliation, a sense of injustice, the death of someone loved, unrequited affection, personal estrangement, boredom, and frustration of many sorts. Indeed emotional suffering, quite unconnected with physical pain, can grip us more inwardly and encroach more inexorably upon the centre of our personal being, and be therefore less endurable, than physical pain.

In the present chapter we shall be dealing with the level and range of suffering that is caused by physical pain, and in the next chapter with the forms of suffering that occur independently of pain.

The distinction between pain as a physical sensation and the psychological reaction to it appears, with variations of terminology, in most of the medical discussions of pain. Thus a standard textbook on *The Relief of Pain* says that 'it is important to remember that there is a vast difference between suffering and pain, either of which may exist without the other, though both are often intimately associated'.[1] And the physiologist Henry Head said that 'it is necessary at the outset to distinguish clearly between "discomfort" and "pain". Pain is a distinct sensory quality equivalent to [i.e. comparable with] heat and cold, and its intensity can be roughly graded according to the force expended in stimulation. Discomfort, on the other hand, is that feeling tone which is directly opposed to pleasure.'[2] However, since for most ordinary purposes the word 'pain' is customarily used to cover both the physical stimulus and the associated emotional reaction, the terminology suggested by Dr. J. D. Hardy is perhaps least likely to cause misunderstandings. He distinguishes between the 'pain sensation' (i.e. physical pain as such) and the 'pain experience', which latter includes the affective state of distress or suffering that is normally produced by physical pain.[3]

[1] Harold Balme, *The Relief of Pain: A Handbook of Modern Analgesia* (London: J. & A. Churchill Ltd., 2nd ed., 1939), p. 2.
[2] Henry Head, M.D., F.R.S., *Studies in Neurology* (London: Hodder & Stoughton Ltd., 1920), vol. ii, p. 665.
[3] J. D. Hardy, 'The Pain Threshold and the Nature of Pain Sensation', in Keele and Smith, op. cit., p. 172.

2. PHYSICAL PAIN

Like other primary sensations, such as those of colour or sound, pain cannot be defined, except ostensively.[1] It is the sensation that you feel when, for example, a pin is stuck into you, or when heat of above about forty-five degrees centigrade is applied to your skin.

The scientific study of pain began approximately a hundred years ago, and the first theory to become widely accepted was that pain was an overloading effect, caused by excessive stimulation of any of the sense organs — for example, excessive heat, cold, pressure, or noise.[2] However, this theory has been generally abandoned in the light of accumulating evidence for the existence of specific pain-receptor nerves, distinct from those that mediate the sensations of touch, heat, and cold. These pain nerves exist virtually all over the surface of the body, but still at definite points (though in some regions grouped as closely as two hundred to the square centimetre[3]) which can be mapped by applying a very fine bristle or a minute point of radiant heat.[4]

As regards the mechanism by which the pain system is excited, it seems that this may occur by a direct impact upon the receptor-nerve membrane triggering off the excitation; or indirectly, through the effect of intermediary chemical substances, which are produced or released (under the influence of the pain-causing stimulus) from tissue cells surrounding the pain receptor. Possibly this latter mechanism commonly operates in the case of deep or visceral pain, with its dull,

---

[1] Cf. Thomas Lewis, *Pain*, p. v; H. K. Beecher, in Keele and Smith, op. cit., p. 161.

[2] However, such is the state of flux of the whole subject that in 1962 W. D. Keele could say of the old intensity theory that 'there is reason to think that in some new guise it may reappear at any time'. ('Some Historical Concepts of Pain', in Keele and Smith, op. cit., p. 24.)

[3] Cf. Thomas Lewis, *Pain*, op. cit., p. 11.

[4] The medical and physiological texts describe various other complications, and questions awaiting further research, which do not, however, enter into the minimal picture required for our present purpose. There are, in particular, the 'free nerve endings' (i.e. the finest terminal branchings out of a nerve cell whose body lies close to the spinal cord) which may apparently mediate both pain and other sensations.

not sharply localizable, quality and its slow build-up and decline, and the former in the case of superficial pain.

So much for the purely physical and neurological event that is called pain. But this is not yet pain as we actually experience it, which normally includes the emotional reaction of suffering. Suffering, however, is not attached to pain in an exact and invariable proportion. The extent to which a given quantity of the pain sensation causes us to suffer, and comes to determine the quality of our consciousness, varies enormously both from person to person and from time to time for the same person.[1] A variety of factors can inhibit the normal feelings of distress when physical pain occurs. After the operations of pre-frontal leucotomy (or pre-frontal lobotomy, as it was formerly called), in which the nerve connections between the receptor areas of the brain and the frontal lobe are severed, pain is still felt but is no longer felt as distressing. 'As is well-known, patients who have been subjected to this procedure "feel" pain, but they do not suffer in the commonly accepted sense of the word.'[2] Again, these operations 'result in a change in the patient's emotional response to pain and to its associated anxiety and anguish. Pain is not reduced but suffering, its emotional response, is.'[3] This evidence underlines the distinction between the pain sensation and the pain experience by showing that 'there are separate anatomical pathways for the experience of pain as a sensory quality and the affective awareness of its unpleasantness'.[4]

Again, it is a matter of common experience that 'Distraction and emotion can block pain'.[5] For example, soldiers and athletes sometimes ignore wounds and injuries to an aston-

[1] There are also quite large variations in the physical pain-threshold (J. D. Hardy, in Keele and Smith, op. cit., pp. 177 f.).
[2] David Bowsher and D. Albe-Fessard, 'Patterns of Somatosensory Organization within the Central Nervous System', in Keele and Smith, op. cit., p. 121.
[3] John Hankinson, 'Neurosurgical Aspects of Relief of Pain at the Cerebral Level', in Keele and Smith, op. cit., p. 139. (Hankinson adds that 'most operations of this type, I think it is fair to say, succeed in their purpose for a few months only at the price of a serious deterioration of personality'.)
[4] Lord Brain in Keele and Smith, op. cit., p. 9.
[5] H. K. Beecher, 'An Inspection of our Working Hypothesis in the Study of Pain and other Subjective Responses in Man', in Keele and Smith, op. cit., p. 163.

ishing extent in the heat of battle or of an athletic contest.
Dr. H. K. Beecher, reporting his observations made during
the Second World War, says :

> At Anzio the wound was often apparently construed as a good
> thing, for it meant release from an intolerable situation. It
> meant the war was over for the individual. There, only one-
> quarter of the severely wounded men (although clear mentally,
> not in shock, with normal blood pressure, and having had no
> analgesics for at least four hours and none at all in many cases)
> had pain enough to want anything done about it — and this
> affirmation was made in response to a direct question which was
> of course suggestive.[1]

Similar phenomena have been described in connection with
the deaths of martyrs.[2] Again, hypnotism can be used to
prevent pain ;[3] placebos are often effective in minimizing
pain ;[4] there is the abnormal masochistic state of mind in
which the pain sensation is experienced as being pleasant ;
and finally, one's conscious interpretation of the significance
of the pain that one feels is also important, as was indicated
by Beecher's Anzio report above. Later in his paper,
Beecher says :

> It hardly seems open to question that the reaction to a stimulus
> is influenced by the subject's concept of the sensation, by its signi-
> ficance, its importance, meaning, or degree of seriousness. (One
> pain, as beneath the sternum, if it connotes sudden death can be
> wholly unsettling ; a pain of the same intensity and duration in a
> finger can be easily disregarded.) The meaning, the reaction to,
> the sensation depends upon information (knowledge of heart dis-

---

[1] Beecher, op. cit., pp. 162–3.

[2] See, for example, the second-century document *The Martyrdom of Polycarp*,
which says of certain Christian martyrs of that time that 'they reached such a
pitch of magnanimity, that not one of them let a sigh or a groan escape them ;
thus proving to us all that those holy martyrs of Christ, at the very time when
they suffered such torments, were absent from the body, or rather, that the
Lord then stood by them, and communed with them'. (Chap. 2, trans. in the
Ante-Nicene Library, vol. i.) For other indications of apparent insensitivity to
pain among the early Christian martyrs, see H. B. Workman, *Persecution in the
Early Church* (London : Charles H. Kelly, 1907), chap. 5 ; e.g.: p. 318 (Per-
petua is in a trance whilst being gored in the arena by a bull, and does not
realize that she has been hurt) ; p. 324 ('The body does not feel when the mind
is wholly devoted to God') ; p. 327 (Dativus, who was 'rather a spectator of
his own tortures than a sufferer').

[3] Beecher, op. cit., p. 162 ; Sauerbruch and Wenke, op. cit., p. 63.

[4] Beecher, op. cit., p. 166 f.

ease, as above), upon past experience and present considerations. Knowledge, memory, meaning, discrimination, judgment, conditioning, all can enter into the process of reaction.[1]

In short,

The sufferer's inner attitude can exercise a powerful formative and transformative effect. How a sufferer supports the pain, how he copes with the pain experience, often depends on this inner attitude. . . . A strong, vital personality will, when in good health, not permit even intense physical pain to take up too much room in his consciousness or to have too large a scope for expression; he will not allow the pain 'to take hold of him'.[2]

It has sometimes been suggested that the progress of civilization and the increase of bodily comforts have softened us and made us more sensitive to pain than were our forefathers. Thus a distinguished French surgeon has written that 'Physical sensibility in men of to-day is very different from what it was in our ancestors', and has also professed to find differences of physical sensibility to pain among different nationalities.[3] Commenting on this kind of view, Sauerbruch and Wenke say, 'The truth is probably that there is greater resentment of pain, and this is usually due to the knowledge that there are now methods by which pain can be alleviated, which is coupled with the belief that the patient is entitled to their use.'[4]

3. Has Pain a Biological Value?

It has long been observed that the pain-receptor system is markedly less sensitive than the other sensory systems, with the result that pain is caused only by stimuli powerful enough to damage, or to threaten to damage, the body. Pain, says J. D. Hardy, 'results from noxious stimulation which indicates the beginning of damage to the pain-fibre ending'.[5]

[1] H. K. Beecher, op. cit., pp. 164–5. Cf. Sauerbruch and Wenke, op. cit., pp. 20–21. On the dissociation between pain perception and pain reaction generally, see also Wolff and Hardy, 'On the Nature of Pain', pp. 192 f.

[2] Sauerbruch and Wenke, op. cit., pp. 19–20.

[3] René Leriche, *The Surgery of Pain*, trans. Archibald Young (London: Baillière, Tindall & Cox, 1939), p. 2 ff.

[4] Sauerbruch and Wenke, op. cit., p. 67.

[5] J. D. Hardy, 'The Pain Threshold and the Nature of Pain Sensation', in Keele and Smith, op. cit., p. 195.

Sauerbruch and Wenke have collected some of the classic affirmations of this view, of which I will quote only two: 'Pain is produced by a strong stimulus; strong stimuli disorganize the tissues and are deleterious to living things and their organic functions' (Charles Richet); 'Pain stimuli are those which in the event of their intensification would soon have deleterious and destructive effects threatening the tissue' (A. Goldscheider).[1]

It has also long been observed that, in contrast to the other sense organs, the pain-receptor system reacts to a very wide variety of forms of stimulation. Whereas, for example, the ears are sensitive only to sound, the pain nerves are sensitive to any kind of environmental impingement upon the organism that is violent enough to damage it.

These considerations suggest that pain has the biological function of a warning signal. Thus Sir Thomas Lewis says, 'The very close adjustment that exists between pain and injury ensures a serviceable response to the former, for cutaneous pain promotes withdrawal, abrupt or more deliberate according to the measure of the threat.'[2] And a standard medical text on pain lists the following three protective functions which it serves:

In the first place, pain serves as a means of alarm, drawing our attention to injury or disease of which we might not otherwise be conscious. . . . It is the symptom of all others which induces a patient to seek the expert advice of a physician or surgeon. Its demands are clamant, and of a nature that cannot usually be ignored. . . . In the second place, pain acts as an invaluable deterrent, preserving us from experiments of a dangerous or injurious nature. . . . In the third place the sense of pain helps to create that condition of voluntary immobilization of an inflamed or injured part which the peripheral sympathetics are attempting to secure, and which is so essential a factor in bringing about a cure.[3]

Once again Sauerbruch and Wenke have assembled some classic statements of this position, of which I quote here only

---

[1] Quoted by Sauerbruch and Wenke, op. cit., p. 30 n. See also Sir Charles Sherrington, *Man on his Nature* (Cambridge: The University Press, 2nd ed. 1951), pp. 224–5.

[2] T. Lewis, op. cit., p. 107.

[3] Balme, *The Relief of Pain*, pp. 5–6.

one : 'Pain, the great torturer of all living things, serves nevertheless as the preserver of life, and we could not do without it. It must therefore be regarded as a benefactor to all living things. Pain is one of those protective arrangements which exist in all organisms, and gives the alarm in times of danger, thereby setting the defensive mechanism in operation' (A. Strümpell).[1]

However, as other writers have indicated, this principle can be stated too sweepingly and unguardedly. Although in general pain is invaluable as a warning sign, and the species probably could not survive without it, the severity of pain is by no means always proportional to the gravity of the danger to which it relates. A toothache can produce a violent pain ; but a toothache involves no serious and imminent threat to life or limb. On the other hand, the pain caused by some very grave and often fatal disorders — such as cancer, peritonitis, coronary thrombosis, and sclerosis of the blood-vessels — comes only when the disease is far advanced and a warning of its existence is no longer needed. Considerations of this kind led Dr. René Leriche to protest against the commonly accepted doctrine of the protective value of pain :

Reaction of defence? Fortunate warning? But, as a matter of fact, the majority of diseases, even the most serious, attack us without any warning. Nearly always, the disease is a drama in two acts, the first of which is played secretly in the silent depths of our tissues, every light extinguished, and not even a candle lit. When pain develops, nearly always the second act has been reached. It is too late. The issue has already been determined, and the end is near. The pain has only made more distressing and more sad a situation already long lost. . . . If nature had any consideration for us, if she had the kindly attributes which we ascribe to her, it is not when a renal calculus can no longer be passed by the natural channels that she would warn us, but rather at the stage when it is no more than fine débris, and could easily be got rid of. One must reject, then, this false conception of beneficent pain. In fact, pain is always a baleful gift, which reduces the subject of it, and makes him more ill than he would be without it.[2]

[1] Quoted by Sauerbruch and Wenke, op. cit., pp. 101–2.
[2] Leriche, *The Surgery of Pain*, pp. 23–24.

Leriche is evidently thinking of pain in its relation to organic disease. And here its function does indeed seem to be ambivalent. On the one hand, it sometimes constitutes a most useful warning sign. Pain in the abdomen may be the warning sign of appendicitis; a stomach-ache may warn of indigestion and cause a temporary healthful abstention from food; ear-ache may indicate an infection of the inner ear, toothache an abscess, and so on. But, on the other hand, there are many cases in which pain serves no useful warning function; and further, even when it does, the warning was often not able to be profited from before the days of modern medicine and surgery, and even now is of use only in those parts of the world where these are available. It was of no constructive use, for example, to feel the intense abdominal pain that warns of an obstruction of the bowel during all the centuries of man's history prior to about 1900. In general, then, we must say that the pain mechanism, considered as a warning system relating to disease, is clumsy and inefficient. Regarded solely from this point of view it is questionable whether pain does more good or more harm.

However, since in general the human organism is rather well adapted to survive, the question naturally arises as to why our pain system should be open to these criticisms. And the answer must surely be that we have been asking the wrong question about it. Perhaps pain has biological value in a different, though adjacent, sphere, and only incidentally and hence in a somewhat haphazard way in relation to disease. Perhaps the primary function of pain relates to the normal state of health rather than to the exceptional state of disease. Perhaps it has primarily to do, not with the internal condition of a diseased organism but with the healthy animal's management of itself within the external environment. There are circumstances here that are so pervasive and familiar that we can easily forget them. Any mobile animal, including man, normally observes certain basic procedures of self-preservation which have been learned in large part by the aid of pain. We have found it painful to have our bodies collide at speed with large, hard, solid objects, such as rocks and trees, or walls and doors;

or to go too close to a burning fire; or to fall from a height on to a hard surface; or to let a limb be cut by a sharp edge or torn by strong teeth. All this, and much more, is invaluable knowledge of the pre-verbal dispositional kind which is common to man and the lower animals. And the two co-operating tutors that have taught it to us are our liability to pain and our capacity for disposition- or habit-formation. The cat avoids going too near the fire because it has been conditioned by the pain of excessive heat to remain at a proper distance. We are careful with sharp knives because at some time we have been cut by them and have found this painful. In these elementary and primitive ways we have all learned how to guide our movements successfully within our material environment.

It is at this quite basic level of utility that our capacity for pain is biologically defensible. Disease is not the normal state of the organism, and pain as a warning system concerning it would be something of a biological luxury; but movement about the material world is our daily occupation and pain, as a teacher of self-preservation amid these movements, is a necessity. The body is adapted by its evolution for survival as a vulnerable fleshy organism inhabiting a fairly fixed and rigid world, and we find that the pain system that has developed to meet this basic need is partly useful but partly merely distressing when the body is diseased. The randomness of the operation of the pain system in these latter circumstances is due to the fact that its primary adaptation is to something else, namely the creature's normal self-direction within a material environment.

However, this theory also must be examined critically in the light of evidence that individuals have been able to live successfully without reliance upon the danger sign of pain. The case has been reported, for example, of a soldier who 'was found to have an almost complete insensitivity to pain'.[1] Dr. Harold G. Wolff, commenting on this case, said, 'I think it is important . . . to realize that pain is not essential to life

[1] E. Charles Kunkle and William P. Chapman, 'Insensitivity to Pain in Man', *Pain*, Proceedings of the Association for Research in Nervous and Mental Diseases, vol. xxiii (1943), p. 105.

A Theodicy for Today

or to reasonably good adjustment. Apparently the individual can use his other equipment to help him in detecting those dangerous situations that would ordinarily indicate the use of pain apparatus.'[1] However, an individual of this kind is something of a freak, and it seems a sound procedure to interpret the abnormal in the light of the normal, rather than vice versa, and to assume that this young man had learned from the normally pain-susceptible people among whom he lived to recognize the sources of pain to them and hence the objects and situations that would be noxious to him also. It may very well be that an individual who is atypically insensitive to pain can maintain himself successfully within human society, being dependent in this respect upon his fellows, but that if the race as a whole were insensitive to pain it would not be able to survive.

Another challenge to the view that our capacity for pain has a positive biological value was formulated by David Hume in part xi of his *Dialogues Concerning Natural Religion* (1779). He there describes 'four circumstances on which depend all or the greatest part of the ills that molest sensible creatures'. Of these four major defects of the world he says that 'None of them appear to human reason in the least degree necessary or unavoidable, nor can we suppose them such, without the utmost license of imagination.' The first of these four ills is the occurrence of pain as well as pleasure to motivate creatures to action. Why should not we (and the other animals as well) be moved by a diminution of pleasure instead of being driven by pain to those actions that are necessary to our survival and well-being?

Now pleasure alone, in its various degrees [he says], seems to human understanding sufficient for this purpose. All animals might be constantly in a state of enjoyment; but when urged by any of the necessities of nature, such as thirst, hunger, weariness; instead of pain, they might feel a diminution of pleasure, by which they might be prompted to seek that object, which is necessary to their subsistence. Men pursue pleasure as eagerly as they avoid pain; at least, they might have been so constituted. It seems, therefore, plainly possible to carry on the business of life

[1] Kunkle and Chapman, p. 106.

338

without any pain. Why then is any animal ever rendered susceptible of such a sensation?[1]

At first hearing, Hume's suggestion sounds as practical as it is benevolent. There is, however, a flaw in it.

We have already distinguished, in the light of the modern scientific investigation of pain, between the physical pain sensation and the psychic pain experience, with its affective quality of suffering. We come most directly to the root of Hume's thought if we understand him, as his examples suggest, as objecting to any experience having the quality of suffering, or of negative hedonic tone. But the flaw begins to appear when we remember that the psychological phenomenon of hedonic tone, which is the pleasant or unpleasant quality of experiences, is relative and variable. A given experience does not have an absolute place on the hedonic scale. Its position consists in its relation to other experiences that are more pleasant or less pleasant than itself. Further, even this relative position varies with circumstances. An experience may rise in the hedonic scale, as when one has become so accustomed to an initially unpleasant situation that it is now less unpleasant than it was at first. Or an experience may fall in the hedonic scale, as when some eagerly sought pleasure palls after a while, and sinks in our scale of delights under the weight of boredom.

If we visualize the hedonic scale, with the pleasant at the top and the unpleasant at the bottom, and a mid-point which is hedonically neutral, Hume's suggestion is in effect to cut off the lower half of the scale. But the purely relative character of the scale makes this impossible. We could not cut off the lower half of the scale and leave the upper half as it was. All that we should be doing is to reduce by half the range of contrast. The lower half of the remaining scale would now be the unpleasant, and the upper half the pleasant; but the pleasant would no longer be *as* pleasant

---

[1] Hume's *Dialogues*, pt. xi, Kemp-Smith's ed., pp. 252–3. The same point had been made by Pierre Bayle when he said of the various threats to animals that pain was not necessary for their avoidance but that creatures 'could avoid them equally promptly and certainly solely by the attraction of pleasures, augmented or diminished according to certain proportions'. *Réponse aux questions d'un provincial*, vol. II (Rotterdam: Renier Leers, 1706), chap. lxxii, p. 104.

as it was, nor the unpleasant *as* unpleasant as it was. We might shrink the scale of hedonic values, but we could only prevent it from being a scale of more pleasant and less pleasant by contracting it down to a single point of hedonic neutrality. Short of that, in the world as Hume suggests that it might have been, the lessened pleasure that is to take the place of the pain, for example, of hunger would be to the inhabitants of that world an unpleasant experience. And in order to fulfil the function of what was formerly called pain or discomfort, it must be sufficiently unpleasant to drive them to eat. There must be sufficient contrast between the greater pleasure which they have lost, and the lesser pleasure remaining, to drive them to take action to remedy the situation. But a condition that we take steps to end is precisely what we mean by an unpleasant state of affairs. Hume's suggestion for abolishing the unpleasant thus amounts either to a proposal to reduce all feeling to a dead level of hedonic neutrality, or else to the semantic recommendation that we describe the unpleasant as a lower degree of the pleasant. But in neither case can it accomplish what Hume had in mind.

4. Pain and the Structure of the World

A general feature of the world has been presupposed in all these discussions, and is singled out in the second of Hume's four complaints. He observes that,

a capacity of pain would not alone produce pain were it not for the *second* circumstance, viz., the conducting of the world by general laws; and this seems nowise necessary to a very perfect Being. . . . If everything in the universe be conducted by general laws, and if animals be rendered susceptible of pain, it scarcely seems possible but some ill must arise in the various shocks of matter and the various concurrence and opposition of general laws.[1]

For animal organisms which are part of the material world are of course subject to its general causal regularities; and these involve, for example, that two solid objects cannot

---

[1] David Hume, *Dialogues*, pt. xi, Kemp-Smith's ed., pp. 253–4.

occupy the same region of space at the same time, and that a certain degree of heat destroys the tissues of the body. But in a world of fixed structure animals are liable sometimes to collide with solid obstacles, or to be submerged in water or burned by fire, or to fall on hard ground, or become entangled with projecting branches and be injured. And if these animal organisms are helped to live out their lifespan by a protective sensitivity to pain, which prompts them to avoid or retreat from dangerous situations, it is inevitable that this mechanism will be used and that they will in the course of their lives experience not a little pain.

However, omnipotence could presumably have created instead a different kind of world in which pain-producing situations would be systematically prevented by special adjustments to the course of nature. Causal regularities could be temporarily suspended so that the pain mechanisms of sentient creatures would never have occasion to be activated. In such a world, animal organisms would not have to learn to move about circumspectly, because all serious hazards would be obviated by a complex system of avoidance or transformation. The effect would be as though each living creature were individually watched over by a miracle-working guardian angel charged with protecting it from pain. Fire, whose heat gives us vital warmth but also burns us if we put our hand in it, would suddenly lose its heat whenever this was about to cause pain. Water, which has certain properties in virtue of which it can both sustain life by slaking our thirst and destroy life by suffocation, would suddenly lose these properties whenever someone was in danger of drowning. Knives, which can cut both bread and flesh, would suddenly become blunt rather than cause a hurt. Food, which is pleasant to taste but hard to digest, and alcohol, which warms and cheers but which, in excess, makes drunken sots and dangerous drivers, would cease to have any undesirable effects; and no amount of tobacco would foster lung cancer. The density and hardness of things, which make it possible to walk and to build houses, but also to be killed or injured by a flying stone or a wielded stick, would be continually adjusted for the avoidance of pain. In short, life would no longer be carried on in a stably-structured environment whose laws we must learn on penalty of pain or death, but would approximate to a prolonged dream in which our experience arranges itself according to our own desires.

Among other far-reaching implications, such a reordering of nature would exclude the whole process of the evolution of the forms of life under the pressure of the struggle to survive within a relatively fixed environment. In a world which lacked a stable nexus of natural law that inflicts pain upon individuals and extinction upon species that are not adapted to its demands, the evolutionary process would scarcely have progressed beyond its earliest stages, and the world would probably still be inhabited mainly by jelly-fish.

If, then, we want to retain man in our reordered world we must suppose that these pain-avoiding suspensions of natural law come into operation only after man has emerged from the long evolutionary struggle. So far as man himself is concerned, one of the most striking features of such a re-arranged world would be the absence of any need to comprehend nature and to learn to predict and manipulate its movements. For in a world of continual 'special providences' the laws of nature would have to be extremely flexible : sometimes gravity would operate, sometimes not ; sometimes an object would be hard and solid, sometimes soft ; sometimes boiling water would be hot, sometimes cool ; and so on. There could be no sciences, for there would be no enduring world structure to investigate. And accordingly the human story would not include the development of the physical sciences and technologies with all that they have meant for the exercise of man's intelligence and the drawing out of his adaptive resourcefulness.

Again , in a painless world man would not have to earn his living by the sweat of his brow or the ingenuity of his brain. For in banishing all pain we banish violent hunger and thirst and excessive heat or cold ; and in excluding these we make needless all those activities — such as hunting, farming, and house-building — by which men have staved off those painful conditions. Human existence would involve no need for exertion, no kind of challenge, no problems to be solved or difficulties to be overcome, no demand of the environment for human skill or inventiveness. There would be nothing to avoid and nothing to seek; no occasion for co-operation or mutual help; no stimulus to the development of culture or the creation of civilization. The race would consist of feckless Adams and Eves, harmless and innocent, but devoid of positive character and without the dignity of real responsibilities, tasks, and achievements. By eliminating the problems and hardships of an objective environment, with its own laws, life would become like a reverie in which, delightfully but aimlessly, we should float and drift at ease. Tennyson's poem 'The Lotus-Eaters' well expresses the desire (analysed by some as a longing to return to the peace of the womb) for such 'dreamful ease'.

Thus an important truth is increasingly borne in upon us as we spell out the implications of the Humean principle that the world would be better if, instead of operating upon general laws, it were specially designed and administered for the avoidance of all pain. This truth is that any such radical alteration in the character of man's environment would involve an equally radical alteration in the nature of man himself. A soft, unchallenging world would be inhabited by a soft, unchallenged race of men.

But let us now ask, Why not? Why should a loving God not have created a world of this luxuriously dreamy kind, and populated it with naturally good people who do not need to be disciplined and challenged 'upon the rack of this tough world'?[1] Under the existing dispensation, each of life's evils may perhaps be necessary to ward off some greater evil, or to attain a good which is not (in our present world) otherwise attainable; but still we must ask the more fundamental question as to why God should have permitted a situation of this kind, in which evils are thus necessary. No doubt our liability to pain teaches us to live successfully in an objective material environment, and no doubt our creaturely vulnerability has been the spur to human culture

---

[1] Shakespeare, *King Lear*, v. iii. 312.

343

and civilization. But to defend the existence of pain on these grounds is only to justify one evil by another, namely the evil of our being able to attain to good only in these arduous and unwelcome ways. Why must there be an objective environment in the first place, with all the hardship that it entails for its inhabitants; and why should not man have been created *ab initio* at the highest condition of human culture and civilization?

It is true that to raise such questions is not merely to propose limited amendments to the world, but rather in imagination to restructure it completely. But why not, if we had the power, 'grasp this sorry scheme of things entire' and 'remould it nearer to the heart's desire'?[1]

At this point I must remind the reader of the nature of our enquiry. We are testing out the Christian claim that the world in which we find ourselves is the creation of an infinitely good and powerful Being. We have just supposed an objector to urge that such a Creator could and should have made a better world. But (as we saw in Chapter XIII) the question whether an omnipotent Creator might have done better is an incomplete question. Better for what? We have to specify the general aim or purpose which has motivated the act of creation, and then ask whether the world could have been better in relation to that purpose. In formulating our conception of the divine intention we have set aside Hume's naturalistic assumption that it must be to arrange a maximum of pleasure and a minimum of pain, and have adopted instead the view (developed especially within the Irenaean strand of Christian thought) that the divine purpose behind the world is one of soul-making.

Now so far as the creation of 'naturally good people' is concerned, I have already urged[2] that the kind of goodness which, according to Christian faith, God desires in His creatures, could not in fact be created except through a long process of creaturely experience in response to challenges and disciplines of various kinds. If this be granted it will, I

---

[1] Edward FitzGerald, *Rubá'iyát of Omar Khayyám*, lxxiii.
[2] See above, pp. 291 f.

think, be further granted that a human environment designed to this end must be similar to our present world at least to the extent that it operates upon general laws and consequently involves at least occasional pains for the sentient creatures within it.[1]

However, even if the general proposition be granted that a place of soul-making must be a world of stable natural law in which sentient creatures sometimes feel pain, the further question will now be asked : in order to further a supposed purpose of soul-making, need the world contain *as much* pain as it does? Need the pedagogic programme include the more extreme forms of torture, whether inflicted by man or by disease? As well as bearable pain, need there be unendurable agony protracted to the point of the dehumanization of the sufferer? Must there be not only salutary challenges but also utterly crushing accumulations of disasters? These very legitimate questions will be faced in the next chapter ; for they are, I think, best taken up in terms not only of physical pain but of suffering in general, especially in its distinctively human forms.

5. Animal Pain

To some, the pain suffered in the animal kingdom beneath the human level has constituted the most baffling aspect of the problem of evil.[2] For the considerations that may lighten the problem as it affects mankind — the positive value of moral freedom despite its risks ; and the necessity that a world which is to be the scene of soul-making should contain real challenges, hardships, defeats, and mysteries — do not apply in the case of the lower animals. These are not moral

[1] Nelson Pike points out very clearly in 'Hume on Evil' (*Philosophical Review*, 1963, reprinted in *God and Evil*, ed. Pike) that in his critique of theism from the point of view of the problem of evil Hume overlooks the possibility that God permits evil for a good and justifying purpose. However, Pike goes altogether too far when he concludes that 'the traditional problem of evil reduces to a non-crucial perplexity of relatively minor importance'. (*God and Evil*, p. 102.)

[2] See, for example, C. C. J. Webb, *Problems in the Relations of God and Man* (London : James Nisbet & Co. Ltd., 1911), p. 268.

personalities who might profit from the hazards of freedom or from the challenges of a rough environment. Why, then, we have to ask, does an all-powerful and infinitely loving Creator permit the pain and carnage of animal life? 'If there are any marks at all of special design in creation', said John Stuart Mill, 'one of the things most evidently designed is that a large proportion of animals should pass their existence in tormenting and devouring other animals. They have been lavishly fitted out with the instruments necessary for that purpose; their strongest instincts impel them to it, and many of them seem to have been constructed incapable of supporting themselves by any other food. If a tenth part of the pains which have been expended in finding benevolent adaptations in all nature, had been employed in collecting evidence to blacken the character of the Creator, what scope for comment would not have been found in the entire existence of the lower animals, divided, with scarcely an exception, into devourers and devoured, and a prey to a thousand ills from which they are denied the faculties necessary for protecting themselves!'[1]

The subject of pain in animals must, of course, remain largely a field for speculation and theoretical interpretation. We cannot enter into the consciousness of the lower species, or even prove demonstratively that they have consciousness. There is, however, sufficient evidence for the presence of some degree of consciousness, and some kind of experience of pain, at least throughout the vertebrate kingdom, to prohibit us from denying that there is any problem of animal suffering.[2] (1) The evidence for man's continuity with other forms of life within a common evolutionary process suggests that man's brain and consciousness differ in degree rather than absolutely from those of the animals;[3] and it would therefore be altogether surprising if man alone experienced

---

[1] J. S. Mill, 'Three Essays on Religion', *Nature*, p. 58.
[2] Such a denial is made, for example, by Dom Illtyd Trethowan in *An Essay in Christian Philosophy* (London: Longmans, Green & Co., 1954), pp. 41 and 92.
[3] For a discussion of the evidence for animal consciousness see W. H. Thorpe, Sc.D., F.R.S., 'Ethology and Consciousness', *Brain and Conscious Experience*, ed. J. Eccles (Berlin: Springer-Verlag, 1966).

pain. (2) We observe that when one of the higher mammals is in a situation in which a human would feel intense pain — for example, the situation of being burned — it behaves in ways characteristically similar to those in which a human would behave : it cries out and struggles violently to escape from the source of heat. This suggests that it is undergoing an experience of pain analogous to our own. Indeed, there is evidence that some of the higher animals not only experience physical pain but also a degree of non-physical suffering, in forms analogous to loneliness, fear, jealousy, and even bereavement. (3) The physical structure — especially the sensory and nervous systems — of the other vertebrates is basically similar to that of man. ' Mammals have all the physical structures which seem to be involved in the production of sensations of pain, and, as far as the physiological evidence goes, these structures seem to work in the same way in other mammals as in man.'[1] (4) The higher vertebrates can be taught by means of their reaction to the pleasure-pain dichotomy. In experimental situations they learn to perform actions that bring a reward and to avoid those which are answered by, say, an electric shock. Indeed, it is the part that pain plays in the learning process, and thus in adaptation to the environment, that constitutes its biological justification. For everything said above about the survival value of the pain mechanism in man applies to those of the lower animals that have a sufficiently developed nervous system to be able to feel pain. Neither the human nor the sub-human animal could survive if it did not quickly learn, under the insistent tutorship of pain, how to guide itself as a vulnerable bodily creature moving about in a relatively hard and rigid world of matter. Given that there are animal organisms with a degree of individual spontaneity, and inhabiting a common environment governed by causal regularities, the liability to pain must be a part of their equipment for survival.

What, however, of the lower vertebrates, such as the fishes and insects? Do they feel pain? Very different answers

[1] G. C. Grindley, M.A., B.Sc., *The Sense of Pain in Animals* (London : The Universities Federation for Animal Welfare, 1959), p. 6.

have been given to this question. At one extreme there is
the assumption that

> The poor beetle, that we tread upon,
> In corporal sufferance finds a pang as great
> As when a giant dies.[1]

At the other extreme we are told that the squashed beetle
feels no more than we do when our nails are pared or our
hair cut. The naturalist Theodore Wood wrote :

> When a crab will calmly continue its meal upon a smaller crab
> while being itself leisurely devoured by a larger and stronger ;
> when a lobster will voluntarily and spontaneously divest itself of
> its great claws if a heavy gun be fired over the water in which it is
> lying ; when a dragon-fly will devour fly after fly immediately
> after its abdomen has been torn from the rest of its body, and a
> wasp sip syrup while labouring — I will not say suffering — under
> a similar mutilation ; it is quite clear that pain must practically
> be almost or altogether unknown.[2]

However, these instances, although accurate, may be misleading. The crab, for instance, has a built-in apparatus
for divesting itself of its claws as an escape mechanism ; but
it does not follow that it is not capable of feeling pain in other
circumstances. Again, insects may be insensitive to what is
happening to them whilst they are intently carrying out some
instinctive operation, such as eating or carrying food, and
yet may perhaps not be similarly insensitive at other times.
It is thus unsafe to conclude that even these relatively primitive creatures have no sensations of pain, however dim or
fleeting. On the other hand, there are lower invertebrates,
such as the sea anemone, which have no central nervous
system at all, and it is therefore extremely improbable that
they have conscious experiences.

I conclude thus far that a realistic response to the problem
of pain in the lower vertebrates and higher invertebrates will
not deny that there is any such problem, but will claim that
in so far as these lower animals do feel pain this occurs
within the general system whereby organic life is able to

---

[1] Shakespeare, *Measure for Measure*, III. i. 89–91.
[2] Theodore Wood, quoted by C. E. Raven, *The Creator Spirit* (London : Martin, Hopkinson and Co., Ltd., 1927), p. 120.

survive, namely by reacting to its environment through a nervous system which steers the individual away from danger by means of pain sensations.

There are, however, two important respects in which the animal's situation, as a being liable to pain, differs from man's. Whereas most human beings die through the eventual wearing out of their bodily fabric and its functions, most animals are violently killed and devoured by other species which, in the economy of nature, live by preying upon them. The animal kingdom forms a vast self-sustaining organism in which every part becomes, directly or indirectly, food for another part. And if we project ourselves imaginatively into this process, and see each creature as a self-conscious individual, its situation must seem agonizing indeed. But to do this is to miss the animal's proper good whilst feeling evils of which it is not conscious. Each individual — or at least each healthy individual — has its own fulfilment in the natural activity of its species, uncomplicated by knowledge of the future or a sense of the passage of time ; and its momentary appreciations of its own physical impressions and activities are totally unaffected by the fact that after this thin thread of consciousness has snapped some other creature will devour the carcase. Death is not a problem to the animals, as it is to us ; and herein lies the second major difference between the quality of human and animal experience. They do not wonder, 'in that sleep of death what dreams may come, When we have shuffled off this mortal coil',[1] or dread 'To lie in cold obstruction and to rot, This sensible warm motion to become, A kneaded clot'.[2] We may indeed say of them that 'Death is not an injury, but rather life a privilege'.[3]

Not only is the animal's experience not shadowed by any anticipation of death or by any sense of its awesome finality ; it is likewise simplified, in comparison with human consciousness, by a happy blindness to the dangers and pains that may lie between the present moment and this inevitable termina-

[1] Shakespeare, *Hamlet*, iii. i. 66–67.
[2] Shakespeare, *Measure for Measure*, iii. i. 119–20.
[3] Lord Samuel, *Memoirs* (London : Cresset Press, 1945), p. 297.

tion; and again by a similar oblivion to the past. Although possibly not total at every level of animal life, these restrictions must render all but the occasional animal genius immune to the distinctively human forms of suffering, which depend upon our capacity imaginatively to anticipate the future. The animal's goods and evils are exclusively those of the present moment, and in general it lives from instant to instant either in healthy and presumably pleasurable activity, or in a pleasant state of torpor. The picture, then, of animal life as a dark ocean of agonizing fear and pain is quite gratuitous, and arises from the mistake of projecting our distinctively human quality of experience into creatures of a much lower and simpler order.

The more fruitful question for theodicy is not why animals are liable to pain as well as pleasure — for this follows from their nature as living creatures — but rather why these lower forms of life should exist at all. Christian theology enables us to understand, within its own presuppositions, why the human creature exists : he is a rational and moral being who may freely respond to his Maker's love and so become a 'child of God' and 'heir of eternal life'. But this explanation cannot cover the lower animals, lacking as they do a rational and moral nature. Their existence remains as a problem.

As we have seen, the Augustinian theodicy resolves this problem by means of the principle of plenitude.[1] The infinite divine nature expresses itself *ab extra* in the creation of every grade of dependent being, from the highest to the lowest, and accordingly the created world includes not only man but also monkeys and dogs and snakes and snails and germs. Each level of life makes its own valid contribution to the harmonious perfection of the whole; and the preying of life upon life is a proper feature of the lower ranges of the animal creation, where individuals are but fleeting ripples in a flowing stream of animate life.[2] Thus the sub-human animals exist because they represent possible forms of being, and therefore of goodness, and because their existence is accordingly necessary to the fullness of the created world.

[1] See above, pp. 76 f.
[2] See above, p. 92.

Pain

Modern theologians have been more troubled than were the ancients by the evil of animal pain and the spectacle of 'nature, red in tooth and claw'; and as a comparatively recent development within the broadly Augustinian tradition the suggestion has been made that either the fall of man or the prior fall of the angels has affected the natural order and perverted the realms of animal life, causing the various species to attack and devour one another.[1] This speculation will be discussed and criticized in a slightly different context in Chapter XVII.[2]

Suppose, however, we find these theories unsatisfying: is there any alternative way of relating sub-human life, in the varied forms that it actually takes, to the creative activity of God? If we conduct our thinking upon the basis of what are, for the Christian theologian, assured data, we shall start from God's purpose for man revealed in the person and life of Jesus, the Christ, and try to work outwards from this centre towards an understanding of animal life. As a possible way of doing this, the concept of epistemic distance developed in the last chapter suggests that man's embeddedness within a larger stream of organic life may be one of the conditions of his cognitive freedom in relation to the infinite Creator. Seeing himself as related to the animals and as, like them, the creature of a day, made out of the dust of the earth, man is set in a situation in which the awareness of God is not forced upon him but in which the possibility remains open to him of making his own free response to his unseen Maker. Protestant theology has generally affirmed that the animals exist for the sake of man, but has interpreted this in terms of man's rule over the lower creation.[3] Possibly, however, another and more fundamental way in which, within the divine purpose, sentient nature supports and serves its human apex is by helping to constitute an independent natural order to which man is organically related and within which

[1] See, for example, C. S. Lewis, *The Problem of Pain*, p. 123. That 'the whole order of nature was subverted by the sin of man' was taught by Calvin. See his *Commentary on Genesis*, chap. 3, p. 177.
[2] See below, pp. 367 f.
[3] See, for example, Calvin's *Commentary on Genesis*. pp. 96, 173, and 290. (See also above, p. 295, n. 2.)

351

CHAPTER XVI

SUFFERING

1. SUFFERING AS A FUNCTION OF MEANING

WE shall, I believe, be using the words to mark an
important distinction within human experience if
we differentiate between pain on the one hand and
suffering, misery, or anguish (three terms that I shall use
synonymously) on the other. Pain is, as we have seen in the
previous chapter, a specific physical sensation.[1] Suffering,
however, is a mental state which may be as complex as human
life itself. The endurance of pain is sometimes, but not
always or even usually, an ingredient of suffering.

Anguish is a quality of experience whose nature could not
be communicated by description to someone who had never
undergone it; and we can in fact conceive of personal
creatures who have always been entirely free from it. However, we human beings are not of such a breed, and have no
great difficulty in communicating with one another about
the forms of human misery. Attempting, then, to define an
all-too-common dimension of our experience, I would suggest
that by suffering we mean that state of mind in which we
wish violently or obsessively that our situation were otherwise. Such a state of mind involves memory and anticipation, the capacity to imagine alternatives, and (in man) a
moral conscience. For the characteristic elements of human
suffering are such relatively complex and high-level modes
of consciousness as regret and remorse; anxiety and despair;[2]
guilt, shame, and embarrassment; the loss of someone loved;[3]

[1] See above, pp. 330 f.
[2] *Constructive Aspects of Anxiety*, ed. by Seward Hiltner and Karl Menninger
(New York: Abingdon Press, 1963), contains valuable psychological and
theological discussions of anxiety.
[3] For a deeply moving diary of meditation upon a bereavement, see *A Grief
Observed*, by N. W. Clerk (C. S. Lewis) (London: Faber & Faber, 1961).

the sense of rejection, of frustrated wishes, and of failure. These all differ from physical pain in that they refer beyond the present moment. To be miserable is to be aware of a larger context of existence than one's immediate physical sensations, and to be overcome by the anguished wish that this wider situation were other than it is.

Suffering, so characterized, is a function of sin. Our human experience can become an experience of suffering to us because we engage in it self-centredly. But in themselves our finitude, weakness, and mortality do not constitute situations from which we should violently wish to escape; if we were fully conscious of God and of His universal purpose of good we should be able to accept our life in its entirety as God's gift and be free from anguish on account of it.

It should be added that suffering can be either self-regarding or other-regarding. My violent and obsessive wish that the situation in which I am involved were different may have in view my own interest or the interest of others. It may spring from self-concern or from sympathy. This distinction is relevant to the theological question as to whether Christ, as one who was sinless, can have experienced suffering. The answer, I suggest, is as follows. In general our human sufferings are self-regarding; we wish *for our own sake* that our situation were otherwise. Christ's suffering, on the other hand — as when he wept in sorrow over Jerusalem — was an other-regarding anguish; he grieved, not for himself, but on account of others. We may say, then, not that Christ, as God incarnate, did not suffer, but that he did not suffer egoistically, as we do.

2. Pain as a Cause of Suffering

As a limiting case, very intense pain may so dominate consciousness as for the time being to shut out the wider context of our existence and itself constitute a situation of suffering — a situation that we violently desire to escape from. But more often pain occurs in interplay with other factors. We saw in the previous chapter that a situation within which we feel

fully in mutual giving and helping and sharing in times of difficulty.[1] And it is hard to see how such love could ever be developed in human life, in this its deepest and most valuable form of mutual caring and sharing, except in an environment that has much in common with our own world. It is, in particular, difficult to see how it could ever grow to any extent in a paradise that excluded all suffering. For such love presupposes a 'real life' in which there are obstacles to be overcome, tasks to be performed, goals to be achieved, setbacks to be endured, problems to be solved, dangers to be met; and if the world did not contain the particular obstacles, difficulties, problems, and dangers that it does contain, then it would have to contain others instead. The same is true in relation to the virtues of compassion, unselfishness, courage, and determination — these all presuppose for their emergence and for their development something like the world in which we live. They are values of personal existence that would have no point, and therefore no place, in a ready-made Utopia. And therefore, if the purpose for which this world exists (so far as that purpose concerns mankind) is to be a sphere within which such personal qualities are born, to purge it of all suffering would be a sterile reform.

At the same time, it is to be noted that we have, in all this, discerned only a very general connection between the kind of world in which we are living and the development of so many of the more desirable qualities of human personality. We have seen that, from our human point of view, this is a world with rough edges, a place in which man can live only by the sweat of his brow, and which continually presents him with challenges, uncertainties, and dangers; and yet that just these features of the world seem, paradoxically, to

[1] As James Hinton argued in his classic meditation on the mystery of pain, 'We could never have felt the joy, never had had even the idea, of love, if sacrifice had been impossible to us.' (*The Mystery of Pain*, 1866, London: Hodder & Stoughton, 1911, p. 51.) Cf. Josiah Royce: 'Even love shows its glory as love only by its conquest over the doubts and estrangements, the absences and the misunderstandings, the griefs and the loneliness, that love glorifies with its light amidst all their tragedy.' (*The World and the Individual*, New York: The Macmillan Co., 1901, vol. ii, p. 409.) Royce presented his own idealist theodicy here and in 'The Problem of Job' (*Studies in Good and Evil*, 1898, reprinted in Walter Kaufmann, ed., *Religion From Tolstoy to Camus*, New York: Harper & Row, 1961).

underlie the emergence of virtually the whole range of the
more valuable human characteristics.

4. EXCESSIVE OR DYSTELEOLOGICAL SUFFERING

But we have now to consider the all-important question of
the *amount* of suffering in the world. The less radical form
of Hume's second suggestion is that God should not interfere
in the workings of nature to such an extent that no objective
order remains, but should interfere only secretly and on
special occasions to prevent exceptional and excessive evils.
'A fleet, whose purposes were salutary to society, might
always meet with a fair wind. Good princes enjoy sound
health and long life: Persons born to power and authority,
be framed with good tempers and virtuous dispositions.' [1]
This suggestion seems more plausible than the previous one.
But nevertheless it is not free from difficulty. For evils are
exceptional only in relation to other evils which are routine.
And therefore unless God eliminated all evils whatsoever
there would always be relatively outstanding ones of which
it would be said that He should have secretly prevented
them. If, for example, divine providence had eliminated
Hitler in his infancy we might now point instead to Mussolini as an example of a human monster whom God ought
secretly to have excised from the human race; and if there
were no Mussolini we should point to someone else. Or
again, if God had secretly prevented the bombing of Hiro-
shima we might complain instead that He could have
avoided the razing of Rotterdam. Or again, if He had
secretly prevented the Second World War, then what about
the First World War, or the American Civil War, or the
Napoleonic wars, and so through all the major wars of
history to its secondary wars, about which exactly the same
questions would then be in order? There would be nowhere to stop, short of a divinely arranged paradise in which
human freedom would be narrowly circumscribed, moral
responsibility largely eliminated, and in which the drama
of man's story would be reduced to the level of a television

[1] *Dialogues*, pt. xi, Kemp-Smith's ed., p. 254.

often so severe as to be self-defeating when considered as soul-making influences? Man must (let us suppose) cultivate the soil so as to win his bread by the sweat of his brow; but need there be the gigantic famines, for example in China, from which millions have so miserably perished? Man must (let us suppose) labour on the earth's surface to make roads, and dig beneath it to extract its coals and minerals; but need there be volcanic irruptions burying whole cities, and earthquakes killing thousands of terrified people in a single night? Man must (let us suppose) face harsh bodily consequences of over-indulgence; but need there also be such fearful diseases as typhoid, polio, cancer, angina? These reach far beyond any constructive function of character training. Their effect seems to be sheerly dysteleological and destructive. They can break their victim's spirit and cause him to curse whatever gods there are. When a child dies of cerebral meningitis, his little personality undeveloped and his life unfulfilled, leaving only an unquenchable aching void in his parents' lives; or when a charming, lively, and intelligent woman suffers from a shrinking of the brain which destroys her personality and leaves her in an asylum, barely able to recognize her nearest relatives, until death comes in middle life as a baneful blessing; or when a child is born so deformed and defective that he can never live a properly human life, but must always be an object of pity to some and revulsion to others . . . when such things happen we can see no gain to the soul, whether of the victim or of others, but on the contrary only a ruthlessly destructive process which is utterly inimical to human values. It seems as though 'As flies to wanton boys, are we to the gods, They kill us for their sport'.[1]

It is true that sometimes — no one can know how often or how seldom — there are sown or there come to flower even in the direst calamity graces of character that seem to make even that calamity itself worth while. A selfish spirit may be moved to compassion, a thoughtless person discover life's depths and be deepened thereby, a proud spirit learn patience and humility, a soft, self-indulgent character be

[1] Shakespeare, *King Lear*, IV. i. 38–39.

made strong in the fires of adversity. All this may happen, and has happened. But it may also fail to happen, and instead of gain there may be sheer loss. Instead of ennobling, affliction may crush the character and wrest from it whatever virtues it possessed. Can anything be said, from the point of view of Christian theodicy, in face of this cosmic handling of man, which seems at best to be utterly indifferent and at worst implacably malevolent towards him?

5. THE TRADITIONAL ANSWER: NATURE PERVERTED BY FALLEN ANGELS

The notion of a pre-mundane fall, whether of human souls (as was taught by Origen, and again in modern times by Julius Müller) or of a life-force that had not yet become individualized (as N. P. Williams suggested), has been rejected above as a speculation which conceals rather than solves the problem of the origin of moral evil.[1] There is, however, another related conception of a pre-mundane fall which differs from these in that it has some basis, however slight, in Scripture; and it has been invoked by several writers as solving in principle our present problem concerning the apparently malevolent aspects of nature. This is the idea of a fall of angelic beings preceding and accounting for both the fall of man and the disordered and dysteleological features of the natural world.[2] It might be suggested that these angels are in charge of the world, perhaps different angels being concerned with different aspects of nature, like the gods of the old Greek pantheon, and that when they fell from grace their apostasy was reflected in distortions of nature and in perverted biological developments of various kinds. That the animal species prey upon one another, that

[1] See p. 287 f. above.

[2] This is to be found in a number of writers; for example, C. C. J. Webb, *Problems in the Relations of God and Man*, pp. 269–71 ; C. S. Lewis, *The Problem of Pain*, pp. 122 f. ; Dom Bruno Webb, *Why Does God Permit Evil?* (London: Burns, Oates & Washbourne, Ltd., 1941), pp. 33–35; Leonard Hodgson, *For Faith and Freedom*, i. pp. 213 f. ; Dom Illtyd Trethowan, *An Essay in Christian Philosophy*, p. 128 ; E. L. Mascall, *Christian Theology and Natural Science* (London: Longmans, Green & Co., 1956), pp. 301 f.

there are microbes and bacteria which cause disease in animal bodies, that there are weaknesses in the earth's crust producing volcanoes and tornadoes, that there are violent extremes of temperature, uninhabitable climates, droughts and blights, may all be due to the malevolence or heedlessness of higher beings who were appointed as nature's guardians but who have become enemies of nature's God.

In support of such a theory it might be claimed that in the New Testament miracles we see the power of God at work reversing nature's tragic departure from the course intended for it in the original divine plan. When Jesus healed the sick and stilled the storm he was overturning the work of fallen spirits who still exercise control over this world. This, it might be added, was a contemporary Jewish understanding of disease which seems also to have been shared by Jesus himself.

Such a speculation has its attractions. It would allow us to recognize unequivocally the inimical character of disease, accident, and decay, even to the extent of seeing a malevolent activity behind them. But, on the other hand, it would revive the fundamental contradiction that lies at the heart of the traditional Augustinian theodicy. It would trace all evil back to a common source in the incomprehensible rebellion of finitely perfect beings who were enjoying the full happiness of God's presence. This would be a return to the unintelligible notion of the self-creation of evil *ex nihilo*, against which I have argued at length above. It must also be objected against this speculation that it would lead to a gnostic rejection of the natural order as evil. If the world is ruled by 'obscure, unfeeling and unloving powers', as Sigmund Freud also believed,[1] we ought to regard it with horror as the sphere of something malevolent and fearful. But for the most part mankind has found it to be otherwise and our minstrels, the poets of all ages, have celebrated the goodness of the earth, the bountifulness of nature, and the infinite delights and ever-changing beauties

[1] *New Introductory Lectures on Psycho-Analysis*, Standard Edition of Complete Psychological Works, vol. xxii (London: Hogarth Press, 1964), pp. 167 and 173.

of earth's seasons. Both Jewish and Christian poets have
been in the forefront of this celebration, directing it upwards
in gratitude to God. Are we, then, to split the seamless coat
of nature and say that the lilies of the field, together with
rich harvests and beautiful sunsets, are ruled by good spirits,
whilst diseases, earthquakes and storms are produced by evil
spirits? This would indeed be a desperate expedient. For
all that the sciences teach us about the workings of nature
tend to emphasize its unity as a single system of cause and
effect exhibiting the same laws throughout.

There are reasons, then, to proceed along a different path
which, whilst not being obliged to reject the traditional
doctrine of the pre-human fall of the angels, does not try to
make use of it to solve the problems of theodicy.

6. SOUL-MAKING AND MYSTERY

The problem of suffering remains, then, in its full force. If
what has been said in this and the previous chapter is valid
the problem does not consist in the occurrence of pain and
suffering as such ; for we can see that a world in which these
exist in at least a moderate degree may well be a better
environment for the development of moral personalities than
would be a sphere that was sterilized of all challenges. The
problem consists rather in the fact that instead of serving a
constructive purpose pain and misery seem to be distributed
in random and meaningless ways, with the result that suffering is often undeserved and often falls upon men in amounts
exceeding anything that could be rationally intended.

Further, I have argued that we ought to reject the traditional theories that would rationalize the incidence of
misery : the theory that each individual's sufferings represent a just punishment for his own sins, whether committed
in this life or in some pre-natal realm ; and the theory that
the world is in the grip of evil powers, so that the dysteleological surplus of human misery is an achievement of
demonic malevolence. Moreover, I do not now have an
alternative theory to offer that would explain in any rational
or ethical way why men suffer as they do. The only appeal

ethical meaning contribute to the character of the world as a place in which true human goodness can occur and in which loving sympathy and compassionate self-sacrifice can take place. 'Thus, paradoxically,' as H. H. Farmer says, 'the failure of theism to solve all mysteries becomes part of its case!'[1]

My general conclusion, then, is that this world, with all its unjust and apparently wasted suffering, may nevertheless be what the Irenaean strand of Christian thought affirms that it is, namely a divinely created sphere of soul-making. But if this is so, yet further difficult questions now arise. A vale of soul-making that successfully makes persons of the desired quality may perhaps be justified by this result. But if the soul-making purpose fails, there can surely be no justification for 'the heavy and the weary weight of all this unintelligible world'.[2] And yet, so far as we can see, the soul-making process does in fact fail in our own world at least as often as it succeeds.

At this point a further, eschatological, dimension of Christian belief becomes importantly relevant, and must be brought into the discussion in the next chapter.

[1] *Towards Belief in God*, p. 234.
[2] Wordsworth, 'Lines composed a few miles above Tintern Abbey'.

CHAPTER XVII

THE KINGDOM OF GOD
AND THE WILL OF GOD

1. THE INFINITE FUTURE GOOD

IT is a commonplace that the prevailing modes of thought in our contemporary Western world are naturalistic. That is to say, the categories of the sciences are today regarded as ultimate, rather than those of religion. In consequence it is widely assumed that God can exist only as an idea in the human mind. Even within the Christian Churches this presupposition is evident in various forms of religious naturalism, or naturalistic religion, according to which statements about God, instead of referring to a transcendent divine Being, are expressions of ethical policies,[1] or of ways of seeing and feeling about the world,[2] or of convictional (as distinct from cognitive) stances,[3] or of one's existential situation.[4] Neither the general naturalistic or positivistic climate of our culture as a whole nor the growing enclave of naturalistic religion within Christianity has any use for the idea of an after-life. From the point of view of secular naturalism, human survival after death is a

[1] R. B. Braithwaite, *An Empiricist's View of the Nature of Religious Belief* (Cambridge: Cambridge University Press, 1955); Peter Munz, *Problems of Religious Knowledge* (London: The S.C.M. Press, 1959); T. R. Miles, *Religion and the Scientific Outlook* (London: George Allen & Unwin, 1959); Paul F. Schmidt, *Religious Knowledge* (Glencoe: The Free Press, 1961).

[2] John Wisdom, 'Gods' (*Proceedings of the Aristotelian Society*, 1944–5. Reprinted in *Essays on Logic and Language*, i, ed. by Antony Flew, Oxford: Basil Blackwell, 1951, and in *Classical and Contemporary Readings in the Philosophy of Religion*, ed. John Hick, Englewood-Cliffs, N.J.: Prentice-Hall, Inc., 1964) and 'The Modes of Thought and the Logic of God' (*The Existence of God*, ed. John Hick, New York: The Macmillan Co., 1964); J. H. Randall, Jr., *The Role of Knowledge in Western Religion* (Boston: The Beacon Press, 1958), chap. 4.

[3] Willem Zuurdeeg, *An Analytical Philosophy of Religion* (Nashville, Tenn.: Abingdon Press, 1958).

[4] Paul Van Buren, *The Secular Meaning of the Gospel* (London: S.C.M. Press, 1964).

Christian theodicy without taking seriously the doctrine of a life beyond the grave. This doctrine is not, of course, based upon any theory of natural immortality, but upon the hope that beyond death God will resurrect or re-create or reconstitute the human personality in both its inner and its outer aspects. The Christian claim is that the ultimate life of man — after what further scenes of 'soul-making' we do not know — lies in that Kingdom of God which is depicted in the teaching of Jesus as a state of exultant and blissful happiness, symbolized as a joyous banquet in which all and sundry, having accepted God's gracious invitation, rejoice together. And Christian theodicy must point forward to that final blessedness, and claim that this infinite future good will render worth while all the pain and travail and wickedness that has occurred on the way to it. Theodicy cannot be content to look to the past, seeking an explanation of evil in its origins, but must look towards the future, expecting a triumphant resolution in the eventual perfect fulfilment of God's good purpose. We cannot, of course, concretely picture to ourselves the nature of this fulfilment; we can only say that it represents the best gift of God's infinite love for His children. But no other acceptable possibility of Christian theodicy offers itself than that in the human creature's joyous participation in the completed creation his sufferings, struggles, and failures will be seen to be justified by their outcome.[1] We must thus affirm in faith that there will in the final accounting be no personal life that is unperfected and no suffering that has not eventually become a phase in the fulfilment of God's good purpose. Only so, I suggest, is it possible to believe both in the perfect goodness of God and in His unlimited capacity to perform His will. For if there are finally wasted lives and finally unredeemed sufferings, either God is not perfect in love or He is not sovereign in rule over His creation.

It is perhaps worth pointing out here the difference between this position and another to which it is in some ways similar, namely the view that the promised joys of heaven

[1] Cf. Emil Brunner, *Man in Revolt*, trans. Olive Wyon (London: Lutterworth Press, 1939), p. 454.

are to be related to man's earthly travails as a compensation or reward. This suggests a divine arrangement equitably proportioning compensation to injury, so that the more an individual has suffered beyond his desert the more intense or the more prolonged will be the heavenly bliss that he experiences. Thus those who have suffered most will subsequently have cause to rejoice most; and presumably, if the just proportion is to be preserved, none will enjoy an endless or infinite bliss, since none will have suffered an unending or unlimited injury. As distinct from such a book-keeping view, what is being suggested here, so far as men's sufferings are concerned, is that these sufferings — which for some people are immense and for others relatively slight — will in the end lead to the enjoyment of a common good which will be unending and therefore unlimited, and which will be seen by its participants as justifying all that has been endured on the way to it. The 'good eschaton' will not be a reward or a compensation proportioned to each individual's trials, but an infinite good that would render worth while any finite suffering endured in the course of attaining to it.

2. Theodicy versus Hell

If this is the true nature of Christian theodicy — a theodicy that is eschatological in character and can be affirmed only by faith — it compels us to question the validity of belief in hell, in the traditional sense of eternal suffering inflicted by God upon those of His creatures who have sinfully rejected Him. For there is a tension within Christian thought between the motives that move towards this doctrine of everlasting punishment and the motives that move towards a theodicy. The sufferings of the damned in hell, since they are interminable, can never lead to any constructive end beyond themselves and are thus the very type of ultimately wasted and pointless anguish. Indeed misery which is eternal and therefore infinite would constitute the largest part of the problem of evil. Further, the notion of hell is no less fatal to theodicy if, instead of stressing the sufferings of the damned, we stress the fact that they are unendingly in sin.

For this is presumably an even greater evil — a greater frustration of the divine purpose — than their misery. Thus in a universe that permanently contained sin, good and evil would be co-ordinates, and God's creation would be perpetually shadowed and spoiled by evil; and this would be incompatible either with God's sovereignty or with His perfect goodness. For the doctrine of hell has as its implied premise either that God does not desire to save all His human creatures, in which case He is only limitedly good, or that His purpose has finally failed in the case of some — and indeed, according to the theological tradition, most — of them, in which case He is only limitedly sovereign. I therefore believe that the needs of Christian theodicy compel us to repudiate the idea of eternal punishment.

Does this mean that we are led to universalism, in the sense of a belief in the ultimate salvation of all human souls? The rejection of the idea of a divine sentence of eternal suffering is not in itself equivalent to universalism, for there remains the third possibility of either the divine annihilation or the dwindling out of existence of the finally lost. In this case there would not be eternally useless and unredeemed suffering such as is entailed by the notion of hell as unending torment; and in working out a theodicy it would perhaps be possible to stop at this point. However, even in such a modified version of a 'bad eschaton' God's good purpose would have failed in the case of all those souls whose fate is extinction. To this extent evil would have prevailed over good and would have permanently marred God's creation. This is accordingly a very dubious doctrine for Christian theism to sponsor, and not one in which we should acquiesce except for want of any viable alternative. But in fact an alternative is available: namely, that God will eventually succeed in His purpose of winning all men to Himself in faith and love. That this is indeed God's purpose in relation to man is surely evident from the living revelation of that purpose in Jesus Christ. In his life we see at work in our human history 'God our Saviour, who desires all men to be saved and to come to the knowledge of the truth'.[1] The

[1] Timothy ii. 3-4.

question, then, is whether God *can* eventually do for the free creatures whom he Has created what He *wants* to do for them. Augustinian and Calvinist theology never doubted that whatever God willed in relation to mankind would finally be brought about. And in this they were surely right; their grave mistake was to suppose that God did not *want* to save all men, but that on the contrary He created some for salvation and others for damnation. Today few would presume to uphold a conception of God so diametrically at variance with the dominant spirit of the gospels. The difficulty that is more often felt about the idea of universal salvation concerns God's power to evoke a right response in personal beings whom He has endowed with the fateful gift of freedom. Can God *cause* them to respond to Him without thereby turning them into puppets?

This difficulty is, I think, often magnified by a failure to distinguish between the logical and the factual questions involved. There would be a logical contradiction in its being, in the strict sense, *predetermined* that creatures endowed with free will shall come to love and obey God. For the thoughts and actions of free beings are in principle unknowable until they occur. They cannot therefore be made the subject of absolute predictions. It would infringe the nature of the personal order if we could assert as a matter of assured knowledge that all men *will* respond to God. It is this logical exclusion of scientific-type predictions that has seemed to many to rule out a doctrine of universal salvation. But, having insisted that it is logically possible that some, or even all, men will in their freedom eternally reject God and eternally exclude themselves from His presence, we may go on to note the actual forces at work and to consider what outcome is to be expected and expected with what degree of confidence.

The least that we must say, surely, is that God will never cease to desire and actively to work for the salvation of each created person. He will never abandon any as irredeemably evil. However long an individual may reject his Maker, salvation will remain an open possibility to which God is ever trying to draw him.

Can we go beyond this and affirm that somehow, sooner or later, God will succeed in His loving purpose? It seems to me that we can, and that the needs of theodicy compel us to do so. God has made us for Himself, and our whole being seeks its fulfilment in relation to Him. He can influence us both through the world without and by the activity of His Holy Spirit within us, though always in ways that preserve the integrity and freedom of the human spirit. It seems morally (although still not logically) impossible that the infinite resourcefulness of infinite love working in unlimited time should be eternally frustrated, and the creature reject its own good, presented to it in an endless range of ways. We cannot say in advance *how* God will eventually free all created souls from their bondage to sin and establish them in love and glad obedience towards Himself; but despite the logical possibility of failure the probability of His success amounts, as it seems to me, to a practical certainty.

William James once used the analogy of two chess players, a novice and a world master, to illustrate the compatibility between divine providence and human freedom.[1] Even though the novice is free at every stage to make his own move, we can predict with complete practical certainty that the master will eventually win. Although we cannot foresee the detailed course of the game, we know that, whatever moves the novice makes, the master can so respond as sooner or later to bring the game to the conclusion that he himself desires. This analogy, of course, breaks down, as all analogies do, if pressed beyond a certain point. It illustrates the way in which, in a process that involves free will and thus precludes logically certain predictions, there can nevertheless be practical certainty as to the outcome. But, on the other hand, the drama of man's salvation is not a contest either of wills or of skills between God and man. God has formed the free human person with a nature that can find its perfect fulfilment and happiness only in active enjoyment of the infinite goodness of the Creator. He is not, then, trying to force or entice His creatures against the grain of their nature,

---

[1] William James, *The Will to Believe and Other Essays* (London: Longmans, Green & Co., 1897), pp. 181–2.

but to render them free to follow their own deepest desire, which can lead them only to Himself. For He has made them for Himself, and their hearts are restless until they find their rest in Him. He is not seeking to subjugate them but to liberate them, in order that they may find in Him their own deepest fulfilment and happiness. Perhaps, then, a less inadequate analogy would be that of a psychotherapist trying to empower a patient to be himself and to cease frustrating his own desires, to face reality and accept his proper place in the affections and respect of others. The divine Therapist has perfect knowledge of each human heart, is infinitely wise in the healing of its ills, has unbounded love for the patient and unlimited time to devote to him. It remains theoretically possible that He will fail; but He will never cease to try, and we may (as it seems to me) have a full practical certainty that sooner or later He will succeed.

In the traditional language of theology this practical certainty is an aspect of the Christian hope. For we do not as Christians hope simply for our own ultimate salvation, but equally for that of any man and therefore of all men. We have no grounds upon which to differentiate in this respect among the objects of God's grace; our Christian hope must accordingly be for the salvation of the whole race. And when we speak of the Christian hope we are not speaking merely of something that is desired and wished for. Christian hope is not parallel to secular hope but is the extrapolation of Christian faith into the future. It is not separable from the Christian's present faith-awareness of the reality and love of God, but is an awareness of that reality as ultimate and of that love as fulfilling itself in the dimension of time. Thus the Christian hope shares the assurance of Christian faith; and within faith and hope we may confidently affirm the ultimate salvation of all God's children.

3. THE INTERMEDIATE STATE

There are, however, certain well-known objections to belief in ultimate universal salvation, and these must now be faced.

The most serious objection is that the New Testament, and

in particular the teaching of our Lord, seems to proclaim an eternal punishment for sinners, in which there is 'weeping and gnashing of teeth'[1] and where 'the worm dieth not, and the fire is not quenched'.[2] These sayings and others in the same category are, I believe, to be taken with the utmost seriousness. An eschatology that merely ignored them, or ruled them out *a priori* as inauthentic, would not be dealing responsibly with the Bible. It is possible, however, to give the fullest weight and urgency to Jesus' sayings about suffering beyond this life without being led thereby to the dualistic notion of an eternal hell.

That Jesus spoke of real suffering and misery, really to be dreaded, which is to come upon men hereafter as a divinely ordained consequence of selfish and cruel deeds performed in this life, is not to be doubted. Nor is it to be doubted that he wanted his hearers to be aware of this inevitable reaction of the moral order upon human wickedness. But that our Lord taught that such misery is to continue through endless time in a perpetual torture inflicted by God cannot safely be affirmed. Between us and Jesus' original meaning there stand the ambiguity of the New Testament word αἰώνιος ;[1] the uncertainty as to how, if at all, the notion of endless duration might have been expressed in the Aramaic language which Jesus presumably spoke; and evidence of the intrusions of Jewish apocalyptic themes into the developing gospel tradition. The textual evidence must accordingly be interpreted in the light of wider considerations drawn from Jesus' teaching as a whole. If we see as the heart of this teaching the message of the active and sovereign divine love, we shall find incredible and even blasphemous the idea that God plans to inflict perpetual torture upon any of His children. If, then, we assume that such sufferings are not eternal and hence morally pointless, but rather temporal and redemptive in purpose, we are led to postulate an existence or existences beyond the grave in which the moral structure of reality is borne in upon the individual, and in which his self-centredness is gradually broken through by a 'godly sorrow'[3] that represents the inbreaking of reality. Such an idea is, of course, not far from

[1] Matthew xxv. 30.　　[2] Mark ix. 48.　　[3] II Corinthians vii. 10.

the traditional Roman Catholic notion of purgatorial experiences occurring (for those who die in a state of grace) between death and entry into the final heavenly Kingdom in which God shall be all in all.[1] Because of the grave abuses that helped to provoke the Reformation of the sixteenth century the word 'purgatory' has a bad sound in many ears; and some of the Protestant theologians who have entertained an essentially similar idea have preferred to use instead the term 'progressive sanctification after death'.[2] The thought behind this phrase is that as sanctification (or perfecting) in this life comes partly through suffering, the same is presumably true of the intermediate state in which the sanctifying process, begun on earth, continues towards its completion, its extent and duration being determined by the degree of *un*sanctification remaining to be overcome at the time of death. For our sanctification — that is to say, our perfecting as persons — is then still radically incomplete. We have not become fully human by the time we die. If, then, God's purpose of the perfecting of human beings is ever to be fulfilled, it must either be brought to an instantaneous completion by divine fiat, perhaps at the moment of death, or else take place through a continued development within some further environment in which God places us. The difficulty attaching to the first alternative is that it is far from clear that an individual who had been instantaneously perfected would be in any morally significant sense the same person as the frail, erring mortal who had lived and died. It would seem more proper to say, not that this previously very imperfect person has now become perfect, but that he has ceased to exist and that a perfect individual has been created in his place. But if we are thus to be transmuted in the twinkling of an eye into perfect creatures, the whole earthly travail of faith and moral effort is rendered needless. For God might by an

---

[1] For a presentation of the moral and rational grounds for this belief by a finely ecumenical Roman Catholic thinker, see Baron von Hügel's *Essays and Addresses on the Philosophy of Religion* (London: J. M. Dent & Sons, Ltd., 1922), vol. i, pp. 201–3.

[2] E.g. Charles A. Briggs, 'Inaugural Address' in *The Edward Robinson Chair of Biblical Theology* (New York: Union Theological Seminary, 1891), p. 54. Cf. I. A. Dorner, *Dorner on the Future State* (New York: Charles Scribner's Sons, 1883), pp. 106–8.

exercise of omnipotence have created us as, or transmuted us into, these perfect creatures in the first place. That He has not done so suggests that the nature of finite personal life requires that man's sanctification takes place at every stage through his own responses and assent, and that the process of man's free interaction with divine grace cannot be by-passed. And this in turn points to the conception of a continued life in an intermediate state.

Lest this notion of purgatory or of progressive sanctification after death should sound offensively Romish to some Protestant ears, let me quote from a distinguished modern Anglican philosopher and theologian, A. E. Taylor. In his Gifford Lectures, *The Faith of a Moralist*, he claims that 'the central thought of the doctrine of Purgatory' is 'really held, in one form or another, by the thoughtful even in communities which nominally repudiate it'. He continues:

I cannot conceive that most of us, with our narrow range of understanding and sympathies, our senseless antipathies and indifferences, and our conventional moral outlook, could ever be fitted by the mere fact of escape from the physical limitations of the body to enter at once into the eternal life of the simply loving souls. I should think it more probable — always with deference to wiser judgements — that death leaves us, as it finds us, still far too much takers and too little givers, and that the process of purgation, begun in this life in all who have made any progress in good, needs, for all but the very few, to be continued and intensified, and that, for most of us, this means severe discipline. It may be well to have got rid of the crude imagination of 'Purgatory-fire' as a 'torment', and still better to have lost belief that one can purchase remission of the torment by cash payment into an ecclesiastical treasury, but the main thought that the hardest part of the work of putting off temporality may, for most of us, lie on the further side of the physical change called death, seems to me eminently sound.[1]

The other main objection commonly offered against the kind of eschatology that seems to be required for a viable Christian theodicy is that by expecting the ultimate salvation of all men it undermines both the Church's missionary enterprise and the individual's sense of a solemn and irrevocable choice confronting him here and now, in this life. As to the

[1] *The Faith of a Moralist* (London: Macmillan & Co. Ltd., 1930), i, pp. 317-18.

first point, everything depends upon the nature of the true missionary motive. If this is to save men from an eternal torment in hell which will befall them if they do not hear and accept the Gospel, then clearly disbelief in such a hell must be fatal to missionary zeal. But surely this is not the motive that blows 'like the rush of mighty wind'[1] through the Acts of the Apostles. The early Church's missionary motive was positive, and twofold : there was both a divine command to preach the gospel to all creatures and also a spontaneous overflowing impulse to share with the world the joy and transforming power of the good news. It was this twofold motive that sent St. Paul and the other early missionaries out upon their great journeys of evangelization. This has always been and always will be the positive motive behind the Church's witnessing activity. And this cannot be injured by abandonment of belief in eternal torment. On the contrary, the good news that is to be shared with the world shines the brighter when it is not clouded by the grim fantasy of unending torment inflicted by God.[2]

The other difficulty, which presents belief in eternal damnation as the foundation of moral and spiritual seriousness in this life, has, I think, already been met in principle by the suggestion that right and wrong actions in the present life do lead to appropriately different experiences beyond this world. Although 'hell' — using this term now as a name for the fact of purgatorial experiences made necessary by our imperfections and sins in this life — is not eternal, it is nevertheless real and dreadful and rationally to be feared. There is thus still a place in Christian preaching for the theme of living righteously as a way of 'fleeing from the wrath to come' — though surely only as a shock tactic to cause grossly selfish minds to look beyond their immediate gratifications. But to assert that the sufferings caused by earthly wrongdoing are eternal is, I believe, to go beyond anything warranted by either revelation or reason, and to fall into a serious perversion of the Christian Gospel.

[1] Acts ii. 2.
[2] One sees the powerful counter-apologetic effect of the traditional doctrine of hell in, for example, J. S. Mill's *Three Essays on Religion*, pp. 113–14.

4. SOME RESIDUAL PROBLEMS

Christian theodicy claims, then, that the end to which God is leading us is a good so great as to justify all the failures and suffering and sorrow that will have been endured on the way to it. The life of the Kingdom of God will be an infinite, because eternal, good, outweighing all temporal and therefore finite evils. We cannot visualize the life of the redeemed and perfected creation, for all our imagery is necessarily drawn from our present 'fallen' world. We can think only in very general terms of the opening up before us of new dimensions of reality 'which eye hath not seen nor ear heard nor the heart of man conceived';[1] a new intensity and vividness of experience; of expanded capacities for fulfilment in personal relationships, artistic and other forms of creativity, knowledge, wonder, the enjoyment of beauty, and yet other goods and kinds of goods at present beyond our ken.

But, having said this, questions and difficulties at once arise. Could even an endless heavenly joy ever heal the scars of deep human suffering? It has been said (by Leon Bloy) that 'Souffrir passe; avoir souffert ne passe jamais.' Physical pain is quickly forgotten; but the memory and the effects of mental and emotional anguish can remain with us throughout our lives and presumably beyond this life so long as there is continuity of personal identity. Would not, then, the recollection of past miseries, shames, crimes, injustices, hatreds, and agonies—including the recollection of witnessing the sufferings of others — destroy the happiness of heaven?[2]

It is very difficult indeed to resolve such a question; for we do not know what is possible, let alone what is probable, in realms of being so far beyond our present experience. We can think only in terms of what Plato called 'likely tales'. It may be that the personal scars and memories of evil remain for ever, but are transfigured in the light of the universal mutual forgiveness and reconciliation on which the life of

---

[1] I Corinthians ii. 9.
[2] For a powerful underlining of this question, see Dostoievski's *The Brothers Karamazov*, pt. II, bk. v, chap. 4.

heaven is based. Or it may be that the journey to the heavenly Kingdom is so long, and traverses such varied spheres of existence, involving so many new and transforming experiences, that in the end the memory of our earthly life is dimmed to the point of extinction. There is no evident ground or need to decide between such possibilities, and I mention them only to suggest that the puzzle that was raised, although not at present soluble, is also not such as to overthrow the theodicy that we have been developing.

Another and even more fundamental difficulty might be formulated as follows: Is there not a fatal contradiction in this use of eschatology to complete theodicy? It is claimed —reducing the argument to its bare bones—that the evils of this life are necessary to prepare us as moral personalities for the life of the future heavenly Kingdom, and that they are justified by the fact that in that Kingdom all evil will have been left behind and unimaginable good will fill our lives. But if only challenges and obstacles and sufferings can evoke the highest moral qualities within us, will not these evils still be necessary in heaven? If, for example, courage presupposes danger on earth, why does it not also presuppose danger in heaven? If love is at its strongest and deepest amidst trials and difficulties, must there not be trials and difficulties in heaven also? And so with all the other virtues.

This again is an exceedingly difficult question to meet; nevertheless I believe that its logical effect is rather to remind us of our ignorance concerning the life of heaven than to forbid us to believe in a heavenly life at all. For once again there are several possibilities between which we are not at present empowered to choose. Perhaps these earthly virtues will become heavenly qualities analogous to courage, perseverance, truthfulness, etc. These analogues would no longer presuppose evils to be overcome and temptations to be resisted, but nevertheless they would be such that they could have been arrived at only via their corresponding earthly virtues. Or perhaps mundane courage, faithfulness, etc., can all be seen as modes of a relationship to God (even when God is not explicitly believed in), which relationship takes other forms in the heavenly environment. Or again, perhaps there

are challenges and tasks, problems and pains in heaven. For
the Christian conception of heaven is not basically that of a
pain-free paradise, but that of life lived in a wholly right
relationship to God. There could not be unhappiness within
such a relationship; but there might (for all that we can now
know) be difficult tasks, not to be performed without the
endurance of even great hardship and pain, and immense
and challenging problems, to be solved only by intense effort.
Arthur Schopenhauer believed that 'after man had trans-
ferred all torments to hell, there then remained nothing over
for heaven but ennui;[1] but in fact the notion that a heavenly
existence must be flat and boring is itself gratuitous and
unimaginative. God's creation is virtually infinite in its
range and complexity, its depths and wonders, and there
may be endless scope for further discovery and exploration,
further adventures in new dimensions of reality, further
experiments in new sciences and experiences in new arts,
further spiritual growth in relation to the infinite divine
plenitude and activity. But our imaginative resources are so
utterly inadequate to whatever the power and wisdom of God
may have in store for 'just men made perfect' that these are
probably all but childish guesses. It is perhaps best merely
to insist that the problem raised does not rest upon any
alleged logical contradiction, but upon the difficulty of conceiving the concrete character of the best possible fulfilment
of human existence; and that such a difficulty does not
render a theodicy impossible.[2]

5. THE BIBLICAL PARADOX OF EVIL

We return in conclusion to the central problem of Christian
theodicy. This problem was posed in Part I,[3] and must have
grown more acute in the reader's mind with each succeeding
chapter: namely, how can we reconcile a serious facing of
evil *as evil* with a full acknowledgement of the absolute

[1] *The World as Will and Idea*, trans. R. B. Haldane and J. Kemp (London:
Trübner & Co., 1883), vol. i, p. 402.
[2] One of the most valuable constructive treatments of the positive character
of the heavenly life is that of Baron von Hügel in *Essays and Addresses on the
Philosophy of Religion*, i, pp. 218–19.   [3] See pp. 21–22.

sovereignty and hence the ultimate omni-responsibility of
God? In the course of this final Part I have suggested that
moral evil is a virtually inevitable result of the epistemic
'distance' from God that is entailed by man's creation as a
morally independent being inhabiting his own world.[1] But
if moral evil is thus a 'virtually inevitable' outcome of God's
own creative work, how can it be truly hateful to Him, truly
at enmity with Him, truly at variance with His purposes and
inimical to all good? Again, I have suggested that pain and
suffering are a necessary feature of a world that is to be the
scene of a process of soul-making; and that even the haphazard and unjust distribution and the often destructive and
dysteleological effects of suffering have a positive significance
in that they call forth human sympathy and self-sacrifice, and
create a human situation within which the right must be
done for its own sake rather than for a reward.[2] But if man's
anguish is thus used to a constructive end, how can it be
truly *evil* and truly contrary to God's will for us?

The problem is not a gratuitous one, created by a theodicy
that needlessly affirms the paradox that God is Himself
ultimately responsible for the existence of evil and yet that
evil is truly evil and truly subject to His condemnation and
rejection. On the contrary, this paradox arises as unavoidably within the scriptural revelation as in rational reflection
upon the theodicy-problem. Both the monistic (or, more
properly, the monotheistic) and the dualist views of evil are
present in the Bible. Their occurrence there should not,
however, be too neatly schematized (as it has sometimes
been)[3] into a contrast between the testaments, as though evil
were depicted in the Old Testament primarily as God's
instrument and servant (as in the prologue to the Book of
Job) and in the New Testament primarily as His enemy (as
in Jesus' expulsion of demons from the sick). In fact the two
conceptions of evil run as intertwined threads throughout
the entire Bible, weaving a pattern that is as complex as the

[1] See pp. 316 f.   [2] See pp. 370 f.
[3] See, for example, Kurt Lüthi, *Gott und das Böse*, pp. 258, 261. However,
there is undoubtedly a development in the biblical conception of Satan, which
is traced, for example, in George B. Caird, *Principalities and Powers* (Oxford:
Clarendon Press, 1956), chap. 2.

complex and sometimes apparently contradictory reactions
of the human spirit to the experience of good and evil. In
tracing this pattern we may find within it a clue to the solution of our problem.

In the Old Testament there are a number of passages in
which evil is directly related to the will of God. There is not
only the picture in the Book of Job of Satan as a functionary
in the divine court.[1] There are also the words of the Most
High, 'I form light and create darkness, I make weal and
create woe, I am the Lord, who do all these things',[2] and the
rhetorical questions of the prophet, 'Does evil befall a city,
unless the Lord has done it?'[3] and of the Wisdom writer, 'Is
it not from the mouth of the Most High that good and evil
come?'[4] Again, in the prophetic interpretation of Old
Testament history God employs the cruelties of the heathen
Chaldeans and Assyrians to discipline and punish His chosen
people.[5] In other contexts too God is depicted as using
suffering for the moral and spiritual training of His children.[6]
But, on the other hand, there are also in the Old Testament
innumerable passages expressing, through the prophets,
God's abhorrence of evil. Such sayings as, 'I will punish the
world for its evil, and the wicked for their iniquity',[7] occur in
almost every book, and there is no need to cite them in detail.

In the New Testament the dualistic strand of thought is
most evident in the gospels. Evil — in the forms both of sin
and of suffering — is depicted here as the enemy of God and
man. Thus Jesus was conscious of temptation as the work of
the devil;[8] he dreaded death as a mortal man;[9] and he saw
disease as a bondage to Satan,[10] and evil-doers as inspired by

[1] Job i. 6–12. Cf. I Kings xxii. 19–22.
[2] Isaiah xlv. 7.   [3] Amos iii. 6.   [4] Lamentations iii. 38.
[5] Jeremiah i. 15; vii. 14; l. 25; and Isaiah x. 5 f.; xxviii. 21. Cf. Lamentations ii.
[6] E.g. Deuteronomy viii. 25; Proverbs iii. 11–12; Job v. 17; Psalm xciv. 12.
On the disciplinary conception of suffering in the Old Testament in general see
J. Alvin Sanders, *Suffering as Divine Discipline in the Old Testament and Post-
Biblical Judaism* (Colgate-Rochester Divinity School Bulletin, 1955).
[7] Isaiah xiii. 11. Cf., for example, lvi. 2; Deuteronomy xix. 19–20;
Jeremiah iv. 4; Micah ii. 1; Malachi ii. 17, etc.
[8] Mark iv. 1–11 = Luke iv. 1–13.
[9] Mark xiv. 33–5 = Matthew xxvi. 38–9 = Luke xxii. 42–6.
[10] Luke xiii. 16.

Satan.[1] But the other and contrasting strand of biblical thought, in which evil is more positively related to the will of God, is also abundantly present in the New Testament. Christ faces the cross as a sacrifice that his heavenly Father desires him to make. 'The Son of man must suffer many things, and be rejected of the elders and chief priests and scribes, and be killed, and on the third day be raised.'[2] A divine purpose overshadows all the events of Passion Week. 'You would have no power over me,' Jesus says to Pilate, 'unless it had been given you from above.'[3] And at the end, as he expires on the cross, he pronounces, 'It is finished.'[4] Like the Suffering Servant of Isaiah liii, he saw the fruit of the travail of his soul and was satisfied. The early Church echoed this conviction that the crucifixion had taken place within the universal divine purpose when it spoke of Jesus as having been 'delivered up according to the definite plan and foreknowledge of God'.[5]

In Christ's attitude to his own sufferings, then, and in the Church's sharing of that attitude, evil is related in a dual way to the divine will. On the one hand, Jesus' death was an experience of agonizing pain and suffering which was emphatically not as such willed by God, but which was worked by human wickedness and moral failure. The avarice of Judas; the blood-lust of the Jerusalem mob; the cowardice of Pilate; the brutality of the soldiers, all contributed to it: and all were contrary to God's will. But, on the other hand, in enduring this Jesus was God's agent overcoming evil with good. The malice which assailed him was part of a solidarity of human evil that was provoked by God's presence in vulnerable human flesh to this violent act of rejection. All this was divinely foreseen, and by enduring 'from sinners such hostility against himself'[6] for the sake of mankind Christ fulfilled the pattern of redemptive suffering. For the Church has always believed as one of its cardinal

---

[1] Matthew xiii. 24–30.
[2] Luke ix. 22 = Mark ix. 31 = Matthew xvii. 22–23.   [3] John xix. 11.
[4] John xix. 30. The overshadowing divine purpose is brought out even more explicitly in the fourth gospel than in the synoptics.
[5] Acts ii. 23. Cf. iii. 18; I Corinthians xv. 3; I Peter i. 20.
[6] Hebrews xii. 3.

doctrines that the judicial murder of the Christ was the focus of God's redemptive work and the turning-point of man's salvation. Here is the paradigm of evil being turned to good by the voluntary endurance of pain and suffering as God's servant and agent. Jesus saw his execution by the Romans as an experience that his heavenly Father desired him to accept, and that was thereby to be brought within the sphere of the divine purpose and made to serve the divine ends.

This view of suffering as capable of a constructive use continues in the remainder of the New Testament documents, expressing the attitude of the members of the earliest Christian communities. There are three main themes. There is, first, a rejoicing in the hardships and persecutions of Christian apostleship as a sharing of both the sufferings of Christ and the joy of his redemptive work. For 'Christ suffered for you, leaving you an example, that you should follow in his steps';[1] and 'as we share abundantly in Christ's sufferings, so through Christ we share abundantly in comfort too'.[2] Again, Paul writes to the Christians at Corinth, 'I rejoice in my suffering for your sake, and in my flesh I complete what is lacking in Christ's afflictions for the sake of his body, that is, the church . . .'.[3]

Closely connected with this theme, there is, second, the vivid expectation of sharing in the joy of Christ's heavenly Kingdom. 'So we do not lose heart. Though our outer nature is wasting away, our inner nature is being renewed every day. For this slight momentary affliction is preparing for us an eternal weight of glory beyond all comparison . . .'.[4]

And third, there is a sense of the 'soul-making' significance of suffering. Even the Lord himself was made perfect in this way. 'For it was fitting that he, for whom and by whom all things exist, in bringing many sons to glory, should make the pioneer of their salvation perfect through suffering.'[5] And the same must be true of us. 'It is for discipline

[1] I Peter ii. 21. [2] II Corinthians i. 5. [3] Colossians i. 24.
[4] II Corinthians iv. 16–17. Cf. I Peter iii. 14, and iv. 13.
[5] Hebrews ii. 10. The redemptive possibilities of suffering are also recognized in Jesus' parable of the prodigal son (Luke xv. 11 f.).

that you have to endure. God is treating you as sons; for what son is there whom his father does not discipline? . . . For the moment all discipline seems painful rather than pleasant; later it yields the peaceful fruit of righteousness to those who have been trained for it.' [1] Therefore, in the words of St. Paul, 'we rejoice in our sufferings, knowing that suffering produces endurance, and endurance produces character, and character produces hope. . .'.[2]

6. Its Source in the Duality of the Christian Life

There are, then, two attitudes to evil within the Bible, one based upon the dualistic view of evil as the irreconcilable enemy of God and man, and the other upon a profound sense of the sole ultimate sovereignty and responsibility of God. These two attitudes reflect, I would suggest, the characteristic duality of the religious and especially of the Christian life, which is lived in conscious relation both to this unparadisal world, with its constant struggle between good and evil, and to the transcendent divine purpose, which sustains and overrules the entire temporal process. Through our immersion in the tensions and pressures, the terrors, pains, and anguishes of our frail, sinful, human life, we see evil as a deadly threat not only to our own welfare but to the very meaning of human existence; and in Christ God Himself shared this temporal standpoint, struggling with us and for us against the powers of darkness. In this perspective we face an uncertain and threatening future in which the evils of life may at any time overwhelm us, and in which we are in any case doomed to an eventual inevitable extinction. And yet we are also and at the same time conscious by faith of the eternal being and love of God, holding both ourselves and the whole creation securely in His grasp and exercising His sovereignty over the entire course of time and history.

These two perspectives come simultaneously to their

[1] Hebrews xii. 7, 11.
[2] Romans v. 3–4. Cf. II Corinthians xii. 7–10.

clearest focus in the New Testament picture of the Christ. On the one hand, he shared our human fear of death;[1] he engaged in spiritual combat with the enemy;[2] and he spoke of the way of destruction which yawns before men,[3] and of Satan as seeking men's downfall[4] and as snatching God's saving truth away from them;[5] and he challenged men to a decision of fateful and eternal significance.[6] And yet, on the other hand, he was utterly confident of God's present rule over His world[7] and saw in the imminent future the defeat of evil and the full establishment of God's Kingdom.[8] In his own person and actions he harmonized these two dimensions of the truth by inaugurating that transformation of our human situation which is God's solution to the problem of evil. So far as moral evil is concerned, this was his work of reconciling men to God in the midst of their sin and suffering. As regards 'natural' evil, the divine power of compassion and self-giving in response to human need, which is finally to banish all illness and pain, was in Jesus in such overwhelming force that in his presence the sick were immediately healed. The end of history was manifested with startling effect in the midst of history; for in Christ the Kingdom or Rule of God had come and at its coming it was seen that 'the blind receive their sight and the lame walk, lepers are cleansed and the deaf hear, and the dead are raised up, and the poor have good news preached to them'.[9]

I suggest, then, that these two contrasting biblical attitudes to evil reflect, on the one hand, our existential involvement in the long, slow, difficult, painful process of 'soul-making' and, on the other hand, our faith that such a process is in fact taking place within and around us; and that both attitudes are wholly valid. The prominence of the monotheistic-instrumental view of evil at several points in the Old

---

[1] Mark xiv. 36 = Luke xxii. 42 = Matthew xxvi. 39.
[2] Matthew iv. = Luke iv. [3] Matthew vii. 13.
[4] Luke xxii. 31–32. [5] Mark iv. 15.
[6] Matthew vii. 21–27 = Luke vi. 46–49.
[7] Matthew v. 45; vi. 25–32; vii. 7–11.
[8] Mark iv. 30–32; Luke xxi. 27.
[9] Matthew xi. 5 = Luke vii. 22. Cf. Matthew xv. 31.

Testament is due to the fact that the faith of the prophets reached beyond our human scene, and looked upward to God, participating by faith in His view of the drama of human history as the outworking of His own sovereign purpose. And the strongly dualistic view of evil is so vividly present at many points in the New Testament because this reflects God's incarnate view of the world and of human life from within. Experienced from within the stresses of human existence, evil *is* a sheerly malevolent reality, hostile alike to God and His creation. It is a threat to be feared, a temptation to be resisted, a foe to be fought. This is the indelible quality of evil as it is seen through human eyes in the midst of time, and felt in the dread power of wickedness, in the anguish of loss or remorse or anxiety, or the weaknesses and smarts of bodily illness. Seen, on the other hand, in the perspective of a living faith in the reality of the great, ongoing, divine purpose which enfolds all time and all history, evil has no status in virtue of which it might threaten even God Himself. It has an interim and impermanent character which deprives it of the finality that would otherwise constitute so much of its terror. Neither its beginning, its course, nor its end lies outside God's ultimate control.

In a coalescence of these two perspectives, or a penetration of our continuing natural awareness by our occasional religious awareness — a coalescence whose fitful and flickering character reflects the precarious quality of our own faith — we are conscious of the divine love, which has ordained our ambiguous human existence, as also actively present within it. This awareness of the divine presence does not negate our agonizing human experience of evil, but sets it within the context of God's purpose of good and under the assurance of the ultimate triumph of that purpose. In virtue of this wider context of meaning there can be a partial but significant transformation of our present encounter with evil. To describe this transformation is a delicate and difficult task, fraught with the danger of several kinds of error. I shall, however, attempt it both in relation to our own sins and sufferings and in relation to our reaction to the sins and sufferings of others.

First, as regards our own sins, the ultimate divine responsibility for the existence of a 'fallen' humanity does not cancel, or even diminish, our individual human moral responsibility. For this latter depends upon the fact that our actions flow from our own responsible choices. They are *our* actions, and we must be judged by them. The ultimate divine responsibility for the universe of which we are a part, and our personal responsibility for our own actions, do not clash with one another. There is thus no mitigation of 'the exceeding sinfulness' of human sin. The sinner cannot shed any of his guilt upon his Maker. We remain responsible for our sins and subject to God's condemnation and revulsion. Nevertheless, our awareness in faith of the universal divine purpose and activity, which is the ultimate context of our life, makes a vital difference. Our sinfulness is no longer a matter for final despair : God's saving purpose continually sets before us the possibility of repentance and a new life. For the divine love, encountered as that which is ultimately real, is able so to reconcile us to reality that we may renounce our own small self-enclosed circles of meaning and receive a place in the universal kingdom of God. Our confrontation with the death of Christ is thus the point at which we have to choose whether to live in a private world of meaning, controlled by ourselves, within which there is place for God only on a man-made cross, or in the illimitable universe of which God is the eternal centre, and in which we exist as His creatures, totally dependent upon Him, and yet at the same time as beloved children in our Father's house.

Again, our own sufferings must be coloured and altered by the conviction that 'neither death, nor life, nor angels, nor principalities, nor things present, nor things to come, nor powers, nor height, nor depth, nor anything else in all creation, will be able to separate us from the love of God in Christ Jesus our Lord',[1] and that 'in everything God works for good with those who love him'.[2] We do not know in what ways or in what scale of time God is bringing future good out of present evil ; but that He *is* doing so, and that we can there-

---

[1] Romans viii. 38–39.   [2] Romans viii. 28.

fore commit ourselves wholly to His providence, is the practical outcome of faith in God's love and sovereignty seen in the life, death and resurrection of the Christ. Even death, that grim reminder of our utter vulnerability as creatures made out of the dust of the earth, loses its sting. To believers, as Martin Luther wrote, death 'is already dead, and hath nothing terrible behind its grinning mask. Like unto a slain serpent, it hath indeed its former terrifying appearance, but it is only the appearance; in truth it is a dead evil, and harmless enough.'[1]

What, however, of the sins and the sufferings of others? When we ask such a question today we almost inevitably think of the Nazi programme for the extermination of the Jewish people, with all the brutality and bestial cruelty that it involved and evoked. What does that ultimate context of divine purpose and activity mean for Auschwitz and Belsen and the other camps in which, between 1942 and 1945, between four and six million Jewish men, women, and children were deliberately and scientifically murdered? Was this in any sense willed by God? The answer is obviously no. These events were utterly evil, wicked, devilish and, so far as the human mind can reach, unforgivable; they are wrongs that can never be righted, horrors which will disfigure the universe to the end of time, and in relation to which no condemnation can be strong enough, no revulsion adequate. It would have been better — much much better — if they had never happened. Most certainly God did not want those who committed these fearful crimes against humanity to act as they did. His purpose for the world was retarded by them and the power of evil within it increased. Undoubtedly He saw with anger and grief the sufferings so wilfully inflicted upon the people of His ancient choice, through whom His Messiah had come into the world.

Our Christian awareness of the universal divine purpose and activity does, however, affect our reaction even to these events. First, as regards the millions of men, women, and

[1] 'The Fourteen of Consolation', trans. by A. T. W. Steinhaeuser in the Philadelphia edition of the *Works of Martin Luther*, vol. i (1943), pt. ii, chap. 2, p. 148.

children who perished in the extermination programme, it gives the assurance that God's good purpose for each individual has not been defeated by the efforts of wicked men. In the realms beyond our world they are alive and will have their place in the final fulfilment of God's creation. The transforming importance of the Christian hope of eternal life — not only for oneself but for all men — has already been stressed above, and is vitally relevant here. Second, within the situation itself, the example of Christ's self-giving for others should have led Christians to be willing to risk their own lives to help the escape of the threatened victims; and here the record is partly good but also, unhappily, in too large part bad. And third, a Christian faith should neutralize the impulse to meet hatred and cruelty with an answering hatred and cruelty. For hatred begets hatred and cruelty begets cruelty in a downward spiral that can be halted only by the kind of sacrificial love that was supremely present in the death of Christ. Such a renouncing of the satisfaction of vengeance may be made possible to our sinful hearts by the knowledge that the inevitable reaction of a moral universe upon cruelty will be met, within this life or beyond it, without our aid. 'Vengeance is mine, I will repay, says the Lord.'[1]

7. Its Eschatological Resolution

The dualistic and instrumental views of evil in the Bible are thus derived from the two perspectives of the religious life, as immersed in the historical process with its uncertainties and threats, and as participating by faith in God's on-going creative purpose, secure in the knowledge of His final triumph. However, the question still remains: What is the relation between these two points of view? Does one of them describe reality and the other appearances? Is evil *really* good but only *seems* from our finite human point of view to be bad? Or is evil *really* bad, but is made to *seem* good in speculative theory? Neither suggestion can be accepted for a moment. We must insist both that evil is

[1] Romans xii. 19.

really evil *and* that God has really willed for a good purpose
a world in which evil, with its demonic quality, arises. For
it is an inevitable deliverance of our moral consciousness,
of which nothing must be allowed to rob us, that evil in all
its forms is to be abhorred and resisted and feared. And it
is — as I have been arguing throughout these chapters of
Part IV — an inevitable theological inference, to which we
must not blind ourselves, that the actual universe, with all
its good and evil, exists on the basis of God's will and receives
its meaning from His purpose. However, these two conclusions do not stand in simple contradiction to one another.
The one says that evil is bad, harmful, destructive, fearful,
and to be fought against as a matter of ultimate life and
death. But the other does not deny this. It does not say
that evil is *not* fearful and threatening, inimical to all good,
and to be absolutely resisted. It says that God has ordained
a world which contains evil — real evil — as a means to the
creation of the infinite good of a Kingdom of Heaven within
which His creatures will have come as perfected persons to
love and serve Him, through a process in which their own
free insight and response have been an essential element.

The bridge between the two standpoints is provided by
the Christian hope of the Kingdom of God. The compatibility between the 'existential' view of evil as utterly
malevolent and harmful and the theological view of it as
divinely permitted and over-ruled, is a form of the compatibility between the temporal process which we know by
our own present immersion within it, and the future completion and transformation of it which we affirm by faith.
We saw in Chapter I that evil must ultimately be defined
as that which thwarts God's purpose for His creation. This
means that if in fact God's purpose of universal good is
eventually attained, then in relation to that fulfilment nothing will finally have been sheerly and irredeemably evil. For
everything will receive a new meaning in the light of the
end to which it leads.[1] What now threatens us as final
evil will prove to have been interim evil out of which good

[1] On the way in which future developments can alter the significance of the
past see William Temple, *Mens Creatrix*, pp. 172–4.

will in the end have been brought. That is to say, there will have been states of the temporal process which might have led, and which indeed by themselves would inevitably have led, to the thwarting of God's purpose and so have been irretrievably evil; but in fact they will, in the retrospect of God's completed work, be seen to have been used as stages in the triumphant fulfilment of the divine purpose of good. Of such moments — which we know as contemporary acts of sin or contemporary moments of suffering — it is true that they really are evil and will remain so until they have been forced to serve God's creative purpose. They are genuinely and unequivocally evil, and at enmity with God and His creatures, unless and until they are turned to an end that is alien to their own character. They do not merely *seem* to threaten us with ultimate destruction; they really are such a threat. And yet precisely because they really do thus threaten our deepest good we may by God's grace so repent of our sins and so bear our sufferings that they become elements within something else, namely our reconciliation with God and our growth into the finite 'likeness' of our Maker. We thus have to say, on the basis of our present experience, that evil is really evil, really malevolent and deadly and also, on the basis of faith, that it will in the end be defeated and made to serve God's good purposes. From the point of view of that future completion it will not have been merely evil, for it will have been used in the creation of infinite good. This duality and paradox is expressed by a sentence which the Christian Church has always cherished even when it has been unable to assimilate it into the prevailing theological framework: 'O felix culpa quae talem ac tantum meruit habere redemptorem' (O fortunate crime which merited such and so great a redeemer). In their far-reaching implications these words are the heart of Christian theodicy.

INDEX

Index

THE END

PRINTED BY R. & R. CLARK, LTD., EDINBURGH